Designing Switch/Routers

This book examines the fundamental concepts and design methods associated with switch/routers. It discusses the main factors that are driving the changing network landscape and propelling the continuous growth in demand for bandwidth and high-performance network devices. *Designing Switch/Routers: Fundamental Concepts and Design Methods* focuses on the essential concepts that underlie the design of switch/routers in general.

This book considers the switch/router as a generic Layer 2 and Layer 3 forwarding device without placing an emphasis on any particular manufacturer's device. The underlying concepts and design methods are not only positioned to be applicable to generic switch/routers but also to the typical switch/routers seen in the industry. The discussion provides a better insight into the protocols, methods, processes, and tools involved in designing switch/routers. The author discusses the design goals and features switch/router manufacturers consider when designing their products as well as the advanced and value-added features, along with the steps, used to build practical switch/routers. The last two chapters discuss real-world 6 switch/router architectures that employ the concepts and design methods described in the previous chapters.

This book provides an introductory level discussion of switch/routers and is written in a style accessible to undergraduate and graduate students, engineers, and researchers in the networking and telecoms industry as well as academics and other industry professionals. The material and discussion are structured to serve as stand-alone teaching material for networking and telecom courses and/or supplementary material for such courses.

Designing Switch/Routers

Fundamental Concepts and Design Methods

James Aweya

CRC Press
Taylor & Francis Group
Boca Raton   London   New York

CRC Press is an imprint of the
Taylor & Francis Group, an **informa** business

First edition published 2023
by CRC Press
6000 Broken Sound Parkway NW, Suite 300, Boca Raton, FL 33487-2742

and by CRC Press
4 Park Square, Milton Park, Abingdon, Oxon, OX14 4RN

CRC Press is an imprint of Taylor & Francis Group, LLC

ISBN: 978-1-032-31769-4 (hbk)
ISBN: 978-1-032-31582-9 (pbk)
ISBN: 978-1-003-31124-9 (ebk)

DOI: 10.1201/9781003311249

Typeset in Times
by SPi Technologies India Pvt Ltd (Straive)

Contents

Preface..xiii
Author xvii

Chapter 1 The Era of High-Performance Networks..1

    1.1   Introduction 1
    1.2   Introduction to IP Routing...1
    1.3   Characteristics of Enterprise and Service Provider
          Networks..2
          1.3.1   Characteristics of Enterprise Networks.......................2
          1.3.2   Characteristics of Service Provider Networks..............3
    1.4   Understanding the Evolution to High-Performance
          Networks..4
    1.5   Introduction to Multilayer Switching and
          the Switch/Router 7
          1.5.1   The Switch/Router as Merging the Best of Layer 2
                 Switching and IP Routing 8
          1.5.2   The Switch/Router as Reducing the Complexity of
                 Three-Tier Networking...9
          1.5.3   Layer 2/Layer 3 Convergence
                 on a Single Platform..11
    Review Questions...12
    References 12

Chapter 2 Introducing Multilayer Switching and the Switch/Router 13

    2.1   Introduction 13
    2.2   Key Terminology and Definitions...13
    2.3   Routing Metric and Administrative Distance
          (or Route Preference) 19
    2.4   Evolution of the Switch/Router 21
    2.5   First-Generation Routing Devices 21
    2.6   Second-Generation Routing Devices.....................................23
    2.7   Third-Generation Routing Devices 25
    2.8   Fourth (Current) Generation Routing Devices 27
    Review Questions...30
    References 31

Chapter 3 OSI and TCP/IP Reference Models..33

    3.1   Introduction 33
    3.2   Development of Network Reference Models 33

3.3   OSI Reference Model 34
          3.3.1   OSI Physical Layer 37
          3.3.2   OSI Data Link Layer 38
          3.3.3   OSI Network Layer 39
          3.3.4   OSI Transport Layer 39
          3.3.5   OSI Session Layer 41
          3.3.6   OSI Presentation Layer 41
          3.3.7   OSI Application Layer 41
   3.4   TCP/IP Reference Model 42
          3.4.1   TCP/IP Link Layer 43
          3.4.2   TCP/IP Network Layer 45
          3.4.3   TCP/IP Transport Layer 46
          3.4.4   TCP/IP Application Layer 48
   Review Questions 51
   References 51

Chapter 4   Mapping Network Device Functions to the OSI
             Reference Model 53

   4.1   Introduction 53
   4.2   Repeater and the OSI Model 53
   4.3   Bridge (or Switch) and the OSI Model 55
          4.3.1   Bridge Logical Reference Model 55
          4.3.2   MAC Relay Entity 58
          4.3.3   Filtering Database (MAC Address Table or
                   Layer 2 Forwarding Table) 61
                   4.3.3.1   Contents of the Filtering Database 61
                   4.3.3.2   Using a Content Addressable Memory
                              for the Filtering Database 63
                   4.3.3.3   Aging Out Dynamic Filtering Entries 63
          4.3.4   Overview of IEEE 802.1D Transparent Bridging
                   Algorithm 65
          4.3.5   Special Focus: Ethernet MAC Address Format 66
                   4.3.5.1   Organizational Unique Identifier and
                              NIC-Specific Identifier 67
                   4.3.5.2   Individual/Group Address Bit 68
                   4.3.5.3   Universally/Locally Administered
                              Address Bit 69
          4.3.6   VLANs – A Mechanism for Limiting Broadcast
                   Traffic 69
   4.4   Router and the OSI Model 71
   4.5   Multilayer Switching and the OSI Model: Integrated
          Layer 2 Switching and Layer 3 Routing 72
   4.6   Layer 4-7 Switching – Going beyond Multilayer
          Switching (Layer 2/3) 76

4.6.1   Benefits of Layer 4+ Switching – Higher
        Infrastructure Return-on-Investment 78
4.6.2   Architecture and Configuration of Layer 4+
        Switching 79
4.6.3   Capabilities and Applications of Layer 4+
        Switching 80
4.7   Generic IP Host Architecture 83
      4.7.1   Types of NICs 84
      4.7.2   High-Level NIC Architecture 87
4.8   Configuring MAC Addresses in Ethernet NICs 88
Review Questions 90
References 91

Chapter 5   Review of Layer 2 and Layer 3 Forwarding 93

5.1   Introduction 93
5.2   Deciding When to Use Layer 2 or Layer 3 Forwarding
      for an Arriving Packet 93
5.3   Layer 2 Forwarding 95
      5.3.1   Layer 2 Forwarding Basics 95
      5.3.2   Lookup Tables Used in Layer 2 Forwarding
              Operations 97
5.4   Internetworking Basics 101
      5.4.1   Bridging (Switching) in Internetworks 101
      5.4.2   Routing in Internetworks 102
      5.4.3   Switching within a Subnet 104
      5.4.4   Routing between Subnets 104
5.5   Control Plane and Data Plane in the Router or
      Switch/Router 106
      5.5.1   Control Plane 106
              5.5.1.1   Basic Control Plane Operations 108
              5.5.1.2   Routing Table or Routing Information
                        Base 108
              5.5.1.3   Forwarding Table or Forwarding
                        Information Base 122
      5.5.2   Data Plane 123
              5.5.2.1   Basic Data Plane Operations 124
              5.5.2.2   Unicast Reverse Path Forwarding 137
              5.5.2.3   Multicast Reverse Path Forwarding 138
      5.5.3   Examining the Benefits of Control Plane and
              Data Plane Separation 140
              5.5.3.1   Scalability and Distributed Forwarding
                        Architectures 140
              5.5.3.2   Control Plane Redundancy and Fault
                        Tolerance 143
      5.5.4   Lookup Tables Used in Layer 3/4 Forwarding
              Operations, QoS, and Security ACLs 144

5.6   Special Focus: Control Plane Management Subsystems 150
        5.6.1   Simple Network Management Protocol 151
        5.6.2   Remote Network Monitoring 154
        5.6.3   Device GUIs and MIB Browsers............................ 160
Review Questions... 161
References 162

Chapter 6   Packet Forwarding in the Switch/Router: Layer 3
              Forwarding Architectures 165

6.1   Introduction 165
6.2   Packet Forwarding in the Router or Switch/Router.............. 165
6.3   Packet Forwarding Architectures.................................... 168
        6.3.1   Traditional Centralized CPU-Based Forwarding
                Architectures: Software-Based Forwarding
                Using the IP Routing Table 169
                6.3.1.1   Packet Forwarding in the Traditional
                          CPU-Based Forwarding Architectures
                          Using the IP Routing Table...................... 171
                6.3.1.2   Limitations of the Traditional CPU-
                          Based Forwarding Architectures.............. 175
                6.3.1.3   Types of Centralized Forwarding
                          Architectures Using the IP Routing
                          Table.. 176
        6.3.2   Forwarding Architectures Using Route Caches 192
                6.3.2.1   Temporal and Spatial Locality of IP
                          Traffic Flows... 194
                6.3.2.2   Handling Exception Packets 196
                6.3.2.3   Route Cache Performance 196
                6.3.2.4   Implementing Architectures with
                          Route Caches 197
                6.3.2.5   Route Cache Maintenance and Timers 206
                6.3.2.6   Exact Matching in IP Route Caches 207
        6.3.3   Architectures Using Topology-Based
                Forwarding Tables 208
                6.3.3.1   Implementing Architectures with
                          Topology-Derived Forwarding Tables 210
6.4   Routing between VLANs 225
        6.4.1   What Is a VLAN?.. 225
        6.4.2   IEEE 801.1Q 226
        6.4.3   Inter-VLAN Routing 229
        6.4.4   Implementing Inter-VLAN Routing........................ 230
                6.4.4.1   Using an External Router.......................... 230
                6.4.4.2   Using a One-Armed Router 230
                6.4.4.3   Using a Switch/Router.............................. 233
Review Questions... 237
References 238

Chapter 7   Review of Multilayer Switching Methods:
Switch/Router Internals 241

    7.1   Introduction 241
    7.2   Multilayer Switching Methods 241
    7.3   Understanding the Front-End Processor Approach with
         Flow-Based Multilayer Switching 243
         7.3.1   Basic Architecture 243
         7.3.2   Flow-Based Multilayer Switching Packet
                Processing Steps 246
         7.3.3   Multilayer Switching Using a Route Cache and
                Access Control Lists 249
         7.3.4   Multilayer Switching Cache Timers 251
         7.3.5   MLS-RP to MLS-SE Communications 252
    7.4   IP Multicast Multilayer Switching 253
         7.4.1   Unicast Multilayer Switching in Cisco Catalyst
                5000, 6000, and 6500 Series Switches 254
         7.4.2   Understanding IP Multicast Multilayer
                Switching 256
              7.4.2.1   IP Multicast Multilayer Switching
                      Network Topology 256
              7.4.2.2   Cisco IP Multicast MLS Components 257
              7.4.2.3   Layer 3 Multicast Multilayer Switching
                      Cache 258
              7.4.2.4   Layer 3-Switched Multicast Packet
                      Rewrite 259
    7.5   The FIB, Fast-Path Forwarding, and Distributed
         Forwarding Architectures 261
         7.5.1   Benefits of Distributed Forwarding 262
         7.5.2   Distributed Forwarding Architecture 262
         7.5.3   Distributed Forwarding with Integrated
                Hardware Processing 265
         7.5.4   Benefits of Multilayer Switching
                with Integrated Hardware Processing 267
    Review Questions 267
    References 268

Chapter 8   Quality of Service in Switch/Routers 269

    8.1   Introduction 269
    8.2   Network Requirements for Real-Time Traffic Transport 270
         8.2.1   Adequate Bandwidth 270
         8.2.2   Network Elements with Non-blocking and
                Improved Switching and Forwarding
                Performance 270
         8.2.3   End-to-End QoS 271

          8.2.4   High Telecom Grade Network Availability 272
          8.2.5   Remark on Network QoS for Real-Time
                  Applications 272
    8.3   Elements of QoS in Switch/Routers 272
    8.4   Traffic Classes 273
          8.4.1   IEEE 802.1p/Q 274
          8.4.2   IETF Type-of-Service 275
          8.4.3   IETF Differentiated Services 276
          8.4.4   Traffic Filtering and Deep Packet Pattern
                  Matching 281
          8.4.5   Real-World Implementations of Classes
                  of Service 281
          8.4.6   Example Applications of Packet Classification at
                  Layer 2 282
    8.5   Traffic Management 285
          8.5.1   Setting the Scenario for Congestion in
                  Internetworks 285
          8.5.2   Output Queue Drops 286
          8.5.3   Buffer Sizing 287
          8.5.4   Packet Discard 288
                  8.5.4.1   TCP Slow Start and TCP
                            Window Sizes 288
                  8.5.4.2   Random Early Detection 289
          8.5.5   Handling Congestion with Multiple QoS
                  Queues 291
    8.6   Packet Marking and Remarking (Rewriting) 293
          8.6.1   IP Precedence Marking 294
          8.6.2   Differentiated Services Code Point Marking 295
          8.6.3   Class of Service Marking Using IEEE 802.1p 295
          8.6.4   MPLS Traffic Class (or EXP) Marking 296
                  8.6.4.1   MPLS QoS Services 297
                  8.6.4.2   MPLS Tunneling Modes 297
          8.6.5   Tunnel Header Marking 298
    8.7   The Need for QoS Mechanisms in the Switch/Router 299
          8.7.1   Important QoS Features 299
    Review Questions 302
    References 302

Chapter 9   Rate Management Mechanisms in Switch/Routers 305

    9.1   Introduction 305
    9.2   Traffic Policing 305
          9.2.1   Metering and Marking with a Policer 306
    9.3   Traffic Shaping 307
          9.3.1   Leaky Bucket Algorithm as a Queue 308

9.4   Traffic Policing and Traffic Shaping Using the Token
       Bucket Algorithm 308
9.5   Traffic Coloring and Marking...311
       9.5.1   Two-Color-Marking Policer 311
       9.5.2   Three-Color-Marking Policer...................................311
9.6   Implementing Traffic Policing..312
       9.6.1   Single-Rate Two-Color Marker – The Token
                Bucket Algorithm 312
       9.6.2   Single-Rate Three-Color Marker............................312
       9.6.3   Two-Rate Three-Color Marker................................314
       9.6.4   Color-Blind versus Color-Aware Policing..............318
9.7   Congestion Management Using Rate Limiters.....................318
9.8   Input and Output Tagging – Packet Reclassification
       and Rewriting 319
9.9   Traffic Management and QoS Features in Crossbar
       Switch Fabrics 320
9.10   Advanced QoS Features in Switch/Routers...........................323
Review Questions...324
References 324

Index...327

Preface

Ethernet switches operate at Layer 2 of the Open Systems Interconnection (OSI) reference model while IP routers operate at Layer 3 of this model. A switch/router, sometimes called the multilayer switch, is a device that is capable of performing the functions of an IP router (i.e., forward packets at OSI Layer 3), and when required or called upon to, the functions of an Ethernet switch (i.e., forward packets at OSI Layer 2). Essentially, a switch/router is a "two-in-one" platform that provides both Layer 2 and Layer 3 forwarding functions.

Combining the functions of Layer 2 and Layer 3 forwarding on a single platform provides an opportunity for developing network devices with smaller form-factor and footprint, and at a reduced cost. Other than deployment in the core parts of some IP networks (where Layer 2 forwarding may not be required), switch/routers can be deployed from the access/edge to the aggregation points of most networks. Also, switch/routers can be designed to support a wide range of network link and interface technology types other than those based on Ethernet technologies.

The cost savings and benefits that can be derived from using switch/routers in terms of equipment room and rack space, wiring, cooling, and power consumption requirements are not hard to see. A single network device playing several important roles obviously brings many benefits. Thus, enterprises and network service providers have long understood the benefits of using switch/routers.

Advances in networking technologies have allowed the switch/router to assume or take on newer features and capabilities like Multiprotocol Label Switching (MPLS), Carrier Ethernet, Provider Backbone Bridges (PBB) which is also known as MAC-in-MAC and specified in IEEE 802.1ah-2008, Shortest Path Bridging (SPB) which is specified in the IEEE 802.1aq standard, and Software-Defined Networking (SDN). These newer features and capabilities, although important in modern networks, are considered out of scope and will not be discussed further in this two-volume book. For Layer 2 operations, this book focuses solely on switch/routers that support the IEEE transparent bridge algorithm and features as specified in the IEEE 802.1D specification (the official standard for transparent bridges).

This book provides an introductory level discussion that covers the functions and architectures of the switch/router as they relate to the OSI and the TCP/IP network reference models. Most practical implementations of routers and switch/routers employ shared-bus, shared-memory, single-stage and multistage crossbar switch fabrics as their internal interconnects. The author's book, *Switch/Router Architectures: Shared-Bus and Shared-Memory Based Systems*, Wiley-IEEE Press, ISBN 9781119486152, 2018, focuses on the design of switch/routers that are based on shared-bus and shared-memory switch fabrics.

The author's companion book, *Switch/Router Architectures: Systems with Crossbar Switch Fabrics*, CRC Press, Taylor & Francis Group, ISBN 9780367407858, 2019, describes the design of switch/routers that are based on crossbar switch fabrics. These two books provide detailed discussions about how switch/routers are designed and provide reviews of the most common architectures, industry practices, and design approaches adopted by switch/router manufacturers. The current book focuses on the fundamental concepts that underline the design of switch/routers in general. The discussion provides a better insight into the protocols, methods, processes, and tools that go into designing switch/routers.

Despite the vast literature available on networking, it is still very difficult to find books that provide a detailed explanation of the inner workings of routers and how they are designed. This two-volume book aims to fill this void by examining the fundamental concepts, discussing in greater detail the various design methods, and describing the well-known architectures associated with switch/routers. **Volume 1** discusses in detail the fundamental concepts underlying IP devices and networks, and the different approaches for designing switch/routers, while **Volume 2** focuses more on the design goals (i.e., the key features of a switch/router), architectures, and practical applications of switch/routers in IP networks.

The continuous growth of enterprise networks, service provider networks, the Internet as a whole, social networking, and eBusiness has ushered in an era of high-performance devices and networks. **Chapter 1** discusses the main factors that are driving the changing network landscape and propelling the continuous growth in demand for bandwidth and high-performance network devices.

Chapter 2 presents some key terminology and definitions that will help the reader understand the role and architectures of switch/routers presented in subsequent chapters. This chapter explains how a routing protocol uses a routing metric to select the best path to a network destination when multiple paths exist (i.e., best path selection within a particular routing protocol). We also explain how a router uses the administrative distance to select which route to install in its routing table when multiple routing information sources provide routes to the same destination (i.e., best path selection amongst routing protocol or routing information sources). This chapter also discusses the evolution of switch/router architectures, and their corresponding internal control plane and data plane architectures and features.

A network reference model is a logical or conceptual model that splits the tasks involved with transferring information from one end-user or network device to another into smaller, more manageable, and realizable task groups called *layers*. The model assigns a task or group of tasks to each of the layers. The goal of the reference model is to make each layer reasonably and sufficiently self-contained so that hardware and/or software developers can implement the tasks assigned to each layer independently. **Chapter 3** discusses the two main network reference models commonly used, namely, the OSI and the TCP/IP network reference models.

Chapter 4 discusses the various network devices (repeaters (also called hubs), Ethernet switches, routers, switch/routers, and web (or content) switches) according to the OSI layer at which they operate. This chapter explains that, unlike an Ethernet switch that forwards packets at OSI Layer 2, and a router at Layer 3, a switch/router is capable of forwarding packets both at Layers 2 and 3. Because the functions of an Ethernet switch running the IEEE 802.1D transparent bridge algorithm are relatively simple and less complex than the routing and forwarding processes in routers and switch/routers, these functions are discussed in greater detail in this chapter, with only limited discussions provided in other chapters only when needed.

Chapter 5 describes how a switch/router performs Layer 2 and Layer 3 forwarding of packets, as well as how a switch/router decides which mode of forwarding to use (Layer 2 or Layer 3) when it receives a packet. This chapter describes the differences between the control plane and data plane functions in a switch/router, the basics of IP routing table structure and construction, and the main data plane operations performed on packets in switch/routers. This chapter describes the input processing of Layer 3 packets, the processes involved in Layer 3 packet forwarding, and the output processing of Layer 3 packets. The chapter describes the main characteristics of the path transit Layer 3 packets take from the inbound Physical Layer (Layer 1) up to the Network Layer (Layer 3) (where the Layer 3 packet forwarding decisions are made), and to the outbound Physical Layer. The discussion includes the key actions involved in Layer 3 packet forwarding.

Chapter 6 highlights that a switch/router is basically the integration of the traditional switch and router on a single platform. This chapter discusses the basic packet forwarding functions in the typical switch/router as well as some of the well-known switch/router architectures used in the industry. The discussion here also lays out the fundamentals for the discussions in other chapters of the book. **Chapter 7** reviews the different methods used for forwarding packets in switch/routers. The discussion in this chapter includes the different Layer 3 forwarding methods and architectures used by switch/routers, including their advantages and disadvantages.

Given that present-day applications and services have different quality-of-service (QoS) requirements (e.g., delay, delay variation, data loss), the discussion in **Chapters 8 and 9** covers the different QoS mechanisms and architectures used in today's switch/routers. **Chapter 9** focuses, in particular, on the fundamental approaches to traffic rate management (e.g., traffic policing and shaping) in switch/routers.

This two-volume book provides an introductory level discussion of switch/routers and is written in a style to appeal to undergraduate- and graduate-level students, engineers, and researchers in the networking and Telecoms industry, as well as academics and other industry professionals. The material and discussion are structured in such a way that it could serve as standalone teaching material for networking and Telecom courses and/or supplementary material for such courses. Each chapter ends with review questions and a list of relevant references. This book is your ultimate guide to switch/router design.

Author

James Aweya, PhD, is a Chief Research Scientist at the Etisalat British Telecom Innovation Center (EBTIC), Khalifa University, Abu Dhabi, UAE. He was a technical lead and senior systems architect with Nortel, Ottawa, Canada, from 1996 to 2009. He was awarded the 2007 Nortel Technology Award of Excellence (TAE) for his pioneering and innovative research on Timing and Synchronization across Packet and TDM Networks. He has been granted 70 US patents and has published over 54 journal papers, 40 conference papers, and 43 technical reports. He has authored six books including this book.

1 The Era of High-Performance Networks

1.1 INTRODUCTION

Advances in electronic and optical technologies coupled with the emergence of electronic Business (eBusiness), social networking, broadband mobile communication, among several others have significantly changed the modern networking landscape. For example, eBusiness, readily available access to mobile networks, and high-speed transport networks have created a fundamental shift in how business is conducted. In the eBusiness world, the traditional business systems and processes are being replaced with internetworked business solutions that exploit the full potential of enterprise and service provider networks as well as the public Internet, allowing organizations to accelerate the attainment of business goals. As eBusiness evolves, organizations continue to evaluate the potential impact on their competitive position and react accordingly.

For most organizations, success in today's competitive business environment hinges very much on the strength of their networks which form the cornerstone of their eBusiness strategy. Businesses now understand that their networks are strategic assets and play a very important role in the increasingly competitive business environment. This chapter discusses the main factors that are driving the changing network landscape and propelling the continuous growth in demand for bandwidth and high-performance network devices.

1.2 INTRODUCTION TO IP ROUTING

Routers are the main network devices that provide the network-wide intelligence for moving information in internetworks, from enterprise networks to service provider networks, and the Internet as a whole. IP routing is a general term used to represent the collection of methods and protocols used to determine the paths across multiple internetworks that a packet can take in order to get from its source to its destination. Each packet is routed hop-by-hop through a series of routers, and across multiple networks from its source to the destination. Each hop represents a routing device (or router).

IP routing protocols are the set of procedures and rules that govern how routers communicate with each other to exchange routing information about how network destinations can be reached [AWEYA2BK21V1] [AWEYA2BK21V2]. The IP routing protocols serve as the brains behind the routers and provide information about how paths can be constructed across internetworks from a packet's source to its destination. Using IP routing protocols, routers pass routing information about network reachability to each other, allowing packets to be delivered successfully to their destinations. The routing information communicated to the routers allows them to calculate best paths to network destinations. The best paths are then installed in the routing tables of the routing

DOI: 10.1201/9781003311249-1

1

devices. Routes can also be configured manually in the routing tables; these routes are called *static routes* and do not adapt dynamically to network changes [AWEYA2BK21V1].

Routers within a routing domain or autonomous system communicate via Interior Gateway Protocols (IGPs) such as Routing Information Protocol (RIP), Enhanced Interior Gateway Routing Protocol (EIGRP), Open Shortest Path First (OSPF), and Intermediate System–Intermediate System (IS-IS). The interior Border Gateway Protocol (or iBGP) may be used in some networks for intra-domain (or intra-autonomous system) routing information exchange (see [AWEYA2BK21V1] [AWEYA2BK21V2]). The inter-autonomous system routing protocol (or exterior gateway protocol (EGP)) used in today's network is BGP (both iBGP and exterior BGP (eBGP) can be used).

1.3 CHARACTERISTICS OF ENTERPRISE AND SERVICE PROVIDER NETWORKS

The *Internet* is a global network that is composed of different interconnected networks and in which devices and the various constituent networks use the Internet Protocol suite (i.e., the TCP/IP suite) to communicate between themselves. The large collection of hosts in the Internet communicates with each other using routers as intermediate packet devices.

A packet originated by a host or router is forwarded through the interconnected system of routers and networks until the packet reaches a router that is attached to the same network as the destination host. The final router then delivers the packet to the specified host on its local network. Each router keeps track of next-hop information that enables it to forward packets through the network to their destinations. A router that does not have a direct connection to a packet's final destination checks its routing table and forwards the packet to another router (i.e., a next-hop router) that is closer to that destination. This process is repeated at each router (hop) until the packet reaches its final destination.

Other than the Internet, networks can broadly be categorized as *enterprise networks* and *service provider networks* [SCHUDSMIT08]. In this section, we describe the main characteristics of these two network categories to set the context of the discussion in this chapter and the rest of the book.

1.3.1 CHARACTERISTICS OF ENTERPRISE NETWORKS

An enterprise network is a network that is built by an organization (e.g., business, company, hospital, school, etc.) to interconnect its computing facilities at various sites (such as departments, offices, production and manufacturing sites, marketing and sales outlets, etc.) in order to share information and computer resources. An enterprise network may be small, medium, or large. The following are the main reasons organizations build enterprise network [SCHUDSMIT08]:

- Provide a network to interconnect applications and users within the enterprise with each other.
- Provide users within the enterprise access to remote sites within the enterprise (e.g., branch offices) and to external sites in the broader Internet.

- Connect external users within the broader Internet to sites and resources within the enterprise (e.g., websites) that are publicly advertised.
- Connect external partners of the enterprise (e.g., business partners) to non-public enterprise computing resources and information.

The following are some of the main characteristics of an enterprise network [SCHUDSMIT08]:

- It has a well-defined network boundary (or network edge) that serves as a demarcation between enterprise facilities and the outside world. The demarcation serves as the boundary that marks the enterprise's end of network ownership and administrative control (i.e., marks the private and public sides).
- It uses a well-defined set of IP routing protocols, for example, IGPs such as RIP, EIGRP, OSPF, and IS-IS, and iBGP [AWEYA2BK21V1] [AWEYA2BK21V2]. The enterprise network also has its own set of network management, control, and configuration protocols (e.g., Simple Network Management Protocol (SNMP), Dynamic Host Configuration Protocol (DHCP), Domain Name System (DNS), Virtual Router Redundancy Protocol (VRRP), syslog), and enterprise client/server applications.
- It has a well-defined architecture typically using the three-layer hierarchical model of access, distribution, and core layers (see Section 1.5.2). In smaller networks, the access, distribution, and core layers are typically consolidated.
- It provides well-defined points for traffic entering and exiting the enterprise (i.e., well-defined entry and exit points). The entry/exit point also serves as security implementation boundaries and present points for filtering and blocking traffic.

Enterprise networks are not used as transit networks, that is, they are not used for transporting traffic arriving at the network with destinations that do not belong to the enterprise network's IP address space.

1.3.2 CHARACTERISTICS OF SERVICE PROVIDER NETWORKS

A service provider network is a network that typically provides access to the Internet, transit services to the Internet, interconnection between two enterprise (or private) networks, web hosting, and access to many other services available on the Internet. A service provider network typically serves as the access point or the gateway to the public Internet. The main reasons organizations build service provider networks are summarized as follows [SCHUDSMIT08]:

- Provide transit connectivity services for enterprise networks (customers) to other enterprise networks and the public Internet.
- Provide Internet access to directly attached customers (e.g., home users).
- Provide access to content and services hosted by the service provider (e.g., web hosting, media broadcasts, conferencing services, etc.).

The main characteristics of a service provider network are summarized as follows:

- It has a well-defined boundary which can be the demarcation between the service provider and the customer (customer-provider boundary), or the demarcation between the service provider and another peer service provider (peering boundary).
- It uses a well-defined set of IP routing protocols which include various IGPs, and BGP (iBGP and eBGP) [AWEYA2BK21V1] [AWEYA2BK21V2]. A service provider network also has its own set of network management, control, and configuration protocols (e.g., SNMP, DHCP, DNS, VRRP, syslog), and client/server applications.
- It has a well-defined architecture typically using the three-layer hierarchical model of access, distribution, and core layers. A service provider may have multiple Point of Presences (PoPs), often located at Internet Exchange points and colocation centers. The PoPs, which serve as local access points for users to connect to the Internet, typically house network switches, routers, servers, and other network transport equipment such as Synchronous Digital Hierarchy/Synchronous Optical Networking (SDH/SONET) and Wavelength-Division Multiplexing (WDM) multiplexers.

It should be noted that IGPs such as RIPv2 and EIGRP [AWEYA2BK21V1] are suitable for small- to medium-size enterprise networks such as those used in companies, universities, hospitals, airports, shopping malls, etc. On the other hand, IGPs such as OSPF and IS-IS [AWEYA2BK21V2] are mostly used in large-scale networks such as those deployed by service providers and telecom companies. BGP is the main EGP used to interconnect all the different and separate enterprise and service provider networks to create the global Internet as we have today.

1.4   UNDERSTANDING THE EVOLUTION TO HIGH-PERFORMANCE NETWORKS

The requirements to protect, optimize, and grow an organization's network have extended from basic connectivity to providing an intelligent service-based infrastructure. The network has evolved to become an even greater asset to the organization and it is called upon to provide greater flexibility, higher reliability, enhanced security, and extensive redundancy. The network is called upon to provide better operational efficiency and faster response to business opportunities today and into the future. Emerging needs such as mobile broadband, social networking, high-bandwidth multimedia applications, non-stop operation, scalability, and IPv6-readiness place new demands on the network.

The consequence of the above is that modern network solutions must be assessed based on a wider set of attributes than earlier generation solutions. In particular, the network must be evaluated on factors that include performance, reliability, scalability, quality of service (QoS), security, and total cost of ownership (TCO). Understanding the TCO of network infrastructure requires analyzing factors such as capital equipment costs, support cost, power consumption, cooling requirements, physical floor space, and future upgrade requirements.

Enterprise and service provider networks are undergoing a number of significant changes with the most important transition stemming from the emergence of IP and Ethernet as the converged technologies for both data networking and real-time communications applications. This convergence is driven not only by the cost reductions achievable through network consolidation but also by the gains that can be achieved through innovative linkages between voice, video, and data applications, resulting in converged applications such as unified messaging.

As application convergence gains further momentum, the enterprise and service provider network infrastructures must also continue to evolve to support a widening range of real-time applications, such as higher definition multimedia broadcasts, high-speed and low-latency bi-directional or multi-directional videoconferencing, high-speed mobile broadband access, massive Internet of Things (IoT), machine-to-machine communications, Virtual Reality (VR)/Augmented Reality (AR), remote surgery, autonomous driving, remote robot management in factory environments, power system control, and real-time traffic information service, smart cities, and so on.

Furthermore, as the breadth of applications supported by the enterprise and service provider networks continues to diversify, the networks will need to accommodate a wider range of attached devices, including voice over IP (VoIP) desk phones and video phones, Wi-Fi access points for connection of laptops, smartphones, digital surveillance cameras, and IP-enabled badge readers. Another major change for the networks comes from the migration of desktop PCs and other edge devices from 10/100 Megabits per second (Mb/s) to Gigabit Ethernet and higher connectivity.

The primary benefit of higher Gigabit Ethernet speeds to the desktop is improved response time for client/server and peer-to-peer applications that may involve burst transfers of large blocks of data. Beyond improved response time, the use of higher Gigabit Ethernet speeds provides the network headroom that can accommodate the diverse bandwidth and latency requirements of the widening range of applications and edge devices.

Current growth trends show that most of the Internet traffic growth is expected to come from wider adoption of high-speed broadband access technologies and services. Advances in networking technologies and the global acceptance of broadband access technologies have greatly increased the availability of low-cost broadband. In many cases, favorable government policies coupled with competition among alternative technologies and service providers have resulted in widespread adoption of broadband. For example, in many developed countries, where broadband has been a national priority, over 94% of Internet subscribers already have broadband connections.

For example, mobile broadband access continues to drive increased usage of the Internet and enables new user services such as high-speed IP telephony and conferencing, video streaming services, digital media distribution, etc. High bandwidth allows service providers to distribute media content and to offer "triple play" services (data, voice, and video). Along with the need for more bandwidth to support these new services, residential and business customers alike are expecting higher levels of network reliability and predictability at lower and lower price per bit levels. The ability to predict and guarantee service levels and maintain QoS will not only add to customer or user satisfaction but also open up new competitive advantages by leveraging the network as a strategic asset.

On the enterprise front, many organizations are currently working to solve the problems created by the explosive expansion of information technology (IT) infrastructure in the late 1990s. In many instances, this rapid expansion to meet business unit demand for IT resulted in highly complex infrastructures that offered suboptimal reliability and security while at the same being overly difficult to manage and expensive to operate.

Furthermore, as this rapid growth in access bandwidth occurred, service providers needed to continue to expand the capacity of their networks while at the same time responding to a number of competitive pressures and challenges such as the following:

- Controlling costs while still offering an attractive mix of broadband access technologies and speeds, and offering new value-added service options
- Increased focus on reliability and manageability because of greater business reliance on the network
- Dynamically distinguishing between types of traffic and assimilation of real-time applications (voice, video, and data)
- Enhancing customer service and customer satisfaction and assuring satisfactory application performance with QoS
- Scaling the network to meet escalating bandwidth requirements
- Ensuring security in the face of a more fluid network environment
- Managing numerous data centers distributed throughout the organization, including ad hoc patchworks of servers located in wiring closets and computer equipment rooms

Today, in order to meet these same challenges, enterprises and service providers are focusing on developing network architectures that are optimized for scalability, robustness as well as simplicity of operations, and manageability. All of these challenges are considerably easier to meet where the underlying network infrastructure has been consolidated to minimize overall network complexity, including the complexity implicit with a diversity of network operating systems, and capabilities for Layer 2 and Layer 3 forwarding.

Organizations with challenges similar to the above are performing internal assessments to determine whether TCO savings and operational efficiencies can be gained by using better network infrastructure designs, and data center and server consolidation options. A TCO approach is needed in order to make the right tradeoffs between the additional capital expenditure required and the savings to be gained by improving resource utilization and reducing operational costs via simplification of the infrastructure. In particular, the TCO assessment has to address the hidden costs of configuring, managing, and supporting distributed and dispersed environments as new technologies are considered for deployment.

Given that a high-availability network is becoming increasingly important to business operations, enterprises and service providers have formed the habit of comparing and contrasting equipment manufacturers' technical capabilities and the various performance and feature capabilities of the available competing platforms. This includes all the operational issues that impact TCO calculations as well as the ability to meet day-to-day performance requirements. Key factors to consider are:

- The length of time the equipment will remain in operation, and what the upgrade cost will be over, for example, a five- to seven-year period.

- The cost in personnel and services required to install, manage, and maintain the network.
- Other costs that must be factored in, such as power, cooling, and physical equipment space. Many organizations also consider factors that impact the environment like materials used, recycling issues, and end-of-life disposal.

In most organizations, network infrastructure, data center and server consolidation, and architectural re-assessment are often planned in the context of an overall IT program to improve the alignment of IT with business goals, reduce operating costs, and improve business continuity, security, and manageability. Organizations tend to balance any potential loss to the business against the cost of building and maintaining a network built for high availability.

With increasing demands on the network, organizations normally look beyond the initial product costs to better understand the scalability, extensibility, and TCO of the network infrastructure. In the current global economic environment, IT capital investments must be backed by a strong business case. IT managers are called upon to demonstrate the benefit realized to the business by their infrastructure purchases, as measured by real cost reduction, improved application support, and better productivity.

Network devices that are deployed to support infrastructure and server consolidation should be selected based on their capabilities to support both near-term and future consolidation. As a result, from equipment acquisition perspective, the emphasis should be on families of switches, routers and switch/routers that offer the right combination of high availability/robustness features, scalability performance, security, and intelligent traffic control. The switch/router, which combines Layer 2 and Layer 3 forwarding on a single platform, has been a key player in this consolidation as will be discussed in this book.

1.5   INTRODUCTION TO MULTILAYER SWITCHING AND THE SWITCH/ROUTER

In addition to the economic, social, and technological factors cited above that are stretching current network designs to their limits, the following factors are also changing the networking landscape:

- Widespread adoption of Ethernet-based LANs, metropolitan area networks (MANs), and wide area networks (WANs)
- Increasing centralization of network servers
- The proliferation of Intranets and IP-based Internet communications
- The emergence of high-bandwidth, lower-cost Ethernet technologies such as 1, 5, 10, 25, 40, 50, 100 Gigabit Ethernet (see Chapter 6 of Volume 2 of this two-part book).

For some time now, enterprise and service provider networks have been transitioning to multi-tiered models based on high-speed Ethernet technologies. As the price of Ethernet bandwidth continues to fall, desktop connections are transitioning from 10/100 Mb/s Ethernet to Gigabit Ethernet, while aggregation and core tiers of the network are transitioning from Gigabit Ethernet links to 10, 25, 40, 50, 100, and 200 Gigabit Ethernet links. In particular, data centers have already evolved to use higher Gigabit Ethernet links to servers and storage. With continuing improvements in cost, there is now high bandwidth available in the LAN to support VoIP, conference room and desktop video conferencing over IP, IP TV, and an increasingly rich array of data applications.

In the following sections, we discuss the factors that are the main contributors to multilayer switching – that is, integrated Layer 2/Layer 3 forwarding on a single platform. There is no industry standard yet on nomenclature for multilayer switching and Telecom equipment manufacturers and vendors, analysts, and Telecom magazine editors do not have a consensus on the specific meaning of terms such as multilayer switch, switch/router, Layer 3 switch, IP switch, routing switch, switching router, and wire-speed router.

Typically, these different terms do not reflect differences in product architecture but rather differing editorial and marketing policies. Nevertheless, the term switch/router or multilayer switch seems to be the best and most widely used description of this class of product that performs both Layer 3 routing and Layer 2 forwarding functions. This is the "definition" we will adopt in this book. We also use interchangeably the terms "switch/router" and "multilayer switch" in the book where appropriate.

1.5.1 THE SWITCH/ROUTER AS MERGING THE BEST OF LAYER 2 SWITCHING AND IP ROUTING

During the mid-1990s, LAN switching performance was greatly increased network by replacing shared media with full-duplex transmission and dedicated bandwidth. Users benefited from direct access to their networks, and the bottlenecks of shared Ethernet disappeared as point-to-point full-duplex switching was deployed. The requirements for proven technology, bandwidth, manageability, and ease of design have made Ethernet a natural fit for networking applications. One key advantage of using Ethernet is the many established protocols and best practices that have evolved over the many years of Ethernet development, including Virtual LANs (VLANs), QoS, Link Aggregation Groups (LAGs), and other features.

But as applications and services are deployed to take advantage of the improved throughput provided by full-duplex Ethernet switching, performance degradations began to emerge in large flat Ethernet networks. These new problems stem from Ethernet switching's roots as a Layer 2 bridging technology. Ethernet switched networks, which had traditionally been flat domains, had to be subnetted to alleviate traffic broadcast overheads. Without the subnetting performed by Layer 3 routing, Ethernet LAN and switching infrastructure do not scale. Large, flat Ethernet switched networks are subject to broadcast storms, network traffic loops, and inefficient addressing limitations. The limitations of traditional Ethernet switching brought routers into bridged networks in the late 1990s.

Routing continues to be as important to Ethernet switched networks as it ever was. At the same time, Ethernet switching allows networks to be designed with greater centralization of servers and other resources in IP subnets or VLANs, helping to streamline network administration and increase overall security. This centralization, in many cases, results in network topologies that have a greater proportion of traffic crossing the network backbone; more traffic being routed beyond a local subnet or VLAN. Corporate Intranets further exacerbate the problem with increased network usage and by granting easy access to resources deployed widely across the enterprise. In such situations, a large proportion of Intranet traffic travels between subnets. Wide-area Internet usage also created a similar effect, as every web session has to be routed to the Internet from the user's local network by an IP router.

The implications of the above are easy to understand – the traditional Ethernet switching and the applications that leverage its performance quickly reached their limits unless the Layer 3 routing between subnets or VLAN is improved. Layer 2 scalability depends on Layer 3 routing, making the throughput of the interconnecting router a great concern to network managers. To address this, the network manager needs a way of handling Layer 2 switching and Layer 3 routing functionality in an integrated manner, which is precisely what the switch/router or multilayer switch provides. This solution is capable of switching traffic within IP subnets or VLANs while at the same time satisfying IP Layer 3 routing requirements at the same. With the right hardware-based architectures, this combination not only solves the throughput problems within the subnet/VLAN but also removes the Layer 3 traffic forwarding bottlenecks between subnets/VLANs. Integrating Layer 2 switching and Layer 3 routing in a single switch/router device simplifies the network topology by reducing the number of separate network devices and network interfaces that must be deployed to implement multi-tiered network designs.

1.5.2 THE SWITCH/ROUTER AS REDUCING THE COMPLEXITY OF THREE-TIER NETWORKING

The network architecture for the typical enterprise or service provider has a three-tier network design as shown conceptually in Figure 1.1. This design typically involves the separation of functional roles, such as access switches connected to the users or servers, aggregation (distribution) switches that aggregate (distribute) the traffic, and core switches that provide connections to the outside world. In the access tier (layer), edge devices (Layer 2 switches or routers) consolidate subscriber connections, possibly over a range of diverse access technologies (Ethernet, Digital Subscriber Line (DSL), Passive Optical Network (PON), Wi-Fi, high-speed mobile access, etc.). Traffic from the edge devices is aggregated by devices in the middle tier (Layer 2 switches or routers) and forwarded to core devices (typically, routers) that provide connections to Internet Exchanges, ISP backbone, other ISPs, and hosting facilities.

This three-tier approach has always been favored by network designers because it has the advantage of facilitating the expansion of capacity and redesign within any tier with minimal disruption of the other tiers. In addition, the three-tier architecture leverages the specialized switching and routing platforms that have evolved in the industry and focus on specific functions:

- **Access**: This layer provides connectivity to end-users and servers; serves as points of entry/exit for end-user traffic. The network designer can deploy edge switching and routing with moderate performance and a diversity of interfaces to support a wide range of access technologies.

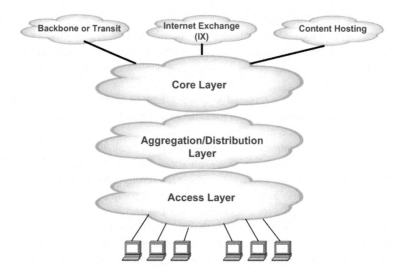

FIGURE 1.1   Typical three-tier networking in the service provider or enterprise.

- **Aggregation/Distribution**: This layer connects the access layer to the core layer, and can provide points for network traffic policy enforcement and control. The aggregation/distribution tier can use cost-effective and high-performance Layer 2 Ethernet switching and IP routing platforms.
- **Core**: This layer serves as the high-speed backbone of the network and can provide connectivity to the public Internet, access points to IP Virtual Private Networks (VPNs), and private IP networks. This tier can use core routing platforms with high performance, Internet-scale routing, and optimized for connection to high-capacity backbones and long-haul optical networks.

The above advantages, however, come at the expense of increase in the network complexity; increase in the total number of devices that must be individually configured and managed. In addition, this architecture may require the configuration of management, traffic control, and security policies at all three levels and at each device. For these reasons, network designers tend to favor platforms with very high switching performance and port density with a small footprint to conserve space in the POP and to minimize the number of devices that need to be managed. Most desirably, the network should be able to support easy upgrades through the use of devices that support simple replacement of line cards and modules. The network must also have the ability to scale bandwidth gracefully with minimal disruption of existing infrastructure.

In view of the above, switch/routers, which support both Layer 2 and Layer 3 forwarding capabilities, have become an effective tool for addressing the requirements of the three-tier architecture. For example, an existing three-tier POP could be upgraded for enhanced performance and robustness by installing switch/routers in the aggregation tier, replacing numerous older aggregation devices, such as ATM switches or earlier generation Layer 2 Ethernet switches. Upgrading the aggregation tier often removes many of the existing performance bottlenecks while preserving the basic architecture of the POP and leaving the existing core and access routers in place.

Furthermore, by exploiting the higher speeds Gigabit Ethernet technologies and Layer 3 routing functionality on the switch/routers, it is possible to collapse the core and aggregation tiers of the POP into a single tier. The POP is simplified through reduction of the number of devices required (i.e., reduced switch count) and the elimination of the need to manage and configure the separate Layer 2 functionality of an aggregation tier. Other benefits of a single layer of aggregation switching/routing within the network include simplified traffic flow patterns, elimination of potential Layer 2 loops and Spanning Tree Protocol (STP) scalability issues, and improved overall reliability.

The POP configuration with switch/routers can be designed to eliminate single points of failure by using redundant inter-tier links between the devices. With the support for the Rapid Spanning Tree Protocol (RSTP), originally specified as IEEE 802.1w, traffic rapidly fails over from primary to secondary paths in the event of link or device failure. With RSTP, failover periods can be as short as milliseconds to hundreds of milliseconds compared to as much as 30 seconds for the original IEEE 802.1d STP.

A multilayer switching fabric with high density of high-speed Ethernet ports can deliver the required switching capacity with a smaller number of devices. This can be done while offering high port capacity for scalable inter-device links based on IEEE 802.3ad (Link Aggregation) trunks using multiple Gigabit Ethernet or even multiple 10 Gigabit Ethernet links. In practice, the switch/router is really an Ethernet-optimized Internet router that also has a complete set of Layer 2 features.

1.5.3 LAYER 2/LAYER 3 CONVERGENCE ON A SINGLE PLATFORM

There are many variations of the Layer 2/Layer 3 convergence or multilayer switching, all focused on relieving router congestion across switched subnets. The approach discussed in this book follows standard Layer 2 (Ethernet) switching combined with standard Layer 3 (IP) routing methods. The book discusses several architectures ranging from high-performance software implementations to architectures that accelerate multilayer switching through integrated hardware and multi-gigabit switching fabrics.

Multilayer switching is a practical evolution of today's enterprise and LAN switching and routing technologies. As an inherent lower-cost solution with integrated Layer 2/Layer 3 functionality, it removes the scalability and throughput restrictions that limit network growth, while also building the foundation for the emerging generation of multi-gigabit-networked applications. In the quickening transition from LAN switching to enterprise and service provider switching/routing over multi-gigabit backbones, simplified management and administration through a Layer 2/Layer 3 device that does not have to be manually configured and has minimal impact on existing topologies makes for the easiest possible migration.

Multilayer switching with its flexible Layer 2 plus Layer 3 architecture, offers investment protection, while enabling vital network applications and capabilities such as multimedia telephony and conferencing, mobile broadband, social networking, wireless LAN access, WebTV, video surveillance, building management systems, and many others. Current switch/routers are cost-effective and have high-performance designs. They deliver the scalability, QoS assurance, resilience, and multimedia communication readiness needed to implement high-value converged network solutions that can scale to meet future growth in traffic.

Additionally, switch/routers can be deployed in MANs and WANs for interconnecting enterprise customers. In this environment, switch/routers may support rich and resilient services like Virtual Router Redundancy Protocol (VRRP), VLAN stacking, and advanced multicast capabilities including Internet Group Management Protocol (IGMP) versions 1/2/3 and Multicast Listener Discovery (MLD) versions 1/2 snooping for controlling multicast traffic, allowing for high-bandwidth content delivery.

Switch/routers can also increase network efficiency and decrease equipment costs through IEEE 802.1Q VLAN tagging. With switch/routers, intra-VLAN traffic can be forwarded at Layer 2 while inter-VLAN traffic at Layer 3. This allows the physical network infrastructure to be shared by multiple subnets, allowing multiple broadcast domains or VLANs to be connected to a single high-speed port on a switch/router (a concept known as one-armed routing). The alternative is to consume one expensive router port for every attached subnet.

All network environments, that require high availability in order to assure service delivery or to guarantee the timely execution of business-critical applications, can derive significant benefits from the features and capabilities that the switch/router provides in networking.

REVIEW QUESTIONS

1. What is the purpose of an IP routing protocol?
2. What is the difference between an Interior Gateway Protocol (IGP) and an Exterior Gateway Routing Protocol (EGP)? Give examples of each.
3. What are the main functions of an Ethernet (Layer 2) switch?
4. What are the main functions of an IP router?
5. What are the main functions of a switch/router (also called a multilayer switch)?
6. Explain briefly the main differences between an enterprise network and a service provider network.
7. Explain briefly the main functions of the access, distribution (also called the aggregation), and core layers in the three-tier network architecture.

REFERENCES

[AWEYA2BK21V1]. James Aweya, *IP Routing Protocols: Fundamentals and Distance Vector Routing Protocols*, CRC Press, Taylor & Francis Group, ISBN 9780367710415, 2021.

[AWEYA2BK21V2]. James Aweya, *IP Routing Protocols: Link-State and Path-Vector Routing Protocols*, CRC Press, Taylor & Francis Group 9780367710361, 2021.

[SCHUDSMIT08]. Gregg Schudel and David J. Smith, *Router Security Strategies: Securing IP Network Traffic Planes*, Cisco Press, 2008.

2 Introducing Multilayer Switching and the Switch/Router

2.1 INTRODUCTION

This chapter presents some key terminology and definitions that help in understanding the role and architectures of switch/routers. This chapter explains the following key terms: network address prefix, network mask, network prefix length, route, path, control plane, data plane, control engine, routing table (also called the Routing Information Base (RIB), Layer 3 topology-based forwarding table (also called the Forwarding Information Base (FIB), Layer 3 route cache, Layer 3 forwarding engine, routing metric, and administrative distance (also called route preference).

We explain how a routing protocol uses a routing metric to select the best path to a network destination when multiple paths exist (i.e., best path selection within a particular routing protocol). We also explain how a router uses the administrative distance to select which route to install in its routing table when multiple routing information sources provide routes to the same destination (i.e., best path selection among routing protocols or routing information sources).

This chapter also traces the evolution of the forwarding features and internal architecture of switch/routers, from the first-generation switch/routers to the current-generation high-performance routers seen in the core or backbone of enterprise and service provider networks.

2.2 KEY TERMINOLOGY AND DEFINITIONS

The terminology and definitions provided here are absolutely necessary for understanding the fundamental concepts discussed here and the subsequent chapters. We focus mainly on the Layer 3 functions of the switch/router here since these are more complex than the Layer 2 functions. Interested readers can refer to [AWEYA1BK18] and [AWEYA2BK19] for more detailed discussion on Layer 2 and Layer 3 functions.

- **Network Address Prefix (or Network Prefix)**: A network address prefix is a contiguous set of the most significant bits in an IP address and represents, collectively, a set of systems within a network. In IPv4 networks, the network portion of an IP4 address is referred to as the "*network part*" which also represents the *network prefix*, while the remaining bits of the address make up the "*host part*" of the address. For a given IPv4 address and network part, the host addresses are selected from the host part. In IPv6

DOI: 10.1201/9781003311249-2

networks, the network portion of an IPv6 address is referred to as the "*subnet prefix*", while the remaining bits represent the "*interface identifier*". The use of Variable Length Subnet Masking (VLSM) [RFC950] [RFC1878], which has replaced the now obsolete classful IPv4 addressing method, allows IPv4 addresses to have arbitrary-length prefixes that do not have to fall on a byte boundary, unlike the classful IPv4 addresses that must fall on an 8-bit boundary (Class A), 16-bit boundary (Class B), or 24-bit boundary (Class C). VLSM allows efficient use of IPv4 addresses since network operators can divide an IPv4 address space into subnets with different IP prefix lengths according to the needs of their networks. In Classless Inter-Domain Routing (CIDR) as discussed below, the network prefix represents the network portion of an IP address and may represent a collection of destinations (i.e., an aggregate or summary address representing a supernet).

- **Network Mask**: A network mask is used to divide an IP address into the network-specific portion and the host- or subnet-specific portion. A network mask, when configured together with an IP address, specifies the boundary between the network prefix and the host or subnet portions of the IP address.
- **Network Prefix Length**: The network prefix length is the number of contiguous bits that make up the network prefix portion of an IP address. The prefix length is equal to the number of contiguous "1" bits in the network mask. For example, the network part of the IPv4 address 192.0.2.130/24 is 24 bits wide, that is, the network prefix length is 24. The network mask is 255.255.255.0, the network prefix is 192.0.2.0, and the host identifier is 0.0.0.130.
- **Route**: A route is defined with respect to a specific next-hop to which a packet may be sent on its way toward the destination. It is defined by a destination network address prefix. A route is the basic unit of information about a specific network destination discovered by the routing protocols and is a candidate for inclusion in the routing table of a routing device. In general, a route is expressed as the n-tuple <*IP address prefix*, *Next Hop* [*...other routing or non-routing protocol attributes...*]> [AWEYA2BK21V1] [RFC4098]. A BGP route is expressed as an n-tuple <*IP Address Prefix*, *Next Hop*, *AS-Path*, [*...other BGP attributes...*]> [AWEYA2BK21V2]. A routing table is a collection of routing entries with each entry representing a route. Simply, a routing table is a collection of routes.
- **Path**: A path is a collection of next-hop data structures each corresponding to a distinct pointer to a destination in the network. Each pointer contains at least one of the following; an outbound interface and an intermediate IP address (e.g., a next-hop IP address). The outbound interface is the interface that should be used to forward packets to a collection of network destinations as described by the network prefix. In this book, we sometimes use the terms route and path interchangeably.
- **Control Plane**: The control plane is a logical component in the routing device that is responsible for exchanging routing information with other routing devices, determining the topology of the network, determining routes to network destinations, and building and maintaining the IP routing

table. The control plane may include signaling protocols such as those used in the Multiprotocol Label Switching (MPLS), for example, Resource Reservation Protocol-Traffic Engineering (RSVP-TE) [RFC3209], and Label Distribution Protocol (LDP) [RFC5036]. In this book, the protocols and tools used to access, configure, monitor, and manage a routing device and its resources (i.e., the housekeeping tools) are considered as part of the control plane and not as a separate logical processing plane usually called the *management plane* [AWEYA1BK18]. Control plane traffic makes up only a small portion of the overall traffic processed by the routing device. The control plane traffic includes exception packets, which are packets that the data plane forwarding mechanisms cannot forward normally because there is not enough forwarding information and/or resources for processing.

- **Data Plane**: The data plane (also referred to as the forwarding plane) is a logical component in the routing device that is responsible for receiving packets, verifying their packet fields, performing destination address lookups, updating packet fields, performing packet field rewrites, and forwarding the packet on its way to the destination. The data plane uses the optimal routing information learned by the control plane and stored in the routing table. A great majority of the traffic processed by the routing device is data plane traffic.

- **Control Engine**: This engine (also referred to as the *routing engine, route processor, control processor, route switch processor*, or *system processor*) runs the routing protocols which construct and maintain the routing tables. It also runs the management and control protocols and routines that manage, monitor, and control device behavior and environmental status, and supports software tools for device management (see Figure 2.1). Routers

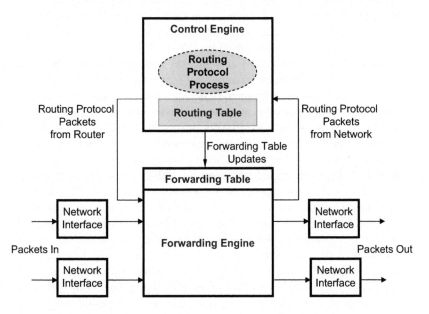

FIGURE 2.1 Routing and forwarding components in a routing device.

use routing protocols to determine routes through internetworks to remote network destinations. Routing protocols allow routers to discover neighboring routers, as well as detect unreachable neighbors and networks. They perform remote network discovery, path computation, and best path selection to reachable remote networks via next-hop nodes.

- **Routing Update**: This is a message sent by a router to other routing devices to indicate the available routes to network destinations and the cost information associated with those routes. Some routing protocols send routing updates at regular intervals (i.e., periodic updates), while others send only after a change in network topology has occurred (i.e., triggered updates).

- **Routing Table**: This database (also referred to as the RIB) is constructed and maintained by a routing protocol running in the router. It describes the network topology from the routing protocol's perspective. The routing table keeps track of routes to all reachable network destinations and the metrics associated with those routes. Different routing protocols use different methods (metrics, algorithms, etc.) to create the contents of their routing tables. Each routing table entry includes at a minimum the following information: destination network prefix address; IP address of the next-hop router; type of network route (such as directly connected route, static route, and dynamic route) and the particular routing protocol that learned the route; routing metric associated with the route and used by the source routing protocol; administrative distance (or route preference) value that is used to select the least-cost route when multiple routing information sources present multiple routes to the same network destination. To keep the size of routing tables, especially, in core routers manageable, address aggregation (or summarization) schemes such as CIDR are used [RFC1517] [RFC1518] [RFC1519] [RFC4632]. IP addresses in CIDR are represented as the tuple *<prefix, length>*, also called aggregates, where *length* indicates the number of bits in the prefix, counting from the left-hand side of the IP address. For example, 192.168.100.0/22, is an aggregate of length 22. Given that some routes such as BGP routes reference only intermediate network addresses (unlike Interior Gateway Protocol (IGP) routes that reference both intermediate network addresses (i.e., next-hop addresses) and corresponding outgoing interfaces), lookups in the RIB may involve recursive lookups to find the outgoing interface associated with an intermediate network address (see the "Routing Table or Routing Information Base (RIB)" section in Chapter 5).

- **Layer 3 Topology-Based Forwarding Table**: This database (simply referred to as the forwarding table, and sometimes called the FIB) is derived/generated from the routing table. It contains the same routing information but structured in a compact and optimized format to be useful for actual packet forwarding (see Figure 2.1). Anytime there is a routing or network topology change, the routing protocol updates the routing table and then those changes are reflected in the forwarding table. The router updates the forwarding table when one of the following occurs: the routing table entry for a network prefix changes, or is removed; the routing table entry for the next-hop changes or is removed; the Address Resolution Protocol (ARP) cache entry for the next-hop times out, changes, or is removed. The routing and forwarding tables contain basically the same information needed for packet forwarding, the forwarding table only removes information that is not directly relevant for packet forwarding (e.g., the routing metric associated with a route is also not listed in the forwarding table). Because of the use of VLSM and CIDR, the address prefixes in the RIB and FIB can be of variable lengths (i.e., contain classless addresses). This means address lookups in the RIB and FIB are based on longest prefix matching (LPM) instead of exact matching. Note that the FIB does not contain recursive routes (and as result does not demand recursive lookups) because all recursive routes in the RIB are resolved before they are installed in the FIB (see the "Resolving Recursive Routes before Installation in the IP Forwarding Table" section in Chapter 5, and "The FIB and Resolved Routes" section in Chapter 6).

- **Layer 3 Route Cache**: This database (also loosely referred to as flow cache, or flow/route cache) maintains forwarding instructions gleaned from recently processed flow or stream of packets that share the same forwarding characteristics (e.g., same destination address). Each entry of the route cache contains at a minimum the destination IP address (i.e., the /32 destination address as carried in the first packet of a flow), the next-hop IP address, and the outbound interface/port. After a successful LPM lookup of the destination address of the first packet of a flow, the actual /32 destination IP address as seen in the packet, the next-hop IP address, the outbound interface, plus other Layer 2 information required for Layer 2 rewrites in the outgoing packet are entered into the route cache. This route cache information is used for faster forwarding of future arriving packets that share the same forwarding characteristics (i.e., subsequent packets of the same flow). Because the route cache entries are based on the actual IP destination addresses of arriving packets which are fixed and not of variable length (i.e., are classful IP addresses), address lookups in the route cache are based on exact (destination address) matching. This exact matching is similar to what is done in Ethernet switches (using the destination MAC address of Ethernet frames), Address Resolution Protocol (ARP) caches (using the destination IP addresses of packets), ATM switches (using the Virtual Path Identifier (VPI) and Virtual Channel Identifier (VCI) of ATM cells), and MPLS Label Edge Routers (LERs) and Label Switch Routers (LSRs) (using the MPLS labels of MPLS packets). Lookups in route caches using exact matching are discussed in detail in the "Exact Matching in IP Route Caches" section in Chapter 6. Recall that lookups in the RIB and FIB are based on LPM.
- **Layer 3 Forwarding Engine**: This engine (simply referred to as forwarding engine) examines specific fields and parameters in arriving packets (including the destination IP address) to determine how packets should be handled as they arrive at the device. It determines the next-hop and outbound interface or port, and updates and performs rewrite operations in packets before they exit the device (see Figure 2.1). The forwarding engine is also responsible for determining which arriving packets are destined for the router itself (local delivery). The process of searching a route cache or

the forwarding table for the next-hop node and outbound port using the IP destination address as the search index or key is referred to as the *address lookup* process.

The control plane and the data plane are discussed further in the "Control Plane and Data Plane in the Router or Switch/Router" section of Chapter 5.

Examples of control engine functions:
The following are examples of the key control engine functions:

- **Routing Protocols**:
  o **Unicast routing protocols**: Routing Information Protocol (RIP), Enhanced Interior Gateway Routing Protocol (EIGRP), Open Shortest Path First (OSPF), Intermediate System to Intermediate System (IS-IS), Border Gateway Protocol (BGP) [AWEYA2BK21V1] [AWEYA2BK21V2].
  o **Multicast Routing Protocols**: Internet Group Management Protocol (IGMP), Multicast Listener Discovery (MLD), Protocol Independent Multicast (PIM) (PIM-Sparse Mode, PIM-Dense Mode, PIM-Source Specific Multicast, Bidirectional-PIM)

- **Network Control Protocols**: Internet Control Message Protocol (ICMP), Address Resolution Protocol (ARP), Dynamic Host Configuration Protocol (DHCP), Domain Name System (DNS), Virtual Router Redundancy Protocol (VRRP)
- **Management Functions**:
  o **Management protocols**: Simple Network Management Protocol (SNMP), Remote MONitoring (RMON), NetFlow, sFlow
  o **Management tools and secure access**: Command-Line Interface (CLI), Graphical User Interface (GUI), Secure Shell (SSH), File Transfer Protocol (FTP), SSH File Transfer Protocol (SFTP), Remote Authentication Dial-In User Service (RADIUS), Terminal Access Controller Access-Control System Plus (TACACS+), syslog

These management functions are discussed in greater detail in Chapters 1 and 2 of Volume 2 of this two-part book.

In all our discussions, the term forwarding table refers to the Layer 3 topology-based forwarding table as explained above. In a more general context, a route cache is considered a type of Layer 3 forwarding table, except it is populated and maintained based on the type of traffic flow mix passing through the router. As noted above, the route cache entries are destination IP address based. The notion of a flow here is a sequence of packets with the same destination IP address (the source IP address of a packet does not matter in this case since forwarding is based solely on the destination address). The traditional definition of a flow is a sequence of packets with the same source and destination IP addresses, the same Transport Layer protocol (TCP or UDP), as well as the same Transport Layer protocol port number.

2.3   ROUTING METRIC AND ADMINISTRATIVE DISTANCE (OR ROUTE PREFERENCE)

A router receives packets from directly attached hosts and other routing devices, and makes forwarding decisions based on the information stored in its routing table. Routers also receive and announce routing information (via routing updates) to allow other routers to maintain network reachability information (in their routing tables). Each router constructs a local routing table containing all the best routes it has learned to all known network destinations.

It is possible for multiple routing information sources to supply routes to the same network destination to a router. The router has to have a way (mechanism) to decide which route among the multiple routes supplied to install in its routing table. The administrative distance (also referred to as the route preference) in the process shown in Figure 2.2 is an integral number that represents the trustworthiness of the routing information source [AWEYA2BK21V1]. The router compares the administrative distance values of all routing information sources to decide which route to prefer.

Administrative distance is often expressed as an integer value from 0 to 255. The higher the administrative distance value, the lower the trustworthiness of the routing information source. A router always prefers a lower value over a higher value. The route with the lowest administrative distance among multiple routes is installed in the routing table. An administrative distance of 0 indicates first priority route, while one

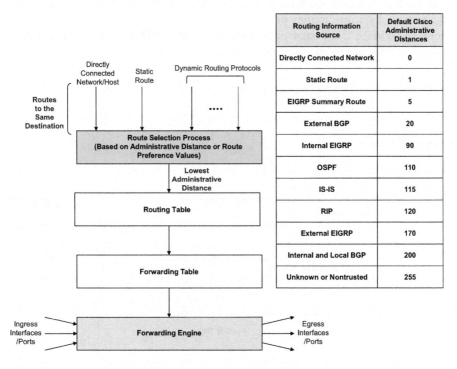

FIGURE 2.2   Routing information sources and route selection.

with a value of 255 indicates no confidence in the routing information source; a router does not install routes with this distance value in the routing table.

As discussed in [AWEYA2BK21V1] [AWEYA2BK21V2], link-state routing protocols such as OSPF and IS-IS are by design, more accurate and trustworthy than distance-vector routing protocols such as RIP. For example, a route supplied by OSPF is preferred over one from RIP. Obviously, routes to directly connected networks and hosts and static route (which are normally configured by a network administrator for obvious reasons even in the presence of dynamic routing protocols) will be preferred over all other routes as indicated by the administrative distances in Figure 2.2.

Directly connected networks and hosts are preferred over all other routes because there is no other better route to get to such destinations other than through the router to which they are directly attached. If there is another route to a directly connected network, then that route must be through other routers and networks; an indirect route that passes through several router hops. This is the main reason why directly connected networks are assigned the default administrative distance value of 0. In Cisco routers, Enhanced Interior Gateway Routing Protocol (EIGRP) routes are generally preferred over all other routes from other Interior Gateway Protocols (IGP).

A router uses the administrative distance when multiple routing information sources supply routes to a given network destination. However, within a routing protocol, the router uses a routing metric to select the best route to any given destination when multiple routes exist to that destination (Figure 2.3). The best route is the one with the lowest cost to that destination. Different routing protocols use different routing metrics. Typically, routing metrics are based on network parameters such as hop count (used in RIP), link bandwidth (typically used in OSPF and EIGRP), delay, data loss, reliability, and so on as discussed in [AWEYA2BK21V1] [AWEYA2BK21V2].

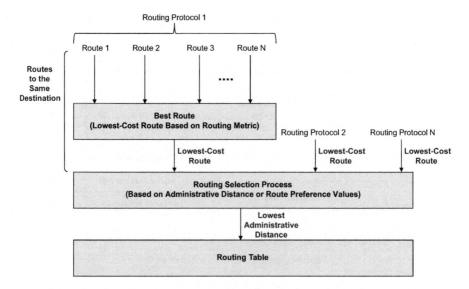

FIGURE 2.3 Route selection within a routing protocol.

Each routing protocol uses its own routing metric type to determine the best route to each destination. This makes comparing routes with different routing metric types not possible. Thus, routers use the administrative distance to address this problem.

In determining the best route to a destination, the routes must consider all available routes to that destination. The router uses a routing protocol to communicate with other routing devices to determine the routes to all known network destinations.

2.4   EVOLUTION OF THE SWITCH/ROUTER

Ethernet continues to evolve as a highly cost-effective and flexible technology for building LANs, MANs, and WANs. The high volume of Ethernet products continues to spur rapidly declining prices and a constant stream of enhancements/innovations. The next sections trace the evolution of the internal architecture and forwarding features of switch/routers [AWEYA1BK18] [AWEYA2BK19] [AWEYA2000] [AWEYA2001]. From a practical marketing and networking perspective, Ethernet technologies have evolved from 10 Mb/s to multi-gigabit speeds. Today, multi-gigabit Ethernet links (10, 25, 40, 100, 200 Gb/s) and technologies are implemented in service provider core networks. Each new generation of Ethernet technology has had a significant impact on the internal architecture and forwarding features of networking devices, as well as the architectures of Ethernet networks.

2.5   FIRST-GENERATION ROUTING DEVICES

The first-generation devices refer to the devices that existed from the late 1980s to the early 1990s (hubs or repeaters, bridges, multi-protocol routers). This period of networking was mostly characterized by the following:

- Earlier in this period, LANs were shared media-based and built using 10 Mb/s Ethernet hubs (resulting in repeated networks). Software-based 2-port bridges were used to partition large repeated segments and to extend the reach of the shared media Ethernet technology. Software-based multi-protocol routers fitted with Ethernet interfaces were also used to segment large networks into subnets. The first-generation routers were basically general computing platforms (workstations) that use *process switching* (Cisco terminology) where a single CPU is solely and directly involved in moving a packet from the inbound interface to the correct outbound interface. As a packet is forwarded through the router, the single CPU is directly involved and responsible for selecting and scheduling the appropriate software processes for moving the packet through the router to the outbound interface. Other than running the routing, control, and management protocols, the CPU is also involved in the following:
  - o Receiving a packet from the inbound interface
  - o Storing the packet in memory
  - o Performing routing table lookups to determine the correct outbound interface and next-hop IP address for the packet which may involve

performing recursive lookups to determine the outbound interface and next-hop IP address (see Chapters 5 and 6)

o Performing IP header updates; IP TTL (Time-to-Live) update and IP header checksum recomputation (see Chapter 5)

o Determining the corresponding Layer 2 address (Ethernet MAC address) of the receiving interface of the next-hop node which may require the router to make an ARP cache lookup or run ARP to determine the Layer 2 address

o Performing Layer 2 rewrites for the outgoing packet; rewrite the MAC address of the outbound interface as the source MAC address of the packet, rewrite the MAC address of the receiving interface of the next-hop node as the destination MAC address of the packet, and recompute the FCS (Frame Check Sequence) of the packet (Ethernet frame)

o Transferring the processed packet to the outbound interface for transmission onto the network toward the next-hop node

• Switched internetworking was ushered in by the mid-1990s with the emergence of Full-Duplex 100 Mb/s (Fast) Ethernet. To exploit the higher bandwidth of 100 Mb/s Ethernet, switch designers decided to take advantage of latest advances in silicon technology. Advances in integrated circuit technology in the form of denser and faster application-specific integrated circuits (ASICs) and field programmable gate arrays (FPGAs) helped set the stage for faster and cheaper switches using Fast Ethernet interfaces. The switches in this era were great improvements over the earlier software-based bridges – the Ethernet switch was born.

• Ethernet hubs (repeaters) in networks were gradually being replaced by the newer 10/100 Mb/s Ethernet switches which were available from a large number of vendors. Within a LAN segment, the ASIC- and FPGA-based Ethernet switches with their 10/100 Mb/s interfaces could operate at wire-speed (line speed). However, as the wire-speed Ethernet switches with Ethernet uplinks became widely deployed, the forwarding capacity of the (software-based) core routers of the day easily became overwhelmed leading to traffic congestion on inter-subnet links; the router's CPU became a packet forwarding bottleneck.

• The software-based routers easily became the bottleneck in the network because they were essentially generally purpose computers running the routing and forwarding software on a centralized CPU (Figure 2.4). This CPU was connected to the network interfaces via a shared bus architecture. The centralized CPU was responsible for forwarding table lookups and packet forwarding, running routing protocols and maintaining routing tables, and performing router management functions.

• Although later in this time period, the packet forwarding process in some routers was enhanced with ASIC hardware assistance, router performance remained bounded by the basic centralized CPU and shared bus architecture. The CPU and shared bus architecture could only handle a few network interfaces and network load.

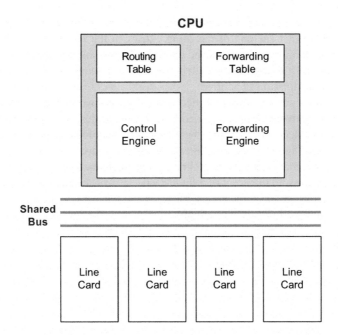

FIGURE 2.4   Routing device with centralized processing.

Many of the network device bottlenecks prevented full utilization of high bandwidth links, resulting in many cases, the devices being incapable of meeting desired service levels. The first-generation switches needed to be upgraded to meet increased network functionality requirements and to eliminate any internal performance bottlenecks.

2.6   SECOND-GENERATION ROUTING DEVICES

The second generation of switches began to appear in the late 1990s with significant improvements to address many of the limitations of the first-generation devices. Advances in silicon technology, ASIC design, and networking technologies ushered in Gigabit Ethernet and the Layer 3 switch or switch/router (which is a switch with the ability to forward both Layer 2 and Layer 3 traffic at wire speed over Fast Ethernet and Gigabit Ethernet ports).

Because of the use of VLSM, the entries in the forwarding table have IP destination address prefixes that have variable lengths. A packet's destination address may match more than one entry in the forwarding table. This means address lookups in the routing table must be based on LPM. In an LPM lookup, the most specific of the matching forwarding table entries, that is, the entry with the longest network prefix (or equivalently, the longest network mask) is selected as the one to be used to forwarding the packet. However, because of the tremendous and continuous growth of networks since the first-generation routing devices were developed, routing table sizes have also increased considerably, making the traditional software-based LPM lookups more time consuming. Thus, designers have always searched for ways to speed up the address lookup process and the packet forwarding rates of routers. It has been recognized that caching of matching entries in the forwarding table after an LPM lookup in a route cache can speed up the lookup of subsequent packets with the same destination IP address; no need to perform LPM lookups repeatedly for packets with the same destination IP address.

The second-generation routing devices typically employed a route cache architecture to hold recently and frequently used IP destination addresses to allow the majority of forwarding decisions to be distributed to ASICs residing on a pool of forwarding engine modules (Figure 2.5), or intelligent line cards (Figure 2.6). Layer 3 forwarding is greatly improved by allowing each line card to maintain a route cache of recently processed IP routes. In the event of a route cache miss or the arrival of a packet with a new destination address, the router forwards the packet via the centralized CPU which maintains the full FIB, also simply referred to as the forwarding table.

The forwarding path through the centralized CPU is sometimes referred to as the "*slow-path*" and that through the route cache as the "*fast-path*". After the slow-path forwarding, during which the device first learns the IP destination address-to-IP next-hop (port) association, the route cache is populated with this association and the fast-path can be used. One example of the architectures in Figure 2.5 is the Cisco 7000 series routers which forward packets via a centralized route cache maintained by a CPU (route processor). An example of the architectures in Figure 2.6 is the Cisco 7500 series routers which forward packets using distributed route caches in line cards called Versatile Interface Processors (VIPs).

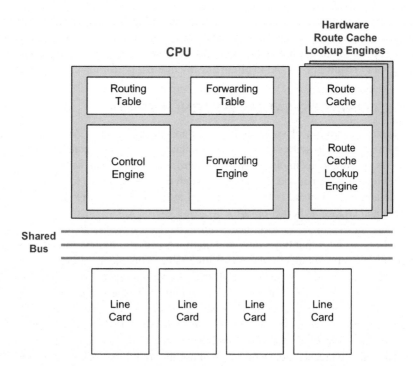

FIGURE 2.5   Routing device with centralized pool of parallel route cache lookup engines.

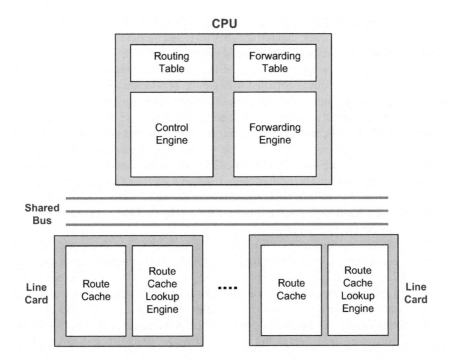

FIGURE 2.6   Routing device with distributed route cache lookup engines in line cards.

The entries in the route cache are periodically aged out to account for changes in the network. The route cache-based switch/router architecture works well in smaller and relatively static networks. However, in large networks and also in core networks, where there are a large number of flows and quite unpredictable and rapidly changing flow patterns, route cache misses can overwhelm the capacity of the centralized CPU (the slow-path). These misses cause many packets to be forwarded via the slow-path resulting in excessive and unpredictable packet delays and delay variations.

With the rapid expansion of the Internet and the World Wide Web (WWW) in the late 1990s, traffic flows in network aggregation points and core networks became continually less predictable and less tolerant for delay, delay variation, and data loss. The chances for service interruptions and catastrophic failure of the network had to be greatly reduced. Network capacities had also moved beyond the Gigabit Ethernet links prevalent in those days. The emergence of 10 Gigabit Ethernet meant that network devices were expected to handle significantly many more flows than a Gigabit Ethernet device.

2.7   THIRD-GENERATION ROUTING DEVICES

By the early 2000s, many networks began carrying high volumes and capacities of real-time traffic (video, voice, interactive media) in addition to the traditional data traffic. Emerging business trends and models created a market need for next-generation Ethernet switch/routers with multi-terabit per second capacity that can allow scalable, full line-rate traffic aggregation. These devices were also required to provide guaranteed low latency, assured data delivery, and the device/network resiliency needed to prevent service interruptions and catastrophic failure.

The early 2000s saw the emergence of the next generation of Ethernet switch/router that employed fully distributed forwarding architectures thereby completely eliminating the slow-path bottleneck created by forwarding packets through the centralized CPU. In the fully distributed switch/router architecture, the centralized route processor CPU is dedicated to running routing protocols and maintaining the FIB, and distributing a copy of the full FIB table to a pool of forwarding engine modules (Figure 2.7) or to each line card (Figure 2.8). These architectures facilitated the development and maintenance of advanced features such as those discussed in the next section. One example of the architectures in Figure 2.7 is the Cisco 10000 with the custom-designed and programmable ASIC called the Toaster that performs Parallel Express Forwarding (PXF).

The architectures in Figure 2.8 typically have dedicated high-capacity control channels between the CPU and the processor or ASIC forwarding engine on each line card. All Layer 2 and Layer 3 packet processing and forwarding decisions are performed by the line card processors or ASIC hardware, allowing all flows to be forwarded with low, consistent latency. In contrast with flow cache-based Layer 2/3 switch architectures, the distributed forwarding architectures do not rely on "slow-path" or software-based forwarding as new flows are initiated. Examples of the

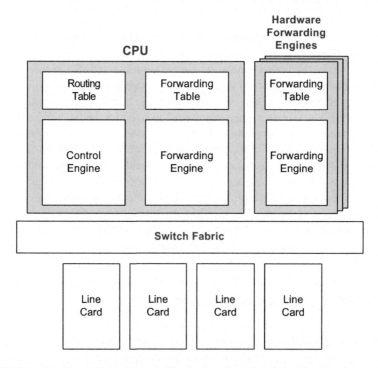

FIGURE 2.7   Routing device with centralized pool of parallel forwarding engines.

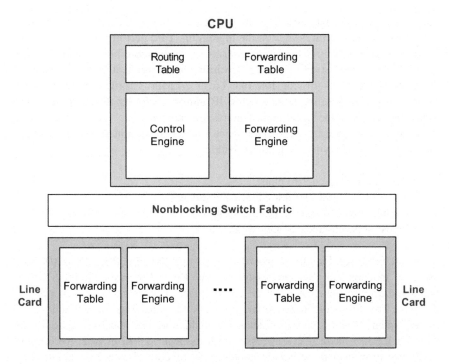

FIGURE 2.8   Routing device with fully distributed forwarding engines in line cards.

architectures in Figure 2.8 are the Cisco 6500 and the Cisco 12000 series routers which use custom-built ASICs in the line cards to perform distributed forwarding.

The fully distributed architectures are well-positioned to handle the next generation of higher capacity Ethernet links. With multi-gigabit Ethernet becoming widely deployed in the core of enterprise and service provider networks, there is still be a high demand of highly reliable Ethernet switch/routers with fully distributed architectures to handle the increasing volumes of network services and applications and also to pave the way for the next-generation higher speeds Ethernet technologies.

2.8   FOURTH (CURRENT) GENERATION ROUTING DEVICES

These next-generation devices were required to both switch and route at full line rate, irrespective of the additional Layer 2/Layer 3 services configured. The integrated Layer 2/Layer 3 features (in the switch/router) considerably reduce network complexity and cost by allowing the consolidation of the legacy access and distribution layers into a single traffic aggregation layer. An important requirement of next-generation switches is their capability to fully exploit 10 and 25 Gigabit Ethernet, while also ensuring a graceful migration path to the next generation of Ethernet at 40 and 100 Gb/s.

High-performance systems like the Cisco 8000, 8300, 8500, and 9000 series routers, Juniper MX and PTX series routers, and Arista 7320X, 7500, and 7800R3 series routers have classification ASICs along with advanced Ternary Content Addressable

Memories (TCAM) that allow line rate lookups within Layer 2 and Layer 3 forwarding tables. Classification ASICs on the ingress line card perform on-the-fly line rate lookups of access control list (ACL) entries for destination, policy, and QoS mappings. This parallelization of packet forwarding and classification processes allows these architectures to provide line-rate Layer 2/Layer 3 forwarding performance independent of forwarding table lengths, IP address prefix lengths, or packet size – even when all ACL, QoS, and other features are enabled.

Systems like those in high-performance routers are designed with a number of high reliability and resiliency features, including:

- multi-processor control plane
- control plane and switching fabric redundancy
- modular switch/router operating system (OS)
- hitless software updates and restarts

Some architectures like the Force10 E-Series [FOR10ESER05] [FOR10TSR06] use three independent microprocessors on each Route Processor Module (RPM) to increase the aggregate capacity of the control plane. This prevents one function (e.g., Spanning Tree Protocol) from depriving processing cycles from other functions (e.g., routing updates), and isolates problems that could otherwise lead to catastrophic failures. Each RPM has a microprocessor with its own pool of error correcting code (ECC)/parity-protected memory to each control function: Layer 2 switching, Layer 3 routing, and system management. The ECC memory also greatly reduces the possibility of parity-related crashes, a fairly common problem in the Internet infrastructure. Control traffic to each microprocessor may also be classified and prioritized, with the lower priority traffic rate-limited as needed to protect critical control tasks.

To ensure that mission critical and other high availability applications continue to work in the event of a network fault, newer generation switch/routers typically support hitless forwarding. The system provides automatic synchronization of configuration information between redundant RPMs in order to minimize recovery time in the case of an RPM or switch fabric module (SFM) failure. With full synchronization and the non-stop "hitless" forwarding feature, should any redundant component fail – whether a line card, switch fabric, RPM, or power supply – the switch/router continues to forward traffic without packet loss. Each line card in the switch/router continues forwarding using its own copy of the FIB, thereby preventing service disruption. Furthermore, the hitless forwarding feature may enable users to perform hitless software upgrades by loading a new version of the router operating system software on a standby RPM, and then initiating an RPM failover to begin operation of the new version.

Current day switch/routers have a variety of serviceability features built into them that are designed to reduce human errors and streamline day-to-day operations for a lower TCO and higher uptime. Human error is a leading cause of network downtime, accounting for a bulk of outages. Switch/routers now include many features for managing, debugging, and troubleshooting the system. Using inline tools, the network operator can easily monitor and diagnose problems without shutting down the switch/router or disrupting application traffic. The system monitors all processes to ensure operations are within normal limits of resource utilization. The system also provides system-wide monitoring for out-of-range environmental conditions and other fault conditions, such as unsynchronized configurations of line cards.

On the issues of robustness, high availability, and security, the next-generation switch/router should be capable of providing non-stop operations even in the face of the full range of software or hardware errors that can possibly occur. Basic high availability features include full hardware redundancy for all major subsystems, including control plane/management modules, switch fabric, power, and cooling. The main features include the following:

- The switch/routers eliminate single points of failure by employing a high degree of subsystem redundancy, including cooling (fans and fan modules), power supplies, switch fabric modules, and RPMs with hitless failovers between redundant RPMs (Figure 2.9). The independence of the data plane allows each line card to continue forwarding using its own copy of the FIB during an RPM failover or software upgrade. As a result, no traffic is lost and all current traffic flows are protected. Additional system robustness is provided by software resiliency features that allow protocol upgrades and restarts without interrupting traffic flow.
- Protocol resiliency includes a number of features to minimize routing table and network re-convergence times and prevent local failures from affecting

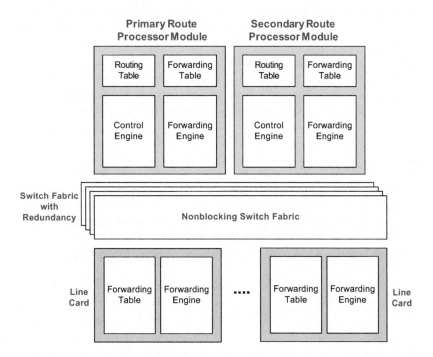

FIGURE 2.9   Routing device with redundant route processors and fully distributed forwarding engines.

the overall network. The switch/router achieves protocol resiliency with the following mechanisms. OSPF and BGP graceful restart mechanisms, for example, allow the data plane to continue forwarding packets while the router's control plane software is reloaded or restarted. The main goal of these restart protocols is to allow a router to restart gracefully without causing a routing flap or transient loops across the network. VRRP eliminates statically defined default gateways as single points of failure in the network. Without VRRP, traffic could be disrupted for a considerable time (seconds or even minutes) while IP configurations are renewed or the default router is restored. Additional network resiliency features include support for IEEE 802.3ad Layer 2 Link Aggregation for trunking and link redundancy, and IEEE 802.1w RSTP or Layer 3 Equal Cost Multi-Path (ECMP) routing to allow for the implementation of highly fault-tolerant networks.

- Software robustness and resiliency can be added by modularization of the control plane to isolate the effects of soft failures. The modularization of switch/router operating system software ensures that all the major functions are separated into distinct processes, with each process supported by its own protected segment of memory. This degree of modularity prevents a fault in one process from affecting or corrupting other processes.
- Wire-speed ACLs can be used to provide restricted access to sensitive data and resources and be able to filter on Layer 2 to Layer 4 information regardless of the forwarding mode of the interface. Security features when implemented contribute to a multi-tiered security solution and to help to protect the switching system from unauthorized access or denial-of-service (DoS) attacks.
- For proactive device management, the switch/router operating system has to constantly perform health checks of switch/router components. When a fault condition is detected, the operating system can alert network operators to take proactive diagnostic action to minimize traffic disruption.

REVIEW QUESTIONS

1. Explain briefly the meaning of the following terms: network prefix, network mask, and network prefix length.
2. Explain briefly the functions that make up the control plane of a routing device.
3. Explain briefly the functions that make up the data (or forwarding) plane of a routing device.
4. Between the routing protocols and management protocols, which of these are absolutely required for a router to function? Explain why.
5. What are the primary functions of the route processor (also called the control or routing engine)?
6. What is a routing update?
7. Explain briefly the main differences between the routing table (RIB) and the forwarding table (FIB).

8. Explain briefly how the entries of a route cache are created.
9. Between the control (or routing) engine and forwarding engine, which of them handles the bulk of the packet traffic arriving at a router and which one requires high-speed packet processing resources? Explain why.
10. Between a route cache and a topology-based forwarding table (FIB), which of these involves relatively more complex lookups during Layer 3 forwarding? Explain why.
11. Between a route cache and a topology-based forwarding table, which table's contents can change more frequently as packets pass through the router from a network? Explain why.
12. Between a route cache based and a forwarding table-based router, which one is more suitable for a network with many short-lived flows? Explain why.
13. What actions will a route cache-based router take when it experiences a route cache miss?
14. What are the advantages of an architecture that decouples the control engine (or control plane) functions from the forwarding engine (or data plane) functions and assigns them to separate processing modules over an architecture that uses a centralized processing unit (CPU) for both functions?
15. Explain briefly the main difference between a routing metric and administrative distance (also called route preference).
16. Explain briefly why directly connected networks and hosts are preferred over all other routes and given an administrative distance value of 0.

REFERENCES

[AWEYA1BK18]. James Aweya, *Switch/Router Architectures: Shared-Bus and Shared-Memory Based Systems*, Wiley-IEEE Press, ISBN 9781119486152, 2018.

[AWEYA2BK19]. James Aweya, *Switch/Router Architectures: Systems with Crossbar Switch Fabrics*, CRC Press, Taylor & Francis Group, ISBN 9780367407858, 2019.

[AWEYA2BK21V1]. James Aweya, *IP Routing Protocols: Fundamentals and Distance Vector Routing Protocols*, CRC Press, Taylor & Francis Group, ISBN 9780367710415, 2021.

[AWEYA2BK21V2]. James Aweya, *IP Routing Protocols: Link-State and Path-Vector Routing Protocols*, CRC Press, Taylor & Francis Group, ISBN 9780367710361, 2021.

[AWEYA2000]. James Aweya, "On the Design of IP Routers. Part 1: Router Architectures," *Journal of Systems Architecture (Elsevier Science)*, Vol. 46, April 2000, pp. 483–511.

[AWEYA2001]. James Aweya, "IP Router Architectures: An Overview," *International Journal of Communication Systems (John Wiley & Sons, Ltd.)*, Vol. 14, No. 5, June 2001, pp. 447–475.

[FOR10ESER05]. Force10 Networks, The Force10 E-Series Architecture, White Paper, 2005.

[FOR10TSR06]. Force10 Networks, Next Generation Terabit Switch/Routers: Transforming Network Architectures, White Paper, 2006.

[RFC950]. J. Mogul and J. Postel, "Internet Standard Subnetting Procedure", *IETF RFC 950*, August 1985.

[RFC1517]. R. Hinden, Ed., "Applicability Statement for the Implementation of Classless Inter-Domain Routing (CIDR)", *IETF RFC 1517*, September 1993.

[RFC1518]. Y. Rekhter, T. Li, "An Architecture for IP Address Allocation with CIDR", *IETF RFC* 1518, September 1993.

[RFC1519]. V. Fuller et al., "Classless Inter-Domain Routing (CIDR): An Address Assignment and Aggregation Strategy," *RFC 1519*, 1993.

[RFC1878]. T. Pummill and B. Manning, "Variable Length Subnet Table For IPv4", *IETF RFC 1878*, December 1995.

[RFC3209]. D. Awduche, L. Berger, D. Gan, T. Li, V. Srinivasan, and G. Swallow, "RSVP-TE: Extensions to RSVP for LSP Tunnels", *IETF RFC 3209*, December 2001.

[RFC4098]. H. Berkowitz, E. Davies, Ed., S. Hares, P. Krishnaswamy, and M. Lepp, "Terminology for Benchmarking BGP Device Convergence in the Control Plane", *IETF RFC 4098*, June 2005.

[RFC4632]. V. Fuller, T. Li, "Classless Inter-domain Routing (CIDR): The Internet Address Assignment and Aggregation Plan", *IETF RFC 4632*, August 2006.

[RFC5036]. L. Andersson, I. Minei, and B. Thomas, Ed., "LDP Specification", *IETF RFC 5036*, October 2001.

3   OSI and TCP/IP Reference Models

3.1 INTRODUCTION

This chapter discusses the two well-known network reference models which describe conceptual or abstract frameworks for communication between entities in communication networks. Network reference models are structured into several layers, with each layer assigned one or more specific networking functions. These functions in turn are implemented as protocols, which provide a system of rules that govern how two or more entities in a communications network can communicate with each other. In real implementations, not every protocol fits perfectly within any specific layer of the reference model; some protocols may spread across two or more layers.

A network reference model provides a conceptual framework or model for communication between network devices, but the model itself is not a method of communication. Actual communication is made possible through the use of communication protocols. Basically, the network reference model presents an abstract model for communication from which real implementations can be created. In data networking, a protocol is a formal set of rules and conventions that governs how network devices exchange information over a network medium. A protocol implements the functions of one or more of the network reference model layers.

Network devices are also designed to adhere to one or more specific layers of the network reference models. It is important to note that network reference models are not physical entities, just abstract models. Devices and protocols operate at specific layers of a reference model, depending on the functions they perform. For example, bridges (or switches) are designed to operate at the Link Layer of the TCP/IP model while routers operate at the TCP/IP Network Layer. Chapter 4 discusses the different network devices and how they align or map to specific layers of the network reference models.

3.2 DEVELOPMENT OF NETWORK REFERENCE MODELS

As computer networks became more widely known and used, the need for a consistent view and standard for implementing end-user and network hardware and software became apparent. The development of a network reference model was undertaken in the 1970s by the International Organization for Standardization (ISO) to address this need. A network reference model serves almost as a blueprint, dictating how communication between entities in the network should happen. This allows network hardware and software products from multiple manufacturers to interoperate when these products are designed to adhere to these models.

As will be discussed below, network reference models are typically structured as a sequence or stack of layers, which are very often collectively implemented and

DOI: 10.1201/9781003311249-3

referred to as a protocol stack. This model of layered network functions is a concep-
tualized view of how network entities should communicate with each other, using
various protocols defined in each layer. Further, each layer is designated or aligned
to a well-defined part of the overall communication system; each layer is designed
for a specific purpose.

Each layer conceptually or logically exists on both the sending and receiving systems involved in a communication (i.e., a "conversation"). Any specific layer on the
reference model on one end-system transmits or receives exactly the same communication object that its remote peer system transmits or receives. Although neighboring
layers in the reference model in a given system can interact with each other, the
activities within any given layer in the system are defined to be self-contained and
operate independently from activities in layers below or above it. Essentially, each
layer on a system is an entity designed to behave independently of the other layers on
the same system. However, each layer on a system is designed to act in concert with
the peer layer on other (remote) systems.

It is not hard to see that without the framework that network reference models
provide, all network hardware and software would have been proprietary. Without
such a common networking framework, the communication industry would have
been littered with many different products, forcing network operators to be locked
into a single vendor's products, and global and worldwide networks like the Internet
as we know it today would have been impractical or even impossible to create.

The network reference models provide flexibility and allow hardware and software vendors to develop products designed for a specific layer of the model. The
segmentation of network functions into well-defined layers splits the overall "communication problem" between networking entities into smaller, more manageable
modules. Each layer of the model is designated clearly understood functions, has
certain responsibilities, and interacts with its neighboring layers and peer layers in
remote systems in a predefined manner.

The following are the two most widely recognized network reference models in
the communications industry:

• The Open Systems Interconnection (OSI) reference model [ISO7498:1984]
• The TCP/IP model (formally called the Department of Defense (DoD)
  model) [DoDARCHM83]

The TCP/IP model was developed by the US DoD and is a more practical model
than the OSI model, and became the basis for the development of the TCP/IP protocol suite (which subsequently, became the collection of protocols that powered the
Internet as we know today).

3.3 OSI REFERENCE MODEL

The OSI Reference Model (ISO standard 7498) was developed by the International
Organization for Standardization (ISO) [ISO7498:1984]. It also uses structured layers for conceptualizing the entities and tasks required for network communication.
The OSI model was developed in the 1970s and published as a standard in 1984 by
the ISO. It was the first network reference model developed for the communication

industry and provided the conceptual framework governing how network entities exchange information across a network. A number of amendments and revisions of ISO standard 7498 have been issued since its first publication in 1984.

As will be shown below, the OSI model describes a structure with seven layers for network activities. Each layer of the OSI model corresponds to a particular set of network functions. One or more protocols are associated with each layer of the model. Each layer represents data transfer functions that can be conveniently grouped together and collectively cooperate to allow entities at that layer and at upper layers, and at remote peer layers to exchange data in a network.

The OSI model structures the protocol layers from the top (Layer 7) to the bottom (Layer 1) as illustrated in Figure 3.1. The OSI model, as a conceptual framework, defines communication operations that are not unique to any particular communication network. The TCP/IP model uses some of the OSI model layers' characteristics or definitions directly or unmodified. TCP/IP model also combines or compacts other layers in the OSI model into more compact or composite layers.

Other network protocol suites, such as Systems Network Architecture (SNA), have eight layers in the model. SNA is IBM's proprietary networking architecture, created in 1974 and is another complete protocol stack for interconnecting computers and their resources (like the TCP/IP protocol suite but is now deprecated).

A range of protocols based on the OSI model were further developed by the ISO; however, these OSI protocols were never widely adopted or implemented by software and hardware system vendors and network operators. Most common modern-day protocol implementations do not align or fit well within the OSI model's layers; thus, making the OSI model has largely become deprecated.

The ISO also developed a complete suite of routing protocols for use in the OSI protocol suite. These include Intermediate System-to-Intermediate System (IS-IS) [ISO10589:2002] [AWEYA2BK21V2], End System-to-Intermediate System (ES-IS) [ISO9542:1988], and Interdomain Routing Protocol (IDRP) [ISO10747:1994]. This chapter does not discuss further any of these protocols.

OSI Model	TCP/IP Protocol Suite	
Application Layer User interface to the communications network	**Application Layer**	Telnet, DNS, TFTP, SNMP, etc.
Presentation Layer Data compression, transformation, syntax, and presentation		
Session Layer Sets up and manages sessions between users		
Transport Layer Creates and manages connections between senders and recipients	**Transport Layer**	TCP, UDP
Network Layer Controls routing of information and packet congestion control	**Network Layer**	IP, ICMP, IGMP
Data Link Layer Ensures error-free transmission by dividing data into frames and acknowledging receipts of frames	**Link Layer**	Network Interface Card and Device Driver
Physical Layer Transmits raw bits over communications channel, ensures 1's and 0's are received correctly		

FIGURE 3.1   OSI model compared to TCP/IP protocol suite.

A given layer in the OSI model (and similarly in the other network reference models) generally communicates with the layer directly above it, the layer directly below it, and the peer layer in other network devices. Each of the seven OSI layers use various forms of control information to communicate with the peer layer in other network devices. A layer exchanges control information consisting of specific requests and instructions with its peer OSI layer. The control information is typically carried in headers and trailers attached to communication messages.

At any given layer, the data that has been passed down from the upper layer has a header prepended to it. A trailer is also appended to the data passed down from the upper layer. An OSI layer is not necessarily required to attach a header or a trailer to data received from an upper layer. The data portion of an information unit at a given OSI layer (see protocol data unit (PDU) below) can potentially contain headers, trailers, and data from all the higher layers.

The information exchange process between a sender and receiver occurs between peer OSI layers in the two systems. As the original user data travels down the layers from the Application Layer, each layer in the source system adds the necessary control information to data. In the destination system, each layer analyzes and removes the control information from that data as it travels up the layers.

The data and control information that is transmitted through the layers of the network reference model and through internetworks, assumes a variety of forms and names. It should be noted that the terms used to refer to these different information formats are not used consistently in the networking industry. These terms are sometimes used interchangeably often leading to confusion about what the terms actually mean. The most common information formats include *frames*, *packets*, *datagrams*, *segments*, *messages*, *cells*, and *data units*:

- **Frame**: A frame is a Data Link Layer information unit that is exchanged between a source and destination that are Data Link Layer entities. A frame consists of a Data Link Layer header and upper-layer data (and possibly a trailer). The header and trailer carry control information that consists of specific requests and instructions for the Data Link Layer entity in the destination system.
- **Packet**: A packet is a Network Layer information unit that is exchanged between a source and destination that are Network Layer entities. A packet is composed of the Network Layer header and upper-layer data (and possibly a trailer). The header and trailer carry control information intended for the Network Layer entity in the destination system. A packet can be generated by Network Layer entities that use a connectionless or connection-oriented service – the type of service does not matter.
- **Datagram**: The term datagram usually refers to a Network Layer information unit whose source and destination are Network Layer entities that use connectionless network service. A datagram has two components, a header and data passed from an upper layer entity. The header contains all the control information sufficient for routing the datagram from the source to the destination without relying on prior exchanges between the two entities and the network. A datagram is also a packet, but a packet that is from a connectionless service.

- **Segment**: The term segment usually refers to a Transport Layer information unit whose source and destination are Transport Layer entities. Typically, a segment consists of a segment header and a data section. The Transport Layer entity accepts data from an upper layer, divides it into chunks, and adds a Transport Layer header creating a segment. The segment is then encapsulated into a Network Layer information unit (a packet), and exchanged with peers.
- **Message**: A message is an upper layer information unit whose source and destination entities exist above the Network Layer (often at the Application Layer).
- **Cell**: In the OSI reference model, the basic information units at the Data Link Layer are generically called frames. In Asynchronous Transfer Mode (ATM), these frames are of a fixed (53 octets or bytes) length and specifically called "cells". A cell is an information unit of a fixed size whose source and destination are Data Link Layer entities. Cells are used in some legacy packet technologies such as ATM and Switched Multimegabit Data Service (SMDS) networks. A cell consists of a header and a payload carrying upper-layer data. The header carries control information to the destination Data Link Layer entity, and is typically 5 bytes long. The payload contains upper-layer data that is typically 48 bytes long.
- **Data Unit**: Data unit is a generic term that does not refer to any particular information unit. A data unit could take the form of a service data units (SDUs) or protocol data units (PDUs). An SDU is a data unit that has been passed down from an upper layer to a lower layer and has not yet been encapsulated into a PDU by the receiving lower layer. The SDU is data that is sent by an entity using the services of a given OSI layer and is transmitted semantically unmodified to a peer service entity at the same layer on another system. A PDU, on the other hand, is the data unit that is sent to the peer protocol layer at the receiving end, as opposed to a lower layer. A PDU can also be a data unit that is generated by a protocol of a given layer and consists of protocol-control information and possibly user data of that layer. For example, Bridge Protocol Data Units (BPDUs) are information units used by the Spanning Tree Protocol (STP), a network protocol that ensures a loop-free topology for Ethernet networks.

Multiplexing is a method by which multiple data streams or channels are combined into a single data stream or channel at the source. The aim is to share the single outgoing channel or medium at the source. The OSI model allows multiplexing to be implemented at any of the layers. Demultiplexing is the process of separating the combined or multiplexed data streams or channels at the destination. One example of multiplexing is when data from multiple applications is combined into a single lower-layer data stream, channel, or packet.

3.3.1 OSI PHYSICAL LAYER

The Physical Layer (Layer 1) defines the characteristics of the network hardware and associated physical transmission medium. This layer defines the mechanical and electrical characteristics of the physical link between the network systems involved in a communication. It also defines the functional and procedural specifications for activating, maintaining, and deactivating the link.

The Physical Layer defines the characteristics of the physical link such as the type of media (copper, optical fiber, wireless), voltage levels, timing of voltage changes, signal encoding, physical data rates, maximum transmission distances, and physical connectors for attaching user systems. Devices such as network interface cards, hubs, repeaters, and cabling types are all considered Physical Layer equipment.

3.3.2 OSI DATA LINK LAYER

The Data Link Layer is responsible for transmitting data across a physical network link. The Data Link Layer provides the functions that work over the Physical Layer and allow the transmission of data across the physical network media (reliable transmission capabilities supported in some specifications). There are different Data Link Layer specifications each defining different network and protocol characteristics, including physical addressing, network topology, error notification, sequencing of frames, and flow control. Some of the most identifying characteristics of the different specifications are described below:

- **Physical Addressing**: Physical addressing defines how devices are addressed at the Data Link Layer. Physical addressing (sometimes called hardware addressing) is different from network addressing. Network addresses (e.g., IP addresses) operate above the physical addressing layer and differentiate between nodes or devices in a network, allowing traffic to be routed through the network (e.g., IP routing). In contrast, physical addressing identifies devices at the link-layer level, differentiating between individual devices on the same physical medium. The primary form of physical addressing is the Media Access Control (MAC) address (e.g., Ethernet MAC addresses).
- **Network Topology**: Network topology specifications identify how devices are linked in a network. Network topology consists of the Data Link Layer specifications that often define how devices are to be physically connected, such as in a bus, star (or hub-and-spoke), or ring topology.
- **Error Notification**: Error notification provides mechanisms for alerting upper-layer protocols that a transmission error has occurred on the physical link. Examples of link-level errors include the loss of a signal, the loss of a clocking signal across serial connections, or the loss of the remote endpoint on a T1 or T3 link.
- **Frame Sequencing**: The frame sequencing capabilities of the Data Link Layer allow frames that are transmitted out of sequence to be reordered (or resequenced) on the receiving end of a transmission. The integrity of the packet may be verified by means of bits in the Layer 2 header, which are transmitted along with the data payload.
- **Flow Control**: Flow control provides mechanisms for regulating the transmission of data so that the receiving device is not overrun with more traffic than it can handle. The Data Link Layer may support flow control

mechanisms the allow devices on a link to detect congestion and notify their downstream and/or upstream neighbors. The neighbor devices may then relay the received congestion information to their higher layer protocols so that the flow of traffic can be adjusted or traffic rerouted.

The Data Link Layer formats the received higher-layer data into frames so that they can be transmitted on the physical wire. This formatting process is referred to as framing or encapsulation. The encapsulation type used is dependent on the underlying data-link/physical technology (such as Ethernet, SONET/SDH, ATM, etc.). Included in this frame is a source and destination hardware (or physical) address.

Hardware addresses usually contain no hierarchy and are usually hard-coded on a device. Each device must have a unique physical or hardware address on the network. The physical address (which is essentially a Layer 2 address) of every network device is unique, fix coded in hardware by its manufacturer and usually never changed.

3.3.3   OSI NETWORK LAYER

In addition to other characteristics, the Network Layer (Layer 3) defines the network addresses, which are logical addresses, and are different from MAC addresses as discussed above. Logical addresses, like IP addresses, are organized as a hierarchy (using a hierarchical address space) and are not hard-coded on devices. A network address is typically based on where a device is in a particular network segment (IP subnet or VLAN), or on logical groupings of network devices that have no particular physical basis. The Network Layer controls the logical addressing of devices, and defines the logical network layout, to allow routers (which are Layer 3 devices) to use this information to determine how to route packets in the network.

The tasks in this layer include the targeted forwarding of packets from a source attached to one network segment (subnet or VLAN) to the destination which is also attached to another network segment (a process called routing). The Network Layer supports the routing functions used for linking individual subnets and networks (facilitating internetworking). Simply, this layer manages data addressing and delivery between individual networks. Routers determine the best path to each destination network and routes the data accordingly.

3.3.4   OSI TRANSPORT LAYER

The Transport Layer (Layer 4) handles the end-to-end transfer of data, that is, from one end-system to another. This layer may include capabilities to ensure that data arrives at its destination without corruption or data loss and in the proper sequence (order). This layer manages the transfer of data and may also assure that the received data is identical to the data transmitted. The Transport Layer in an end-system receives data from the Session Layer, and segments it for transport across the network to another peer Transport Layer in another end-system.

The multiplexing feature of the Transport Layer enables data from several applications (upper-layer protocols) to be combined and transmitted onto a single physical link. Logical or virtual circuits between the communicating entities (applications) are established, maintained, and terminated by the Transport Layer.

There are two types of Transport Layer communication:

- **Connection-oriented**: Connection-oriented service involves three phases; connection establishment, data transfer, and connection termination. Appropriate connection parameters must be agreed upon by both end-systems before a connection is established. This results in connection-oriented network services having more communication overhead than connectionless services.
- **Connectionless**: In a connectionless service, any end-system can send data without the need to establish a connection first. No connection parameters are established before data is sent. An end-system can simply send the data without the added overhead of creating and tearing down a connection.

The parameters that are negotiated by connection-oriented protocols include:

- **Flow Control (Windowing)**: Flow control is the process of regulating the rate of data transmission to ensure that a transmitting device does not overwhelm a slower receiving device with too much data. Flow control mechanisms manage data transmission between the devices so that the sending device does not transmit more traffic than the receiving device can process. Windowing (similar to what is used in TCP) is one example of a flow-control scheme. In windowing, the source device sends data but requires an acknowledgment from the destination after a certain number of packets have been sent. The windowing dictates how much data the source can send between acknowledgments from the destination.
- **Congestion Control**: Congestion control is about controlling the traffic sent into a network so as to avoid resource oversubscription and congestive collapse. This is done by allowing an end-system to take steps to reduce resource overconsumption, such as reducing the rate of sending data.
- **Error-Checking**: Error checking is different from error recovery. Error checking involves utilizing various mechanisms for detecting transmission errors, while error recovery involves taking action, such as requesting that data be retransmitted, to resolve any errors that occur during transmission. A common error-checking scheme is the Cyclic Redundancy Check (CRC), which when applied to received data, detects errors, allowing the receiver to discard corrupted data. Generally, the Transport Layer is responsible for making sure that the data is delivered error-free and in the proper sequence, but the Data Link Layer can also support such capabilities

The Transport Layer segments data into smaller units (called *segments*) for transport. Each segment is then assigned a sequence number, so that the receiving device can reassemble the segments in the right order on arrival to create the original data.

3.3.5   OSI SESSION LAYER

The Session Layer (Layer 5) which functions as the communication control layer, establishes, maintains, and ultimately terminates connections (sessions) between end-systems. Sessions can be full-duplex (i.e., send and receive simultaneously), or half-duplex (i.e., communication in a single direction at a time, send or receive, but not simultaneously). A simplex session sends data in only one direction. The Session Layer provides the necessary controls required for full-duplex or simplex sessions.

Communication sessions between applications located in different end devices involve the use of service requests and service responses. The service requests and service responses exchanged between the communicating entities are coordinated by protocols implemented at the Session Layer. The Session Layer controls the dialog between the two communicating entities including the establishment of security functions such as authentication of entities.

3.3.6   OSI PRESENTATION LAYER

The Presentation Layer (Layer 6) ensures that data is delivered to the receiving system in a format that the system can understand. This layer controls the formatting of user data, whether it is text, video, sound, or an image. The Presentation Layer ensures that data from the sending device can be understood by the receiving end-system.

The Presentation Layer receives Application Layer data and provides a variety of data coding and conversion functions that can be applied to the received data. These Presentation Layer functions ensure that data sent from the Application Layer of the sending system would be readable by the Application Layer of the receiving system. Examples of Presentation Layer coding and conversion schemes include common data representation formats, conversion of character representation formats, common data compression schemes, and common data encryption schemes.

Additionally, the Presentation Layer may be responsible for the compression and encryption of data. The data compression functions at the Presentation Layer ensures that data that is compressed at the source end-system can be properly decompressed at the destination end-system. The data encryption functions ensure that data encrypted at the source end-system can be properly deciphered at the destination end-system.

3.3.7   OSI APPLICATION LAYER

The Application Layer (Layer 7) consists of standard communication services and applications that end-user applications can use. This layer provides the actual interface between the user application (which falls outside the scope of the OSI model) and the lower OSI layers and the network. The user directly interacts with this layer. Some examples of Application Layer services and applications include Telnet, File Transfer Protocol (FTP), and Simple Mail Transfer Protocol (SMTP).

The functions of the Application Layer typically include identifying communication partners on other systems, determining resource availability needed for communication, and synchronizing communication between communication partners. To identify communication partners on remote systems for an application on the local system that has data to transmit, the Application Layer determines the identity and availability of communication partners.

To determine resource availability for communication, the Application Layer must decide whether sufficient network resources exist for the requested communication. To synchronize communication between applications, communication between the entities require cooperation that is managed or coordinated by the Application Layer.

3.4 TCP/IP REFERENCE MODEL

The acronym "TCP/IP" is commonly used for the set of communication protocols that make up the Internet Protocol suite (often also called the Internet model). TCP/IP provides end-to-end connectivity and specifies how data should be packetized, addressed, and transmitted by an end-system, routed in a network, and received at the destination. In the networking literature, many texts use the term "Internet" to describe both the protocol suite itself and the global wide area network that allows for worldwide connectivity and communication. In this book, we use "TCP/IP" to refer specifically to the Internet Protocol suite and "Internet" to refer to the global wide area network and the bodies that govern it.

As described above, the OSI model describes essentially only an idealized network communications framework with only a limited family of protocols developed, a majority of which are not widely used or not used at all. The seven-layer OSI model was the first network reference model developed. However, the OSI model has been largely deprecated, replaced with a more practical model like the TCP/ reference model. TCP/IP which was developed by the US DoD does not directly map to the OSI model. TCP/IP was developed as a more compact model where several OSI layers are combined into a single layer, or certain layers not used at all.

Figure 3.1 shows the four layers of the TCP/IP protocol suite (or reference model). Figure 3.1 shows the layers from the topmost layer (Application Layer) to the bottommost layer (Link Layer). The figure shows the TCP/IP protocol layers and the OSI model equivalents. Figure 3.2 shows examples of the protocols that are available at each layer of the TCP/IP protocol stack.

As data is passed from the user application down the layers of the network reference model (OSI model and TCP/IP model), each of the lower layers adds a header (and sometimes a trailer) containing protocol information specific to that layer. The header plus user data (and possibly, plus trailer) is called a PDU of the layer, and the process of adding the header (and trailer) is called encapsulation. For example, the Transport Layer adds a header containing flow control and sequencing information (in the case of TCP). The Network Layer header adds logical addressing information, and the Link Layer header contains physical (or hardware) addressing and other hardware-specific and layer-specific information.

Each layer on a sending device communicates with the corresponding layer on the receiving device. For example, on the sending device, hardware addressing is placed

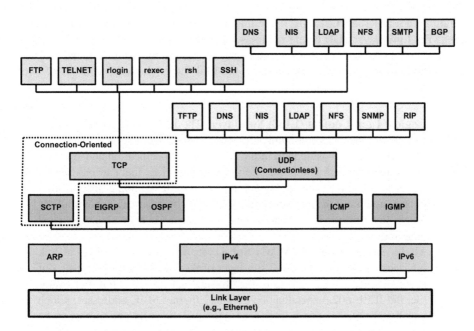

FIGURE 3.2 TCP/IP protocols.

in a Link Layer header. On the receiving device, that Link Layer header is processed and stripped away before it is sent up to the Network Layer and other higher layers.

Specific devices are often identified by the OSI layer or TCP/IP model layer the device operates at, or, more specifically, what header (or PDU) the device processes. For example, Ethernet switches are usually identified as Layer 2 devices, as they process hardware (usually MAC) address information carried in the Link Layer header of a frame. Similarly, routers are identified as Layer 3 devices, as routers examine and process IP addressing information in the Network Layer header of a packet.

3.4.1 TCP/IP LINK LAYER

The Link Layer specifies the characteristics of the hardware participating in the network connectivity in addition to its physical (hardware) addresses. For example, the Link Layer specifies the physical characteristics of the communications media such as those in the IEEE 802.3 specification for Ethernet-based media. The Link Layer performs functions such as adding a packet header (which carries the source and destination physical link addresses and control information) and trailer when a packet is ready for transmission, and transmitting the packet over the physical medium.

The Link Layer also identifies the network protocol type of the data encapsulated in the packet (e.g., IPv4, IPv6, ARP). The Link Layer provides error control and "framing" procedures for the encapsulated data. Examples of Link Layer protocol framing are Ethernet IEEE 802.3 framing and Point-to-Point Protocol (PPP) framing. The Link Layer PDU is referred to as a *frame*.

For protocols such as Ethernet (and legacy ones such as Token Ring and FDDI), in addition to the physical link interface, the Link Layer (Layer 2) consists of two sub-layers:

- **Logical Link Control (LLC) Sub-layer**: The LLC sub-layer of the Data Link Layer manages communications between devices over a physical link of a network. LLC is defined in the IEEE 802.2 specification and supports both connectionless and connection-oriented services that can be used by the Network Layer and higher-layer protocols. IEEE 802.2 specification defines a number of fields in Data Link Layer frames that enable multiple higher-layer protocols to share (or be multiplexed over) a single physical data link.
- **Media Access Control (MAC) Sub-layer**: In addition to physical link interface addressing, the MAC sub-layer of the Data Link Layer consists of protocols that manage access to the physical network medium. For example, the IEEE 802.3 specification defines Ethernet MAC addresses, which enable multiple devices or interfaces to uniquely identify one another at the Data Link Layer.

The LLC sub-layer serves as the intermediary between the physical link and the Network Layer and all higher layer protocols. It ensures that Network Layer protocols like IPv4 and IPv6 can function and interact with peer entities regardless of what type of physical link is being used (Ethernet, Token Ring, FDDI, etc.). Additionally, the LLC sub-layer can use flow-control and error-checking to enhance data transfer reliability, either in conjunction with a Transport Layer protocol with flow-control such as TCP, or with a Transport Layer protocol without flow control such as UDP.

The MAC sub-layer controls access to the physical medium by serving as an arbitrator if multiple devices are competing for transmission on the same physical medium. The IEEE 802 technologies have various methods of accomplishing this. For example, Ethernet can use carrier sense multiple access with collision detection (CSMA/CD) (which is now deprecated and rarely used), and Token Ring utilizes a token mechanism.

An Ethernet frame header has a 2-byte EtherType field that indicates which protocol is encapsulated in the payload of the frame. The receiving end uses the EtherType field value to determine how the payload should be processed. IPv4 packets are carried in Ethernet frames with the EtherType field in the Ethernet frame header set to 0x800 (hexadecimal) which is equivalent to 2048 (decimal). For IPv6 packets, the EtherType field is set to 0x86DD (hexadecimal) which is equivalent to 34525 (decimal). When ARP messages are carried in Ethernet frames, the EtherType field is set to 0x0806 (hexadecimal) which equivalent to 2054 (decimal). When Multiprotocol Label Switching (MPLS) packets are carried in Ethernet frames the EtherType field is set to 0x8847 (for unicast) and 0x8848 (for multicast) which are equivalent to 34887 and 34888 (decimal), respectively.

3.4.2 TCP/IP Network Layer

The Network Layer, also known as the Internet Layer or IP Layer, sends IP packets at the source device, routes packets at the intermediate network devices (i.e., IP routers), and delivers IP packets for the logical network (IP subnet) on which the destination device is attached to. These logical networks are designed based on a hierarchical address structure using identifiers (IP address prefixes) within the address space that identify logical network segments (subnets) and also nodes or interfaces within those logical networks.

The Network Layer is responsible for exchanging IP packets across the logical network boundaries. It defines the addressing and routing structures used in the TCP/IP protocol suite. The primary protocol in this protocol suite is the powerful Internet Protocol (IP) [RFC791], which defines the IP header format, IP addressing, IP routing, and related capabilities as described below.

The Network Layer also includes the Address Resolution Protocol (ARP), the Internet Control Message Protocol (ICMP), and Internet Group Management Protocol (IGMP). The Internet Layer is not only agnostic to the data structures at the Transport Layer, but it also does not distinguish between operation of the various Transport Layer protocols (TCP, UDP, and SCTP). The Network Layer PDU is referred to as a *packet*.

- **IP**: IP and its associated routing protocols are definitely the most significant and highly recognized part of the entire TCP/IP protocol suite. IP is responsible for the following:
  - **IP addressing**: IP performs host and network device addressing and identification. IP addressing conventions (which are different for IPv4 and IPv6) are part of the TCP/IP protocol suite.
  - **Host-to-host communications**: IP via the routing protocols determines the best path a packet must take through a network (consisting of a collection of network segments or subnets) to its destination. The routing is based on the receiving system's IP address carried in the IP packet.
  - **Packet formatting**: IP assembles data from upper protocol layers into units that are known as *datagrams*. These data units are most often commonly referred to simply as *packets*.
  - **Fragmentation**: If a packet is too large for transmission over a network interface, IP on the sending end system (for IPv6), and IP on the sending system or an intermediary network node (for IPv4), divides the packet into smaller fragments, each sent as a complete IP packet in its own right. IP on the receiving system then reassembles the received fragments into the original packet.

When the term "IP" is used in this book, in many cases it applies to both IPv4 and IPv6.

- **ARP**: The Address Resolution Protocol (ARP) [RFC826] is used in discovering the Ethernet MAC address of a remote host interface when the IP address of that interface is known. It assists IP in directing packets to the appropriate receiving system by encapsulating them in Ethernet frames with the correct destination Ethernet MAC addresses. ARP allows for the mapping of a known IP address (32 bits long) to a corresponding Ethernet MAC address (48 bits long).

- **ICMP**: The Internet Control Message Protocol (ICMP) [RFC792] is responsible for detecting and reporting network error conditions on the path between a sender and a receiver. In general, ICMP can report the following conditions:
  - o **Dropped packets**: These are packets that arrive too fast at a device to be processed and are discarded
  - o **Connectivity failure**: This is a condition where a destination system cannot be reached and communication cannot be established
  - o **Redirection**: This is a condition where a network device redirects a sending system to use another router

  ICMP messages are sent over IP using protocol number 1 in the IP header. An IP packet header has a 1-byte Protocol field that indicates which upper-layer protocol is encapsulated in the payload of the IP packet. The receiving end uses the Protocol field value to determine the layer above IP to which the payload should be passed.

- **IGMP**: This protocol is used by hosts and adjacent routers on IP networks to establish multicast group memberships [RFC1112]. It is an integral part of IP multicast. IGMP messages are sent over IP using protocol number 2 in the IP header.

3.4.3 TCP/IP Transport Layer

The TCP/IP Transport Layer is responsible for end-to-end message transfer (independent of the underlying network), error control, segmentation, flow control, congestion control, and establishing process-specific transmission channels for applications (called ports). The Transport Layer establishes a basic data channel that an application uses when it is to exchange data with another application. This layer provides end-to-end connectivity and services that are independent of the structure of user data and the underlying physical network infrastructure used for communicating the data.

The Transport Layer may provide additional features that ensure that packets arrive at a receiving system in sequence and without error, by exchanging acknowledgments of data reception, and retransmitting lost packets. Transport Layer protocols at this level are Transmission Control Protocol (TCP), User Datagram Protocol (UDP), and Stream Control Transmission Protocol (SCTP). TCP and SCTP provide reliable, end-to-end service. UDP provides unreliable datagram service. The Transport Layer PDU is referred to as a *segment* for TCP, and the *datagram* for UDP.

- **TCP**: TCP [RFC793] enables end applications to communicate with each other by providing reliable end-to-end data delivery as if the end applications were connected by a dedicated loss-free circuit. The transmission between the communicating entities consists of the following phases:

- o **Connection establishment**: During connection establishment, the end-systems may reserve resources for the connection. The end-systems also may negotiate and establish certain parameters for data transfer, such as the TCP window size.
- o **Data transfer**: The data transfer phase occurs when the actual data is transmitted over the connection. During data transfer, TCP monitors the transmission for lost packets and handles how they are retransmitted. The protocol is also responsible for arranging the received packets in the right sequence at the receiving end-system before passing the data up the protocol stack.
- o **Connection termination**: When the transfer of data is complete, the end-systems participating in the data exchange terminate the connection and release all resources reserved for the connection.

TCP defines a common Transport Layer header format which includes destination and source port numbers. TCP uses a three-way handshake mechanism to establish a connection-oriented session between two end-systems that want to exchange information. It uses a sliding window for flow control between the systems and supports its own congestion control mechanism.

When TCP receives data from an upper layer, it attaches a header onto the data and transmits it. This header contains many control parameters that help processes on the sending system connect and exchange data reliably with the peer processes on the receiving system. TCP ensures that a packet has reached its destination by establishing an end-to-end connection between sending and receiving systems and ensuring that the packet is actually received (using acknowledgment sent by the receiving system). TCP is therefore considered a "reliable, connection-oriented" Transport Layer protocol. TCP *segments* are sent over IP using protocol number 6 in the IP header.

- **UDP**: UDP [RFC768] also defines a common Transport Layer header format which includes destination and source port numbers. However, unlike TCP, UDP provides a connectionless session between two end systems and provides no reliability or flow control between the systems. UDP provides datagram delivery service and does not establish or verify connections between sending and receiving systems. UDP eliminates the processes of establishing and verifying connections between sender and receiver, which is the main reason why some applications that send small amounts of data use UDP. Also, real-time streaming applications such as voice and video over IP (which will suffer performance degradation under the handshaking process of TCP) use UDP. UDP *datagrams* are sent over IP using protocol number 17 in the IP header.
- **SCTP**: Unlike UDP, SCTP [RFC4960] is a reliable, connection-oriented Transport Layer protocol and provides similar services to applications like TCP. However, SCTP is message-stream-oriented not byte-stream-oriented like TCP, and allows multiple streams to be multiplexed over a single connection. Furthermore, SCTP can support connections between systems that have

more than one IP address, or are *multihomed*. The SCTP connection established between the sending and receiving system is referred to as an *association*. Data sent over the association from sender to receiver is organized in manageable chunks (messages). By SCTP's ability to support multihoming, certain applications, particularly applications used by the telecommunications industry, preferably are designed to run over SCTP, rather than TCP. SCTP messages are sent over IP using protocol number 132 in the IP header.

3.4.4 TCP/IP APPLICATION LAYER

The Application Layer defines standard network applications and Internet services that can be used by end-users. These applications and services work with the Transport Layer protocol described above to exchange data with peer applications and services on other end systems. The Application Layer PDU is referred to as a *message*. The widespread adoption of TCP/IP has resulted in many Application Layer protocols being developed and many more are still being developed. The following are some examples of Application Layer protocols:

- **Standard TCP/IP services such as the `ftp`, `tftp`, and `telnet` commands**:
  - **FTP and Anonymous FTP**: The File Transfer Protocol (FTP) [RFC959] is a utility program that can be used to transfer files to and from a host on a remote network. The protocol includes the well-known `ftp` command and the `in.ftpd` daemon that work on UNIX systems as well as a variety of other (non-UNIX) systems. FTP has features that enable a user to specify the (Domain Name System (DNS), see discussion below) name of the remote host and file transfer command options on the local host's command line. The `in.ftpd` daemon on the remote host then receives and processes the requests from the local host. Unlike the related protocol **rcp**, `ftp` works even when the remote host does not run a UNIX-based operating system. A user login is required to the remote system in order to make an `ftp` connection, unless the remote system has been configured to allow anonymous FTP (where no login is required). FTP uses TCP port number 21 to establish the connection between the two hosts (control/command channel) and TCP port number 20 to transfer data.
  - **Telnet**: The Telnet protocol [RFC854] enables a user using a terminal and terminal-oriented process to access a remote host. Telnet allows the user to log on to the remote host as a regular user and with whatever privileges may have been granted to the specific application and data on that host. The Telnet protocol is a client-server protocol and is implemented as the `telnet` program on local systems and the `in.telnetd` daemon on remote machines. Telnet provides a user interface through which the user can communicate with the remote host on a character-by-character or line-by-line basis. Telnet includes a set of commands that are fully documented in the `telnet(1)` manual (man) page. A man page is an online software documentation for commands and utilities usually found on a UNIX or UNIX-like operating system. The Telnet

client establishes a connection over TCP port number 23, where the Telnet server application (`telnetd`) is listening.

o **TFTP**: The Trivial File Transfer Protocol (TFTP) [RFC1350] is another protocol that allows a client to copy files to and from a remote host. TFTP provides functions that are similar to `ftp`, but `tftp` does not establish `ftp`'s interactive connection. As a result, a user cannot list the contents of a directory or change directories on the remote host. Also, when using `tftp` a user must supply the full name of the file to be copied. A user can access the `tftp(1)` man page to view the `tftp` command set. TFTP uses UDP and always initiates transfer request on UDP port number 69, but the UDP ports used for data transfer are chosen independently by the sender and receiver during the transfer initialization.

- **UNIX "r" commands, such as `rcp` (remote copy), `rlogin` (remote login), `rsh` (remote shell), and `rexec` (remote execution)**: The UNIX "r" commands enable a user to issue commands on a local host that will in turn run on a remote host. The description of these commands is given in the `rcp(1)`, `rlogin(1)`, and `rsh(1)` man pages in the operating system. The rcp command enables a user to copy files to or from a remote host or between two remote hosts (using TCP port number 514 similar to `rsh`). The `rsh` command enables a user to execute a single command on a remote host without having to log in to that host (using TCP port number 514). The `rlogin` command enables a user to log in to another UNIX host on a network (using TCP port number 513). The `rexec` command enables a user to run shell commands on a remote host (using TCP port number 512), but unlike `rsh`, the `rexec` server (`rexecd`) requires login. The `rexec` server authenticates the user using the username and password (unencrypted). The capabilities of the "r" commands have been largely replaced by the Secure Shell (SSH) Protocol [RFC4251] (see Chapter 2 of Volume 2 of this two-part book).
- **Name services, such as NIS and the Domain Name System (DNS)**:
  o **DNS**: The Domain Name System (DNS) [RFC1034] [RFC1035] is the name service provided by the Internet and it is used for mapping more memorizable hostnames to the numerical IP addresses needed for locating and identifying computing devices and services. Simply, the DNS provides hostname to IP address mapping services. DNS focuses on making communication simpler by allowing the use of hostnames instead of numerical IP addresses. DNS works over UDP or TCP. DNS uses UDP port number 53 for a majority of DNS messages, but when the UDP message is greater than 512 bytes, DNS uses TCP port 53 (e.g., for zone transfers). See details on DNS in Chapter 2 of Volume 2.
  o **NIS**: Network Information Service (NIS) [SUNNIS90] is a client–server directory service protocol for maintaining and distributing system configuration information between computing systems on a network such as user and group information, hostnames and addresses, e-mail aliases, and other important information about the network itself. The focus of NIS is to make network administration more manageable by providing centralized control over a variety of network information. Lightweight Directory

Access Protocol (LDAP), discussed below, is now a more widely used directory service and has largely replaced NIS. The NIS server listens on port number 111 (for TCP and UDP), 714 (for TCP), and 711 (for UDP).

- **Directory services (LDAP)**: Similar to NIS, LDAP [RFC4510] is an application protocol that allows a user to access and maintain distributed directory information services over an IP network. It allows the sharing of information about users, systems, networks, services, and applications in a network. LDAP is widely used for directory information access and is the underlying protocol for directory access at many businesses and organizations; it is in a variety of systems such as email systems, Web systems, and enterprise applications. LDAP focuses on simplifying directory management and reducing the cost of network administration by enabling central management of users, groups, devices, and other data. LDAP works over TCP or UDP using the well-known port number 389. An LDAP client starts an LDAP session to an LDAP server on TCP or UDP port 389 for insecure sessions or port 636 for secure sessions; LDAP server listens on these ports.
- **File services, such as the NFS service**: The Network File System (NFS) [RFC1813] is an Application Layer protocol that provides file services for a TCP/IP network. It is a client/server application that allows a computer user to view and optionally store and update files on a remote computer as if those files were stored on the user's own computer. The user's system needs to support an NFS client and the remote computer needs the NFS server. The NFS server works over port number 111 (for TCP and UDP) and 2049 (TCP and UDP).
- **Simple Network Management Protocol (SNMP), which enables network management**: SNMP [RFC 1157] is an application protocol that enables a network manager, a variety of tools, and applications, monitors, collects information, and manages devices in an IP network. For example, SNMP enables a user to view the topology of a network and the status of key devices in the network. Organizations mostly use SNMP in network management systems to monitor network devices for conditions that require administrative attention. SNMP also enables the user to obtain statistics about the performance of a network and to use appropriate software to display the information on a graphical user interface (GUI). Most network management packages implement SNMP as a standard feature. SNMP messages are carried in UDP datagrams, and an SNMP agent (in a managed device) receives SNMP requests on UDP port 161. An SNMP manager may use any available UDP port (i.e., source port) to send SNMP requests to UDP port 161 on the SNMP agent. See details on SNMP in Chapter 5 of this volume and Chapter 2 of Volume 2.
- **SMTP for electronic mail transmission**: SMTP [RFC5321] is used by mail servers and other message transfer agents to send and receive mail messages over TCP port number 25 for plaintext and 587 for encrypted communications. User-level email clients typically use SMTP over TCP port 587 or 465 for sending messages to a mail server for relaying, but use

Internet Message Access Protocol (IMAP) [RFC9051] for retrieving messages from the mail server using TCP port 143 for insecure sessions and 993 for secure sessions; IMAP server listens on these ports.

• **Routing protocols such as Routing Information Protocol (RIP)**: A routing protocol specifies how routers communicate with each other, disseminating information that enables them to determine the best to network destinations. Each routing protocol uses appropriate algorithms and routing metrics to determine the best routes. Each router has a priori knowledge only of networks directly attached to it. A routing protocol shares this information first among immediate neighbors, and then this information is propagated throughout the network. This way, routers acquire knowledge of the topology of the network. Both RIPv1 and RIPv2 send and receive messages over UDP well-known port number 520 [RFC1058] [RFC2453]. Border Gateway Protocol (BGP) sends traffic and communicates with neighbors over TCP port 179 [RFC4271]. Note that, unlike RIP and BGP, both Enhanced Interior Gateway Routing Protocol (EIGRP) and (OSPF) operate directly over IP. EIGRP packets are carried (encapsulated) directly in IP using protocol number 88 in the IP header [RFC7868]. OSPF operates directly over IP using protocol number 89 [RFC2328].

REVIEW QUESTIONS

1. What is a network reference model?
2. What is a communication protocol?
3. Explain briefly why the TCP/IP reference model is considered a more practical model than the OSI reference model.
4. Name the main layers of the TCP/IP reference model and explain briefly the main functions at each layer.
5. Explain the main differences between a physical address (also called a hardware address) and a Network Layer address.
6. What is the purpose of flow control in a network?
7. Explain briefly what windowing is as a flow control mechanism.
8. Explain the difference between connection-oriented service and connectionless service.
9. Explain briefly the three phases of TCP data transfer.
10. Why is UDP not suitable for real-time streaming data transfers like streaming voice and video?

REFERENCES

[AWEYA2BK21V2]. James Aweya, IP Routing Protocols: *Link-State and Path-Vector Routing Protocols*, CRC Press, Taylor & Francis Group, ISBN 9780367710361, 2021.
[DoDARCHM83]. Vinton G. Cerf, and Edward Cain, "The DoD Internet Architecture Model", Computer Networks, Vol. 7, North-Holland, 1983, pp. 307–318.
[ISO7498:1984]. ISO 7498:1984 – Information Processing Systems – Open Systems Interconnection – Basic Reference Model, October 1984.

[ISO9542:1988]. ISO 9542:1988 – Information Processing Systems – Telecommunications and Information Exchange Between Systems – End system to Intermediate System Routing Exchange Protocol for use in Conjunction with the Protocol for Providing the Connectionless-Mode Network Service (ISO 8473), August 1988.

[ISO10589:2002]. ISO/IEC 10589:2002 – Information Technology – Telecommunications and Information Exchange between Systems – Intermediate System to Intermediate System Intra-Domain Routing Information Exchange Protocol for use in Conjunction with the Protocol for Providing the Connectionless-Mode Network Service (ISO 8473)", International Organization for Standardization (ISO). November 2002.

[ISO10747:1994]. ISO/IEC 10747:1994 – Information technology – Telecommunications and Information Exchange Between Systems – Protocol for Exchange of Inter-Domain Routing Information Among Intermediate Systems to Support Forwarding of ISO 8473 PDUs, October 1994.

[RFC768]. *IETF RFC*, User Datagram Protocol, August 1980.

[RFC791]. *IETF RFC 791*, Internet Protocol, September 1981.

[RFC792]. *IETF RFC 792*, Internet Control Message Protocol, September 1981.

[RFC793]. *IETF RFC 793*, Transmission Control Protocol, September 1981.

[RFC826]. David C. Plummer, "An Ethernet Address Resolution Protocol", *IETF RFC 826*, November 1982.

[RFC854]. J. Postel and J. Reynolds, "Telnet Protocol Specification", *IETF RFC 854*, May 1983.

[RFC959]. J. Postel and J. Reynolds, "File Transfer Protocol (FTP)", *IETF RFC 959*, October 1985.

[RFC1034]. P. Mockapetris, "Domain Names - Concepts and Facilities", *IETF RFC 1034*, November 1987.

[RFC1035]. P. Mockapetris, "Domain Names - Implementation and Specification", *IETF RFC 1035*, November 1987.

[RFC1058]. C. Hedrick, "Routing Information Protocol", *IETF RFC 1058*, June 1988.

[RFC1112]. S. Deering, "Host Extensions for IP Multicasting", *IETF RFC 1112*, August 1989.

[RFC1157]. J. Case, M. Fedor, M. Schoffstall, and J. Davin, "A Simple Network Management Protocol (SNMP)", *IETF RFC 1157*, May 1990.

[RFC1350]. K. Sollins, "Trivial File Transfer Protocol (Revision 2)", *IETF RFC 1350*, July 1992.

[RFC1813]. B. Callaghan, B. Pawlowski, and P. Staubach, "NFS Version 3 Protocol Specification", *IETF RFC 1813*, June 1995.

[RFC2328]. J. Moy, "OSPF Version 2", *IETF RFC 2328*, April 1998.

[RFC2453]. G. Malkin, "RIP Version 2", *IETF RFC 2453*, November 1998.

[RFC4251]. T. Ylonen, and C. Lonvick, Ed., "The Secure Shell (SSH) Protocol Architecture", *IETF RFC 4251*, January 2006.

[RFC4271]. Y. Rekhter, T. Li, and S. Hares, Ed., "A Border Gateway Protocol 4 (BGP-4)", *IETF RFC 4271*, January 2006.

[RFC4510]. K. Zeilenga, Ed., "Lightweight Directory Access Protocol (LDAP): Technical Specification Road Map", *IETF RFC 4510*, June 2006.

[RFC4960]. R. Stewart, Ed., "Stream Control Transmission Protocol", *IETF RFC*, September 2007.

[RFC5321]. J. Klensin, "Simple Mail Transfer Protocol", *IETF RFC 5321*, October 2008.

[RFC7868]. D. Savage, J. Ng, S. Moore, D. Slice, P. Paluch, and R. White, "Cisco's Enhanced Interior Gateway Routing Protocol (EIGRP)", *IETF RFC 7868*, May 2016.

[RFC9051]. A. Melnikov, and B. Leiba Ed., "Internet Message Access Protocol (IMAP) - Version 4rev2", *IETF RFC 9051*, August 2021.

[SUNNIS90]. Sun Microsystems, "System and Network Administration", March 1990.

4 Mapping Network Device Functions to the OSI Reference Model

4.1 INTRODUCTION

As discussed in Chapter 3 of this volume, network reference models split the tasks involved with exchanging information between communicating entities in a network into smaller, more manageable task groups or modules called layers. A task or group of tasks is assigned to each of the layers. Each layer is defined to be reasonably self-contained so that the tasks (i.e., network functions) assigned to it can be implemented independently without being constrained by the other layers. This allows the services offered by any one layer in the model to be updated or modified without adversely affecting the other layers. This chapter discusses the various network devices (repeaters (also called hubs), Ethernet switches, routers, switch/routers, and web (or content) switches), according to the OSI layer at which they operate. Hubs and repeaters are used interchangeably, similar to bridges and switches.

Traditionally, the networking industry has categorized network devices such as bridges (or switches) and routers by the OSI model layer at which they operate and the role they play in a network. As discussed below, bridges and switches operate at Layer 2, and they are used for forwarding traffic within LANs or virtual LANs (VLANs). Traditionally, bridges and switches operated by forwarding traffic based on Layer 2 addresses. Bridges and switches and the LAN protocols they run, operate at the physical and data link layers of the OSI model and provide communication over the various LAN media.

Routers operate at Layer 3 and perform route calculations and packet forwarding based on Layer 3 addresses. They are used for interconnecting separate IP subnets (or VLANs) and route packets across internetworks in a hop-by-hop manner. Traditionally, routers operated solely on Layer 3 addresses. As discussed in previous chapters, a multilayer switch (or switch/router) is simply the integration of the traditional Layer 2 switching and the Layer 3 routing and forwarding capabilities into a single product, usually through a hardware implementation to allow for high-speed packet forwarding. This chapter discusses the various network devices according to the OSI layer at which they operate. The main difference between a repeater, switch, router, and switch/router, is how many OSI layers are implemented in each of these devices.

4.2 REPEATER AND THE OSI MODEL

The discussion here is limited to Ethernet repeaters or hubs. A repeater, commonly called a hub, allows multiple end-user devices to be connected to the same physical LAN segment. It is a device for connecting multiple Ethernet devices together and

DOI: 10.1201/9781003311249-4

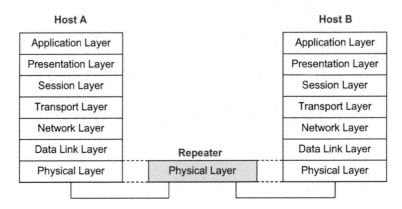

FIGURE 4.1   Repeater (Hub) and the OSI model.

making them act as a single physical LAN segment. The devices on that segment share the total available bandwidth among themselves; for example, a 100 Mb/s hub will have 100 Mb/s bandwidth available to all user devices connected to the hub. In reference to the OSI model, a hub is considered a Layer 1 (Physical Layer) device (Figure 4.1).

A hub senses the electrical signal (on the wire attached to the port) on the LAN segment it is connected to and passes this signal along to the other ports [KANDJAY98] [SEIFR1998]. A hub has multiple input/output (I/O) ports, such that, a signal introduced at the input of any port appears at the output of every port except the original source port. Ethernet hubs also participate in Ethernet collision detection, allowing a hub to forward a jam signal to all other ports if it detects media access collision.

A multiport hub works by repeating bits received from one of its ports to all other ports. It can detect Physical Layer "packet" start (preamble), idle line (interpacket gap), and sense access collision which it also propagates to other ports by sending a jam signal. A hub cannot further examine or manage any of the traffic that comes through it – any packet entering any port is rebroadcast on all other ports. Essentially, a repeater provides signal regeneration by detecting a signal on an incoming port, cleaning up and restoring this signal to its original shape and amplitude, and then retransmitting (i.e., repeating) this restored signal on all ports except the port on which the signal was received.

The simple hub/repeater has no memory in which to store any data – a packet must be transmitted while it is received or is lost when a collision occurs (the sender should detect this and retry the transmission). Due to this property, hubs can only run in half-duplex mode. Consequently, because of the larger collision domain they create, packet collisions are more frequent in networks connected using hubs than in networks connected using more sophisticated devices. A network built using switches does not have these limitations.

Hubs are now largely obsolete, having been replaced by network bridges (switches) except in very old installations or specialized applications. Hubs have been defined for Gigabit Ethernet but commercial products have failed to appear due to the industry's transition to switching. There were other repeater types such as full-duplex buffered repeaters and managed repeaters [CUNNLAN99], which can all be considered as earlier attempts at developing Layer 2 forwarding devices (bridges or switches) as discussed in the next section.

4.3   BRIDGE (OR SWITCH) AND THE OSI MODEL

The bridge, which has been developed to address the limitations of the hub, can physically replace the hub in a network. The terms bridge and switch are used interchangeably in many parts of this chapter and rest of the book. A bridge allows multiple devices to be connected to the same network, just like a hub does. But a bridge allows each connected device to have dedicated bandwidth instead of shared bandwidth. The bandwidth between the bridge and the device is reserved for communication to and from that device alone.

An Ethernet LAN is traditionally implemented as a broadcast domain. This means, all hosts that are connected to the LAN will receive all broadcast transmissions on the LAN. Each host on the LAN uses the destination Media Access Control (MAC) address of the Ethernet frame to determine which frame it should receive and process. Hosts (i.e., end-systems) learn the MAC addresses on a LAN segment through the Address Resolution Protocol (ARP) [RFC826] while bridges learn and filter MAC addresses on the fly.

Unlike a hub, a bridge can greatly increase the available bandwidth in the network, which can lead to improved network performance. A hub just passes electrical signals along (other ports), while a bridge assembles the signals (bits) into a (Layer 2) frame and then makes a decision on how to forward the frame. After a bridge has determined how to forward a received frame, it passes the frame out the appropriate port (or ports).

Bridging is the simplest and earliest form of packet forwarding in LANs. Bridges operate at the Data Link Layer of the OSI network model (see Figure 4.2) and make their forwarding decisions based on the MAC address information carried in Ethernet frames. By reading the source MAC address for each packet received on a port, bridges learn where users are located on the network. The MAC address-to-port correlation is then stored in a local address table for future reference. This *Layer 2 forwarding table* is also referred to as the *Filtering Database* or *MAC address table*.

Because the functions of a bridge (or switch) are relatively simple and less complex than the routing and forwarding processes in routers and switch/routers, these functions are discussed in greater detail in this chapter, with only limited discussions provided in other chapters only when needed. The concepts presented here provide the foundation for all the Layer 2 forwarding operations in switch/routers discussed in later chapters and in Volume 2 of this two-part book.

4.3.1   BRIDGE LOGICAL REFERENCE MODEL

A bridge may have two or more ports, with each port having a MAC and Logical Link Control (LLC) layer. The MAC and LLC layers are essentially the same as in routers, switch/routers, end-systems, and other network devices. Each port of a

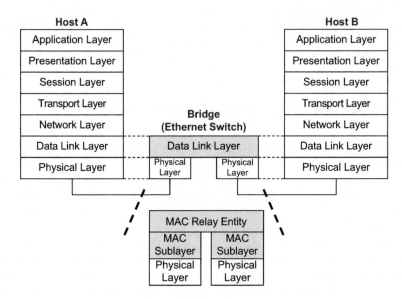

FIGURE 4.2   Bridge (Ethernet Switch) and the OSI model.

bridge has its own MAC address. The specifics of the Physical Layer (PHY) of an Ethernet bridge are defined by one of various Ethernet PHY standards (see Chapter 6 of Volume 2).

Figure 4.3 shows the main elements of the logical reference model of a bridge (or switch). This model consists of a MAC and LLC for each bridge port (see MAC and LLC in Chapter 3), a MAC Relay Entity, and higher layer entities (that include the Bridge Protocol Entity, Spanning Tree Protocol Entity, Bridge Management Entity, etc.) [IEEE802.1Q05]:

- **MAC Relay Entity**: This entity is responsible for transferring (relaying) Ethernet frames at the MAC level between any two Ethernet LAN segments. This entity handles the relaying of frames between bridge ports (i.e., it interconnects bridge ports), frame filtering, and MAC address learning and information storage.
- **Bridge Protocol Entity**: This entity supports management operations for controlling the states of each bridge port.
- **Spanning Tree Protocol Entity**: This entity handles the operation of the Spanning Tree Algorithm and the Spanning Tree Protocol, the reception and transmission of Bridge Protocol Data Units (BPDUs), and the modification of bridge port state information during the process of determining the active Spanning Tree topology of the Ethernet LAN.
  - ○ The Spanning Tree Protocol (which has various versions) is responsible for building a logical loop-free topology for an Ethernet LAN to prevent bridging loops and endless propagation of broadcast traffic in the LAN. The Spanning Tree Protocol Entity is responsible for the notification of

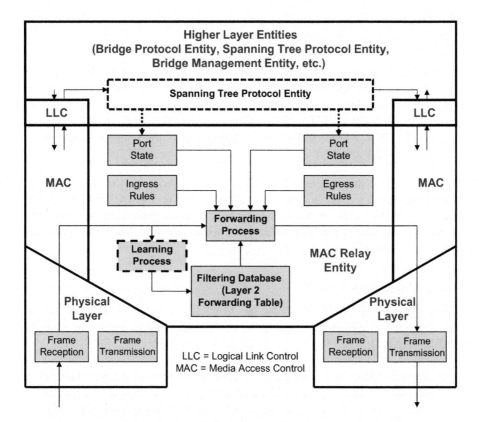

FIGURE 4.3   Layer 2 forwarding (relaying) function.

the Filtering Database of changes in the active topology of the Ethernet
LAN as signaled by the Spanning Tree Protocol.

o BPDUs are special Ethernet frames that contain information about the
Spanning Tree Protocol and are transmitted across the Ethernet LAN to
allow bridges to detect loops in the network topology. A BPDU contains
information regarding bridges in the LAN, bridge ports, port priorities,
and MAC addresses. The information contained in the BPDUs is used
to configure and maintain the Spanning Tree topology for the LAN. A
bridging loop occurs when there is more than one active bridged path
from a node on one LAN segment to another node on another LAN
segment.

• **Bridge Management Entity**: This entity identifies the objects that com-
prise the managed resources of a bridge, namely, the bridge configuration
and the port configuration for each bridge port.

As shown in Figure 4.3, the two main paths through a bridge are the path through
the MAC Relay Entity and the path through one of the higher layer entities (e.g., the

Spanning Tree Protocol Entity). The particular path an arriving frame takes is determined as follows [CUNNLAN99]:

- If the destination MAC address of the received Ethernet frame is *different from* the MAC address of the receiving bridge port, the frame is sent to the MAC Relay Entity for processing.
- If the destination MAC address of the received Ethernet frame is *the same as* the MAC address of the receiving bridge port, the frame is sent to the LLC for processing.

Received frames that require higher layer protocol processing in a bridge (e.g., SNMP messages) are addressed to the receiving bridge port's MAC, which forwards them via the LLC to the appropriate higher layer protocol entity for processing. This means frames carrying the MAC address of a bridge port (e.g., SNMP messages) are acted upon within the bridge itself.

4.3.2   MAC Relay Entity

Figure 4.3 further shows details of the MAC Relay Entity which contains all the components needed for Layer 2 forwarding of Ethernet frames in a bridge [IEEE802.1Q05]:

- **Ingress Rules**: These rules are used for the classification of received frames to determine which specific VLANs they belong to. These rules are used to determine which frames are untagged, priority-tagged, or VLAN-tagged as well as determine which frames should be discarded or accepted for processing.
- **Forwarding Process**: This consists of rules that decide whether to filter or forward a received frame between bridge ports. This process implements the forwarding decisions/rules governing how each frame should be forwarded as determined by the current VLAN topology, end-station location information, and configured management controls. The Forwarding Process enforces a loop-free active topology for each VLAN, performs filtering of frames using their VLAN IDs and destination MAC Addresses, and forwards/relays received frames that satisfy forwarding rules to other bridge ports. The Forwarding Process compares the destination MAC address and VLAN ID of a received frame with the entries in the Filtering Database, and if a matching entry is found, it conditionally forwards the frame to the indicated bridge port(s).
- **Learning Process**: This process examines the source MAC addresses of Ethernet frames received on each bridge port, and updates the Filtering Database. The main purpose of this process is to determine which bridge port a particular source MAC address and VLAN ID is associated with and to update the Filtering Database accordingly. The Learning Process monitors the source MAC address of each frame received on a bridge port and conditionally updates the Filtering Database depending on the state of the port (see port states below).

- **Filtering Database**: This database (also called the Layer 2 forwarding table or MAC address table) holds the filtering information that is used by the Forwarding Process to decide whether a frame with a given VLAN ID and destination MAC address can be forwarded to a particular bridge port. The entries of this database can be either *static* and explicitly configured by management action (the network administrator), or *dynamic* and automatically entered by the normal operation of the bridge and the protocols it supports. This database contains both *individual* and *group* MAC addresses of end-stations and bridge ports, and *logical* IDs (VLAN IDs). The Learning Process is responsible for the creation, update, and removal of dynamic entries. Addresses can also be inserted into or deleted from the Filtering Database by management action (i.e., manually).
- **Egress Rules**: These are rules that decide if a frame passed by the Forwarding Process must be sent untagged or tagged. These rules determine how frames should be queued for transmission through the selected device ports, management of queued frames, how frames are selected for transmission (scheduling policy), and determination of the appropriate format type for outgoing frames (untagged or VLAN-tagged).
- **Bridge Port States**: The *port state* of each bridge port is an operational state that governs whether the port forwards frames on the basis of classifying them as belonging to a given VLAN and whether the port learns from received source MAC addresses. Simply, the port state identifies the operational state of each port. The following are the primary port states in [IEEE802.1Q05] (and also in the Rapid Spanning Tree Protocol (RSTP) which is standardized in IEEE 802.1w):
  - **Discarding**: In this state, the port is not allowed to participate in the forwarding (relaying) of frames. Frames received on a port that is in the *Discarding* state are discarded to prevent duplicate frames from propagating and circulating endlessly in the LAN. The Spanning Tree Protocol determines which bridge ports should be placed in the *Discarding* port state in the LAN when calculating the active Spanning Tree topology. A *Discarding* port is one that would cause a switching loop if it were placed in the *Forwarding* state. To prevent looped paths, a *Discarding* port does not send or receive frames; it is excluded from forwarding and learning from MAC frames. However, BPDUs are still received on a port in *Discarding* state. When a bridge port is freshly connected to a LAN, the port will first enter the *Discarding* state.
  - **Learning**: In this state, the port does not yet forward frames, but it does learn source MAC addresses from received frames and adds them to the Filtering Database. The port has MAC address learning enabled but frame forwarding disabled.
  - **Forwarding**: In this state, the port is part of the active Spanning Tree topology and is participating in the forwarding/relaying of frames. In this state, the port is both learning and forwarding frames. A port in this state is in normal operation and is receiving and forwarding frames. The port still monitors incoming BPDUs that may indicate if it should return

to the *Discarding* state in RSTP (or *blocking* state in STP) to prevent a switching loop from occurring.

o **Disabled**: In this state, the port has been disabled by network management action and is not considered in the active Spanning Tree topology calculations. The *Disabled* state represents exclusion of the port from the active topology. In this state, no frames are sent over the port.

IEEE 802.1D [IEEE802.1D04] and the Spanning Tree Protocol (STP) define five STP bridge port states: *disabled, blocking, listening, learning,* and *forwarding.* The *learning* and *forwarding* states here correspond exactly to the *Learning* and *Forwarding* port states described above and specified in [IEEE802.1Q05] and RSTP. The *disabled, blocking,* and *listening* in STP all correspond to the *Discarding* port state in [IEEE802.1Q05] and RSTP; these mainly serve to distinguish reasons for discarding frames in IEEE 802.1D and STP.

It is important to note that the operation of the Forwarding and Learning processes is the same in both [IEEE802.1Q05] and RSTP, and IEEE 802.1D and STP. RSTP has only four port states (*Disabled, Discarding, Learning,* and *Forwarding*), while STP has five port states (*disabled, blocking, listening, learning,* and *forwarding*). The first letters of the RSTP port states are capitalized here only to aid readers distinguish them from those of STP (note that the learning and forwarding states are exactly the same in both protocols). The STP *blocking* and *listening* states are defined as follows:

- *Blocking* represents exclusion of the port from the active topology by the Spanning Tree Algorithm. When a bridge port is connected to a LAN segment, it will first enter the *blocking* state.
- *Listening* represents a port that the Spanning Tree Algorithm has selected to be part of the active topology (i.e., to be part of computing a Root Port or Designated Port role) but is temporarily discarding frames to guard against loops or incorrect learning. In this state, the bridge will listen for and send BPDUs. In the listening state, the bridge port processes BPDUs and waits for possible new information that would cause it to return to the *blocking* state. The port does not send information to populate the Filtering Database and it does not forward frames.

When a bridge is first attached to a LAN segment, it will not immediately start to forward frames. It will instead go through a number of port states while it processes BPDUs and determines the topology of the Ethernet LAN. RSTP (defined as IEEE 802.1w), which has introduced new convergence behaviors and bridge port roles, provides significantly faster Spanning Tree convergence after a LAN topology change, and has rendered the original STP obsolete.

Every time the bridge reads the destination MAC address of a packet, it checks the MAC address-to-port table for a match. When a match is found, the bridge forwards the packet to the port indicated in the Filtering Database (or MAC address table). Bridges also ensure that packets destined for MAC addresses that lie on the same port as the originating station are not forwarded to the other ports or transmitted back on the same source port.

4.3.3 FILTERING DATABASE (MAC ADDRESS TABLE OR LAYER 2 FORWARDING TABLE)

The Forwarding Process uses the Filtering Database to decide whether an arriving Ethernet frame with given destination MAC address and VLAN ID is to be forwarded through a potential bridge port. Although the Filtering Database description here includes extensive information, this book focuses on only the basic behavior of the Ethernet bridge and not the extended behaviors as described in [IEEE802.1Q05] and related IEEE standards that defined more advanced features and behaviors.

4.3.3.1 Contents of the Filtering Database

The Filtering Database contains information in the form of filtering entries that are either *static filtering information* or *dynamic filtering information* as discussed above and in [IEEE802.1Q05].

4.3.3.1.1 Static Filtering Information

Static filtering information is represented using two entry types, *Static Filtering Entry* or *Static VLAN Registration Entry*.

- **Static Filtering Entry**: This entry type represents static information for *individual* and for *group* MAC addresses. This entry type allows administrative control (i.e., manual control) of the forwarding of Ethernet frames with specific destination MAC addresses and the inclusion in the Filtering Database of dynamic filtering information associated with Extended Filtering Services that will use this static entry information.
- **Static VLAN Registration Entry**: This entry type represents all static information in the Filtering Database for VLANs. This entry type allows administrative control of forwarding of Ethernet frames with specific VLAN IDs, the inclusion/removal of VLAN tag headers in forwarded frames, and the inclusion in the Filtering Database of dynamic VLAN membership information that will use this static entry information.

Static filtering information is inserted into, modified, and removed from the Filtering Database only under explicit management control (i.e., manually). Static entries are not automatically removed by any aging mechanism. The management of static filtering information may be carried out through the use of the remote management capability provided by the Bridge Management Entity.

4.3.3.1.2 Dynamic Filtering Information

Dynamic filtering information is represented using three entry types, *Dynamic Filtering Entries*, *Group Registration Entries*, and *Dynamic VLAN Registration Entries*:

- **Dynamic Filtering Entries**: These entries specify the bridge ports on which individual MAC addresses have been learned. These entries are created and updated by the Learning Process and are subject to aging and removal by the Filtering Database.

- **Group Registration Entries**: These entries contain information about the registration of group MAC addresses. They are created, updated, and deleted by the Generic Multicast Registration Protocol (GMRP) in support of Extended Filtering Services.
- **Dynamic VLAN Registration Entries**: These entries specify the bridge ports on which VLAN membership has been dynamically registered. They are created, updated, and deleted by the Generic VLAN Registration Protocol (GVRP) in support of automatic VLAN membership configuration.

4.3.3.1.3   Filtering Entry Parameters

Each Static Filtering Entry and Group Registration Entry comprises the following parameters:

- A MAC address specification (individual or group MAC address)
- A VLAN ID, and a Port Map with a control element for each outbound bridge port that specifies filtering for that MAC address specification and VLAN ID.

Each Dynamic Filtering Entry comprises the following parameters:

- A MAC Address specification
- A locally significant Filtering Identifier (ID)
- A Port Map with a control element for each outbound bridge port that specifies filtering for that MAC address specification in the VLAN(s) allocated to that Filtering ID.

The *Filtering ID* is an identifier that is assigned by the bridge to identify a set of VLAN IDs for which no more than one Dynamic Filtering Entry can exist for any individual MAC address. A Filtering ID identifies a set of VLANs among which shared VLAN learning takes place. Any pair of Filtering IDs identifies two sets of VLANs between which independent VLAN learning takes place. The allocation of VLAN IDs to Filtering IDs within a bridge determines how the bridge uses learned individual MAC address information in forwarding/filtering decisions. That is, whether such learned MAC address information is confined to individual VLANs, shared among all VLANs, or confined to specific sets of VLANs.

A *Port Map* identifies to which single bridge port an Ethernet frame with a particular destination MAC address and Filtering ID should be forwarded. A Port Map is equivalent to specifying a single port number in the bridge.

Dynamic Filtering Entries are created and updated by the Learning Process and are automatically removed from the Filtering Database after a specified time (i.e., the *Aging Time*). The Aging Time is the time elapsed since the entry was created or last updated. The Filtering Database must not contain more than one Dynamic Filtering Entry for a given combination of MAC Address and Filtering ID. Dynamic Filtering Entries cannot be created or updated by management action (i.e., manually).

Each Static and Dynamic VLAN Registration Entry comprises the following parameters:

- A VLAN ID
- A Port Map with a control element for each outbound bridge port that specifies filtering for the VLAN.

4.3.3.2 Using a Content Addressable Memory for the Filtering Database

When a bridge receives a frame, it must make a decision on how to handle that frame. The bridge could discard (or filter) the frame, it could forward the frame out one other port, or it could forward the frame out many other ports (i.e., when handling multicast and broadcast traffic). To determine how to handle the frame, the bridge learns the MAC addresses of all devices on the LAN segment and the ports on which these devices (i.e., MAC addresses) are located. This MAC address-to-port mapping information is typically stored in a CAM (Content Addressable Memory) table for fast address lookups because it supports exact matching of the destination MAC address of Ethernet frames with CAM table entries [SEIFR2000] [SEIFR2008].

The CAM table stores, for each device, the device's MAC address, the bridge port on which that MAC address can be found, and the VLAN to which the port or MAC address is associated with. The bridge continually performs this learning process as frames are received into the bridge and the bridge's CAM table is also continually being updated.

The MAC address-to-port mapping information in the CAM table is used to determine how a received frame should be handled (Figure 4.4). To determine where to forward a frame, the bridge examines the destination MAC address in a received frame and then looks up that destination MAC address in the CAM table. The CAM table indicates which port the frame should be forwarded on to reach the specified destination MAC address.

In Ethernet LANs that support VLANs, the forwarding decision lookup in the MAC address table is usually done using the destination MAC address and VLAN ID fields of the arriving Ethernet frame. These fields are passed through a hash function to generate a search or lookup key which is then used for the MAC address table lookup as shown in Figure 4.5 (i.e., exact matching in the CAM using the lookup key generated by the hash function as explained in Chapter 5 of this volume). In the basic implementation where the CAM does not support VLAN ID entries, the search key is simply the 48-bit destination MAC address (i.e., exact matching lookup in the CAM using the destination MAC address).

4.3.3.3 Aging Out Dynamic Filtering Entries

The Filtering Database (Figure 4.4) contains an *Age* for each Dynamic Filtering Entry in addition to several other fields. Aging out Dynamic Filtering Entries is to ensure that end stations that have been moved to a different part of the Ethernet LAN will not be permanently prevented from receiving frames. Aging also allows the Filtering Database to take into account changes in the active Spanning Tree topology of the

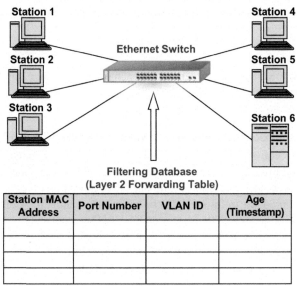

FIGURE 4.4   Ethernet switch with Layer 2 forwarding table.

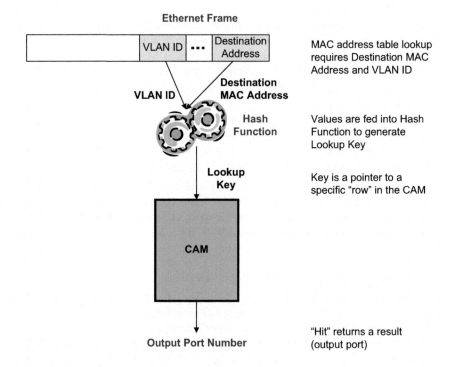

FIGURE 4.5   MAC address lookup using a CAM.

Ethernet LAN that can cause end stations to appear to move from the point of view of the bridge, that is, the path to those end stations subsequently lies through a different bridge port in the LAN.

Although the actual implementation of aging in the Filtering Database is done differently for efficiency reasons (see [SEIFR2000] [SEIFR2008]), all implementations are conceptually the same. When a bridge learns the MAC address, it records the MAC address itself, the inbound bridge port (plus other relevant information), and also sets the age of the entry to zero. Once each second, the age of the entry is incremented and if the age reaches a specified value called the *Aging Time* (or *Age Limit*), the entry is removed from the Filtering Database. The Age Limit has a range of applicable values (10 to 1000000 seconds) but the recommended default value is 300 seconds (5 minutes) [IEEE802.1Q05].

An entry is reset to zero anytime the bridge sees the associated MAC address on the same inbound bridge port. This is done to keep the MAC addresses of active stations always up to date in the Filtering Database. Bridges have a maximum size for their Filtering Databases and if the database fills up, no new entries can be added. The aging process ensures that if a particular station is totally disconnected from the LAN segment, its MAC address entry will be removed from the database.

Also, if a station is moved to another part of the LAN, the database will be updated to reflect this move. In the latter, the bridge will relearn the MAC address of the moved station and the database will be updated accordingly. Because bridge ports operate in promiscuous mode and most end stations are typically very "chatty" in various ways and send user data and protocol information regularly, a bridge will learn the MAC addresses on its connected LAN segments very quickly.

Generally, a network administrator may set the MAC address entries for devices that are directly attached to bridge ports such as networked printers and servers as static filtering entries in the Filtering Database. Such entries are never aged out of the Filtering Database unlike dynamic filtering entries as discussed above.

4.3.4   OVERVIEW OF IEEE 802.1D TRANSPARENT BRIDGING ALGORITHM

The traditional Ethernet bridge determines how a received frame should be handled using transparent bridge algorithm [IEEE802.1D04]. Throughout this book, we assume that an Ethernet switch or a switch/router that supports Layer 2 forwarding uses the transparent bridging algorithm. The following are the basic rules that Ethernet bridges (switches) use when running the *fundamental* transparent bridge algorithm about Ethernet frame handling (the basic forwarding decision process for every received frame):

- If the destination MAC address is found in the Filtering Database, then the bridge will forward the frame out the port that is associated with that destination MAC address in the Filtering Database. This process is referred to as *forwarding*. A frame is said to be *forwarded* when it is received on one port of a bridge and transmitted on another port.
- If the port associated with the frame's destination MAC address in the Filtering Database is the same port on which the frame originates, then there

is no need to forward the frame back on that same port, so, the frame is ignored. This process is referred to as *filtering*. A frame is said to be *filtered* when it is received on one port of a bridge and is discarded (not transmitted on another port).

- If the destination MAC address is not in the Filtering Database (the address is unknown), then the bridge will forward the frame on all other ports that are in the same VLAN as the received frame. This process is referred to as *flooding*. The bridge will not flood the frame on the same port on which the frame was received. A frame is said to be *flooded* when it is forwarded from its incoming port to all other ports.

- If the destination MAC address of the received frame is the broadcast address, then the frame is forwarded on all ports that are in the same VLAN as the received frame. This process is also referred to as *flooding*. The frame will not be forwarded on the same port on which the frame it was received. The broadcast MAC address is a special address placed in an Ethernet frame with all 48 bits set to a 1 value (FF.FF.FF.FF.FF.FF in hexadecimal). A broadcast frame is an Ethernet frame with a broadcast MAC address. All broadcast frames sent on the LAN are copied and processed by the MACs of all network interfaces attached to the LAN.

- Multicast frames are also forwarded on all bridge ports except the inbound port. A bridge forwards a multicast frame to all other ports (just like broadcast frames) because it has no way of knowing which end stations are listening for that multicast address.

Each port of a bridge operates in *promiscuous mode* (also called the *monitor mode*); it receives and examines every frame transmitted on the connected LAN segment. In promiscuous mode, the MAC copies all received frames regardless of a frame's destination address. This behavior is key to the operations of a bridge where it receives frames on a bridge port and decides whether to filter or forward them. This decision-making process is possible because a bridge learns the MAC addresses that are on the LAN segment connected to each port. The promiscuous mode is also called the monitor mode because this is the mode used by network traffic analyzers to monitor and record all received network traffic.

Figure 4.6 describes the fundamental transparent bridge algorithm. The downside to bridging is when the bridge reads a packet with a destination MAC address that has not yet been learned. When this occurs, the bridge forwards the packet on all ports, a condition known as flooding. Flooding can spell trouble from a network bandwidth usage and security standpoint, as data may be sent to users that should not be allowed to receive it.

4.3.5   Special Focus: Ethernet MAC Address Format

Each Ethernet interface on a network has a 48-bit MAC address with format as shown in Figure 4.7 (see Appendix A of [AWEYA1BK18]). This address is programmed in the MAC of the network interface; the reason it is called a MAC address. Every network interface card (NIC) or module manufactured is assigned a unique MAC

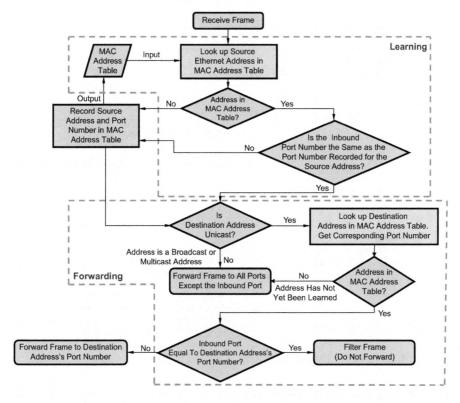

FIGURE 4.6   Basic transparent bridge algorithm.

address as explained below. This prevents any two NICs manufactured from having the same MAC address.

4.3.5.1   Organizational Unique Identifier and NIC-Specific Identifier

The Organizational Unique Identifier (OUI) is the basic mechanism the IEEE uses to administer MAC addresses globally. It occupies bytes 0 to 2 of the 6-byte Ethernet MAC address. In the OUI field, bits 0 and 1 of byte 0 are used as the I/G bit and the U/L bit, respectively. The IEEE assigns one or more OUIs to each manufacturer of Ethernet network interfaces. Using OUIs, the IEEE has to keep track of fewer (often only one) OUIs per manufacturer instead of individual MAC addresses.

Each manufacturer is then responsible for assigning a unique organization-specific identifier for each network interface module manufactured using the IEEE assigned OUI. The NIC-Specific Identifier is the part of the 48-bit MAC address that is assigned to the network interface module by the manufacturer. It occupies bytes 3 to 5 of the MAC address. The organization (manufacturer) is only responsible for keeping track of the NIC-Specific Identifiers it assigns to manufactured network interface modules. The 24-bit NIC-Specific Identifier field allows a manufacturer to build 16 million network interface modules with unique MAC addresses before it needs another OUI.

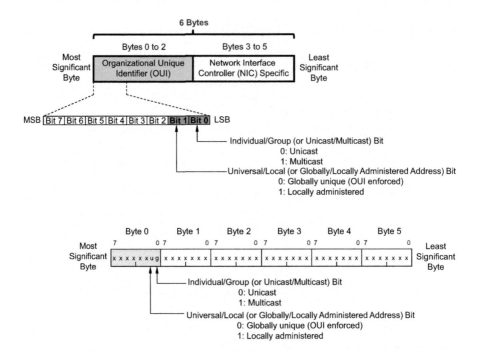

FIGURE 4.7 Ethernet MAC address format.

4.3.5.2 Individual/Group Address Bit

The first bit of byte 0 (i.e., bit 0) in the 48-bit Ethernet MAC address is used to indicate if the address is an *individual* (I) or *group* (G) address. The first bit is usually called the *Individual/Group* (I/G) bit or flag. *Individual MAC addresses* are assigned to one and only one interface in an Ethernet network. MAC addresses with the I/G (i.e., first) bit set to 0 are normally called MAC addresses and are always unique to one and only one interface on the network.

When the I/G bit is set to 1, the MAC address is a *group address*, and normally called a *multicast MAC address*. A group address is one that can be assigned to one or more network interfaces on an Ethernet LAN. Ethernet frames sent to a group address are received and copied by all network interfaces in the Ethernet LAN configured with that group address. Multicast MAC addresses allow Ethernet frames to be sent to a subset of network interfaces on an Ethernet LAN.

- If the I/G bit is set to 1, then all the bits 1 through 47 of the MAC address are treated as the multicast address.
- If bits 1 through 47 are all 1s, then the MAC address is the well-known Ethernet broadcast MAC address, (FF.FF.FF.FF.FF.FF).

Multicast MAC addresses are similar to broadcast MAC addresses except that multicast frames can be received by none, one, some, or all nodes on the LAN. A multicast MAC address has the I/G bit set to 1 and at least one of the other bits is a 0. If all the other bits are 1s, then we get a broadcast address. A network node can elect to listen for and copy frames carrying only certain multicast addresses. A node can listen for the multicast frames in which it is interested and copy only those frames. The MAC of the network interface is responsible for filtering and copying multicast frames. Note that a broadcast or multicast MAC address can only be carried in the destination MAC address field of an Ethernet frame and not in the source MAC address field. The source MAC address of a frame can only be a unicast MAC address.

4.3.5.3   Universally/Locally Administered Address Bit

The second bit of byte 0 (i.e., bit 1) in the 48-bit Ethernet MAC address is used to indicate if the MAC address is globally administered or locally administered. An Ethernet MAC address is a *globally* or *universally administered MAC address* if the U/L bit is set to 0. If both the I/G and the U/L bits are 0, then the MAC address is unique to a single network interface. The MAC address is universally unique because the address is assigned to the network interface by the manufacturer using a combination of the IEEE assigned OUI and the organization's NIC-Specific Identifier.

An Ethernet MAC address with the U/L bit set to 1 is a *locally administered MAC address*. Setting this bit to 1 means that someone other than the NIC manufacturer has set the MAC address. For example, the network administrator of an organization may set the MAC address on a network interface to a value having only local significance by setting the U/L bit to 1, and then bits 2 through 47 to the locally chosen value. In this case, the organization would have to keep track of the locally administered addresses to make sure that there are no duplicate addresses. However, since all NICs are delivered with universally administered addresses, locally administered addresses are rarely used.

4.3.6   VLANs – A Mechanism for Limiting Broadcast Traffic

Another instance where bridges are deficient is with regard to broadcast packets. A broadcast packet uses a fairly primitive technique to force all users on a LAN or VLAN to read the broadcast packet. This packet uses the reserved destination address (FF.FF.FF.FF.FF.FF) which does not actually exist for a host. Since this destination address belongs to no particular port or host, its location can never be learned, so, bridges have no other choice but flood broadcast frames. Under some conditions, large amounts of broadcast frames may be sent on the network, reaching almost all corners of the LAN and resulting in a condition called *broadcast storm*. Furthermore, the broadcast MAC address causes frames from any connected segment (not separate by routers) to be flooded onto all other segments in the same broadcast domain, regardless of IP address or other logical address assignments.

Broadcast storms are the primary reason network architects began to move from networking with simple bridges to switches with VLANs capabilities [IEEE802.1D04] and routers [RFC791]. A bridging loop in the LAN can cause bridges to endlessly forward broadcast and multicast frames. VLANs, just like IP subnets, divide broadcast domains in a network into smaller more manageable or controlled broadcast segments. Furthermore, VLANs allow a network administrator to create logical connectivity of nodes that are separate from the physical network connectivity;

something like a software patch panel. Recall that a broadcast domain is a group of nodes in a bridged network that receive each other's broadcast frames. A broadcast domain can be made up of one or more LAN segments.

Whenever hosts in one VLAN need to communicate with hosts in another VLAN, the traffic must be routed between them (i.e., via a router). This is known as *inter-VLAN routing*. Routers are used in an internetwork of VLANs to provide broadcast filtering, inter-VLAN traffic flow management, security, and IP address summarization. During inter-VLAN routing, the traditional traffic filtering and security functions of a router can be used.

A switch running the transparent bridging algorithm floods unknown and broadcast frames on all the ports that are in the same VLAN as the received frame. The flooding of unknown and broadcast frames causes a potential problem in a VLAN. If the LAN switches running this algorithm so happen to be (mis)connected in a physical loop in the network, then flooded frames (such as broadcasts) will be forwarded continuously and endlessly around from switch to switch. Depending on the topology of the LAN segment the switches are located in, the number of frames may actually multiply exponentially in volume as a result of the flooding algorithm, which can cause serious network problems or even network collapse.

There is a benefit, however, to having physical loops in the LAN as the links creating the loop can actually be used to provide redundancy in the network. The only requirement here is to logical block some switch ports, thereby creating a logical tree topology (i.e., a Spanning Tree) in the LAN even though it contains physical loops. If one link fails, the LAN topology is logically rearranged, allowing the other link(s) to still provide alternative paths for the traffic to reach its destination. To derive benefits from this sort of redundancy without creating problems like broadcast storms in the network because of flooding, the Spanning Tree Protocol (STP) was created and standardized in the IEEE 802.1D specification [IEEE802.1D04]. Newer versions of STP are the Rapid Spanning Tree Protocol (RSTP) standardized in IEEE 802.1w and Multiple Spanning Tree Protocol (MSTP) standardized in IEEE 802.1s.

The purpose of STP and its newer variants (RSTP and MSTP) is to identify and temporarily block the ports creating loops in a network segment or VLAN to prevent the flooding problem described above. The switches run STP which supports loop prevention mechanisms. Part of the loop prevention mechanisms involves electing a root bridge or switch. STP (and its variants) creates a logical loop-free topology of the LAN, called a spanning tree, from the root bridge. The other switches in the LAN segment measure their distance from the root switch and if there is more than one path to get to the root switch, then it can safely be assumed that there is a loop. The switches use STP to determine which ports should be blocked to break the loop and create a Spanning Tree for the LAN.

STP is dynamic and responds to network topology changes, creating a new Spanning Tree as network changes occur. If a link in the segment fails, then ports that were originally blocking may possibly be changed to forwarding mode. The Spanning Tree Algorithm and Protocol monitor, evaluate, and configure (or reconfigure) the Ethernet LAN topology to ensure that there is only one active network path at any given time between any pair of end stations.

The bridges in the LAN use BPDUs to communicate with each other to discover the topology of the LAN and detect switching loops. If the bridges discover loops, they cooperate with each other to place selected bridge ports in the *Discarding* (or *blocking*) mode in order to prevent the loops, while still maintaining a Spanning Tree that reaches all stations. The Spanning Tree Algorithm and Protocol allows a bridged network to be intentionally built with switching loops to provide redundant backup paths between LAN segments. Once the bridges have computed and built the Spanning Tree, they monitor the network to ensure that all the links are functioning as intended (*Discarding*, *Learning*, or *Forwarding*).

4.4   ROUTER AND THE OSI MODEL

Routing was designed to remedy many of the earlier problems associated with bridges or switches. Routers operate at Layer 3 or Network Layer of the OSI network model (Figure 4.8). The Network Layer uses a hierarchical and logical addressing scheme and routing protocols to divide a network into smaller broadcast domains (subnets of VLANs) that are interconnected by routers. Routers do not forward or propagate broadcast traffic from one broadcast domain to the other and allow network operators to create more manageable networks. As a result, the forwarding logic of a router is much more complex than that of a bridge.

The relatively more complex decision processes provided by routers increase the design flexibility and capabilities of networks but tend to make the typical router much more expensive than a bridge of equal performance. Given that the routing and forwarding processes in routers are relatively more complex and involving, these are discussed in greater detail in Chapters 5, 6, and 7 of Volume 1, and Chapters 3, 4, and 5 of Volume 2.

Routers run routing protocols that specify how routers in a network communicate with each other to disseminate information about the network topology and conditions, and how to forward (route) packets to their destinations. The routing protocols have routing algorithms built into them that allow routers to determine the best routes to network destinations. Using a routing protocol, a router shares routing information

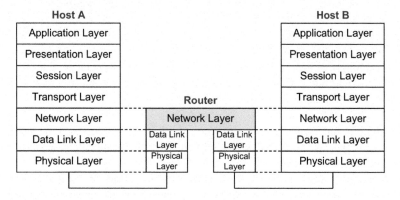

FIGURE 4.8   Router and the OSI model.

first with immediate neighbors, and this information then gets propagated to other routers in the network. Basically, routing protocols allow routers to gain knowledge of the topology of the network as well as prevailing conditions. Some of the common IP routing protocols are the following:

- Interior Gateway Protocol (IGP): Routing Information Protocol (RIP), Enhanced Interior Gateway Routing Protocol (EIGRP), Open Shortest Path First (OSPF), Intermediate System to Intermediate System (IS-IS)
- Exterior Gateway Protocol (EGP): Border Gateway Protocol (BGP)

An IP router performs a lookup of the IP destination address of an incoming packet in its forwarding table or forwarding information base (FIB) to determine the IP address of the next-hop node and outbound interface, and then forwards the packet through that interface on its way to the final destination. This process is generally called the IP forwarding table lookup operation, and is performed on every arriving packet by every router on the path that the packet takes from its source to the destination. The two parameters of next-hop IP address and outbound interface identified at each router determine the path a packet takes through a network to its final destination.

The adoption of Classless Inter-Domain Routing (CIDR) **[RFC4632]** means that a routing lookup needs to perform a "longest prefix match" search in the forwarding table. A router maintains a set of IP destination address prefixes (of variable length) in a forwarding table, each associated with a next-hop IP address and outbound interface (plus, possibly, other important information as discussed in Chapters 2 and 5). Given a packet, the "longest prefix match" search involves finding the longest prefix in the forwarding table that matches the first contiguous set of bits of the IP destination address of the packet.

4.5   MULTILAYER SWITCHING AND THE OSI MODEL: INTEGRATED LAYER 2 SWITCHING AND LAYER 3 ROUTING

Switch/routers or multilayer switches implement both Layer 2 and Layer 3 functionalities; the functions of an Ethernet switch and a router are combined on a single platform (Figure 4.9). The high-end, high-performance switch/router, for example, combines the functionality of a full-featured Layer 2 switch with that of an Internet core router. In addition to having switching/forwarding functionalities (including high switching/forwarding capacity, port density, and redundancy features), the high-end switch/router normally supports comprehensive routing functionality that is required for Internet core routing such as the following:

- Routing and forwarding functionality with a full forwarding information base (FIB), and in some designs, with the entire FIB replicated on each line card on the platform.
- Standards routing protocols such as RIP, EIGRP, OSPF, IS-IS, and BGP.
- Graceful restart mechanisms of the routing protocols to allow the data plane to continue forwarding packets while the router's control plane software is reloaded or restarted.

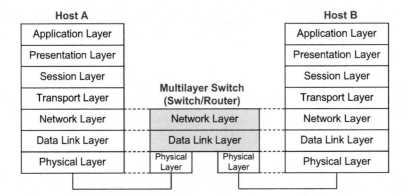

FIGURE 4.9   Switch/router and the OSI model.

- Standards multicast routing protocols including IGMP, MBGP (Multiprotocol BGP), and the various variants of Protocol Independent Multicast (PIM)to support IP multicast applications.
- Equal Cost Multi-path routing (ECMP) to allow load sharing within logical groupings of, for example, up to 16 equal-cost IP links and active-active redundancy.
- Support for the Virtual Router Redundancy Protocol (VRRP) to eliminate a single point of failure in networks.
- Support for Transport Layer protocol (UDP/TCP) five-tuple and IETF Differentiated Services (DiffServ) traffic classification to allow traffic prioritization (QoS), traffic shaping, policing, and rate limiting.
- Support of priority queuing per port and congestion management mechanisms such as buffer and bandwidth management based on Weighted Random Early Detection (WRED), Weighted Fair Queuing (WFQ), strict priority queuing, and so on.
- Support of access control lists (ACLs) to classify and control traffic.

By providing fully integrated switching and routing functionality on every port, switch/routers provide great network design flexibility and can be deployed in the access, aggregation, and core layers of any multi-tiered network architecture. Modern switch/routers are feature-rich and support many capabilities as discussed in Chapters 1 and 2 of Volume 2 of this two-part book.

Many switch/router vendors now offer a diverse range of platforms that meet the requirements of the Layer 2/Layer 3 edge, aggregation, or small-network backbone connectivity. These devices typically support intelligent QoS and security features, predictable performance, comprehensive management tools, and integrated resiliency features.

Integrating Layer 2 switching and routing in a single device (the switch/router) simplifies the network topology by reducing the number of systems and network interfaces that must be deployed to implement multi-tiered network designs. For example, in a collapsed access/aggregation network configuration to a data center

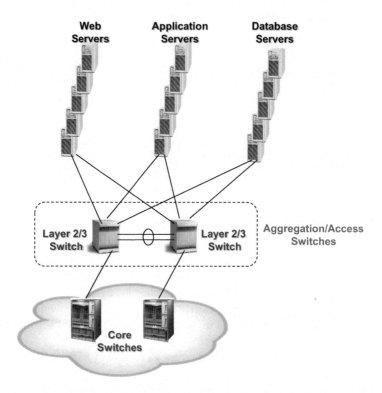

FIGURE 4.10   Switch/router in a data center application.

(Figure 4.10), the switch/router would play the dual role of a Layer 2 switch for server access connections and a Layer 3 router (with VRRP functionality) for distribution function and connections to the core of the enterprise LAN.

In addition to providing robustness for large network designs, routing allows optimum load balancing using ECMP routing of traffic over the redundant paths. Redundant paths are prevalent in highly meshed networks including very large data centers designed for both highest performance and maximum availability. For example, from the core switches to the data center switch/routers, there could be at least two equal cost routes to the server subnets. This permits the core switches to load balance Layer 3 traffic to each switch/routers using, for example, OSPF ECMP routing.

It is important to note that Layer 3 routed redundant paths differ from Layer 2 redundant paths. For example, routed redundant paths do not have to be blocked to break a physical loop, even though the switch/routers also run loop resolution protocols like STP and RSTP.

In Figure 4.10, the dual redundant Ethernet switch/routers are configured for Layer 2 switching between nodes in each cluster VLAN/subnet, and Layer 3 forwarding (IP routing) among the cluster subnets and between the data center and the remainder of the enterprise network. In order to avoid single points of failure in the access portion of the network, servers can be connected with dual-port network interface cards (NICs) to each of the switch/routers using Gigabit Ethernet, Gigabit

Ethernet trunks, or 10 or higher Gigabit Ethernet links depending on the performance level required.

NIC teaming or bonding can also be used to allow each physical server to be connected to two different access/aggregation switches. For example, a server with two teamed NICs, sharing a common IP address and MAC address, can be connected to both switch/routers as shown in Figure 4.10. The primary NIC can be placed in the active state, and the secondary NIC in standby mode, ready to be activated in the event of a failure in the primary path to the data center. NIC teaming can be used for bonding several Gigabit Ethernet NICs to form a higher speed Link Aggregation Group (LAG) connected to one of the switch/routers. However, in a LAG, all the NICs are active and none are in standby mode.

For server connectivity, one of the switches can be configured as the IEEE 802.1D primary root bridge for the Layer 2 connectivity within each group of clustered servers and as the VRRP primary router for connectivity outside the cluster subnet. The second switch could be configured to act as the secondary root bridge and the secondary VRRP router. This way, each access/aggregation switch plays a consistent role relative to the servers in each subnet at both Layer 2 and Layer 3.

VRRP creates a virtual router interface that can be shared by two or more routers. The interface consists of a virtual MAC address and a virtual IP address. Each database server is configured with the VRRP virtual interface as the default gateway. Each of the routers participating in the virtual interface is assigned a priority that determines which is the primary router.

The VRRP routers multicast periodic "hello" messages that include the priority of the sending router. In the event the primary router fails, the secondary router detects the missing hello, determines that it has the highest remaining priority, and begins processing the traffic addressed to the virtual interface. VRRP, in conjunction with STP and dual-ported server NICs, assures that server connectivity will be maintained even if one of the switch/routers fails.

Layer 3 routing between the clusters and the remainder of the network maximizes the degree of control over traffic entering the data center and within the server tiers of the data center. Layer 3 extended ACLs, rate limiting, and policing can all be enabled on the Layer 3 interfaces to implement tight control over the traffic entering each subnet. In order to provide the highest level of service for critical data center traffic, such as latency-sensitive IPC (inter-process communication) traffic in the database subnet, the basic goal could be to give high priority to general data center traffic, while ensuring that IPC traffic and other more critical traffic does not incur any added latency by spending time queued up in a buffer behind less critical traffic.

Highly scalable and resilient switch/routers (particularly with 10 and higher Gigabit Ethernet interfaces) provide the network operator the opportunity to greatly simplify the network design of the data center core. Leveraging VLAN technology together with switch/router scalability and resiliency allows the distinct access and aggregation layers of the legacy data center design to be collapsed into a single access/aggregation layer, providing both Layer 2 switch and routing, as shown in Figure 4.10.

As illustrated in Figure 4.10, the integrated access/aggregation switch/router (i.e., the multilayer switch) becomes the basic network data forwarding element upon which the data center is built. The benefits of using a single layer of switching/routing within the data center network include reduced network device count, simplified traffic flow patterns, elimination of Layer 2 forwarding loops and associated scalability issues, and improved overall reliability.

The high density, reliability, and performance of most of today's high-end switch/routers maximize the scalability of the network design across the data center core. The scalability of these high-end switch/routers often enables network consolidations with a significant reduction in the number of data center switches. This high reduction factor is due to the combination of the following factors:

- Elimination of the access switching layer (made of mostly Layer 2 switches).
- More servers per data center aggregation switch, resulting in fewer aggregation switches.
- More data center aggregation switches per core switch, resulting in fewer core switches.

4.6   LAYER 4-7 SWITCHING – GOING BEYOND MULTILAYER SWITCHING (LAYER 2/3)

Today's data centers are expanding as demand for data and storage continues to grow exponentially. Moreover, requirements such as application hosting, non-stop operation, scalability, high availability, and power efficiency are placing even greater demands on the network infrastructure. To meet this challenge, today's data center network solutions must provide a broad set of capabilities, including higher levels of performance, reliability, security, and QoS, as well as low total cost of ownership (TCO).

A switch/router works at both Layers 2 and 3 – performs the functions of a LAN switch and router on one chassis. Recently the term "Layer 4 switching" has emerged in the networking literature and the Telecom market, adding to the confusion of many in the Telecom industry who are still trying to understand the many terms around. This term is mostly a marketing term used by vendors rather than a precise technical description of a networking device.

Layer 4 switching generally refers to a product's ability to make various traffic handling decisions based on the contents of OSI Layer 4 (the Transport Layer) information in packets where the end-station application is identified (via the TCP or UDP port number). Layer 4 switching refers broadly to capabilities that augment the Layer 2 and 3 functions of a network device rather than to some new type of switching.

Layer 4 and Layer 4-7 switches (all referred to here as Layer 4+ switches) use detailed application message information beyond the traditional Layer 2 and 3 packet headers, directing client requests to the most available servers. Layer 4+ switches are sometimes referred to as Web switches, content switches, content service switches, or application switches.

These intelligent (Layer 4-7) application switches transparently support any TCP- or UDP-based application by providing specialized acceleration, content caching, firewall load balancing, network optimization, and host offload features for Web

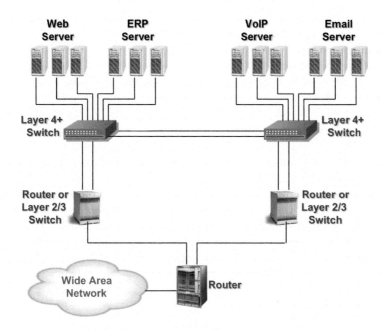

FIGURE 4.11   Layer 4+ switch in an internetwork.

services. Figure 4.11 provides an overview of the architecture of a consolidated data center based on Ethernet switch/routers providing an integrated layer of access and aggregation switching and Layer 4–Layer 7 services provided by standalone appliances (Web switches).

Layer 4+ switches provide efficient traffic distribution among infrastructure devices such as firewalls, DNS servers and cache servers. Layer 4+ switches also provide a reliable line of defense by securing servers and applications against many types of intrusion and attack without sacrificing performance. Layer 4+ switches forward traffic flows based on Layer 4-7 definitions, and provide high performance for higher-layer application switching functions. Intelligent content switching capabilities include customizable rules based on Uniform Resource Locator (URL), and other HTTP headers, as well as cookies, Extensible Markup Language (XML), and application content.

Layer 4+ devices may support a number of advanced features (integrated functionality) as a standalone device. For example, server load balancing, Secure Sockets Layer (SSL) termination/acceleration, Virtual Private Network (VPN) termination, firewall, and packet filtering functionality may be integrated within a single device, reducing box count, while improving the reliability and manageability of a data center.

Furthermore, Layer 4+ switches may provide hardware-assisted, standards-based network monitoring for all application traffic flows, improving manageability, and security for network and server resources. To enable real-time problem detection, extensive and customizable service health check capabilities can be included to monitor Layer 2 to 7 connectivity along with service availability and server response.

If a problem arises, client requests are automatically redistributed to other servers capable of delivering optimum service. This approach helps to keep applications up and running smoothly.

To optimize application availability, Layer 4+ switches may also support many high-availability options, using real-time session synchronization between two Layer 4+ switches available to protect against session loss during outages. As one device shuts down, the second device transparently resumes control of client traffic with no loss to existing sessions or connectivity. Organizations can use advanced synchronization capabilities to simplify the management of two Layer 4+ switches deployed in high-availability mode, minimizing network downtime caused by configuration errors.

Similar to switch/routers (Layer 2/3 switches), Layer 4+ switches can be configured and managed using a Command Line Interface (CLI) or browser-based Graphical User Interface (GUI). The CLI could use well-known industry-standard syntax for fast, error-free configuration. The switches may support Simple Network Management Protocol (SNMP) to allow device management through well-known industry applications such as HP OpenView (now called HP BTO (Business Technology Optimization) Software). Moreover, organizations can use network management tools to monitor traffic, chart traffic patterns, and perform comprehensive configuration management.

4.6.1 BENEFITS OF LAYER 4+ SWITCHING – HIGHER INFRASTRUCTURE RETURN-ON-INVESTMENT

With their intelligent application-aware load balancing and content switching capabilities, Layer 4+ switches significantly improve application and server farm performance while increasing availability, security, scalability, and resource utilization. Layer 4+ switches perform highly customizable real-time health checks, dynamically monitoring the ability of servers in order to optimize performance and transparently reacting to server farm congestion by distributing client traffic loads to the most available servers. Intelligent content switching maximizes server utilization and performance by eliminating the need to replicate content and application functions on every server.

Layer 4+ switches provide very high scalability to IP-based applications and server farms in a cost-effective manner. They allow the use of multiple servers with load balancing and failover, eliminating complete overhauls of the server farms and disruption to applications. The switches provide high return on investment (ROI) for server and application infrastructure in a short timeframe, and support significantly higher application traffic and user loads on existing infrastructure by maximizing server resource utilization.

With support for server connection offload, Layer 4+ switches reduce connection management overhead, freeing up resources for application processing and improving overall server farm capacity and performance. The on-demand and high server farm scalability provided by Layer 4+ eliminates the need for complete upgrades and dramatically improves server farm infrastructure ROI.

Particularly, Layer 4+ switches simplify server farm management and application upgrades by enabling organizations to easily remove resources and insert them into the pool, helping to minimize TCO. The switches provide a single platform that can reduce network device footprint and extend server farm network design and scalability. They can accomplish this by combining high-performance Layer 4-7 packet processing architecture with multi-gigabit Gigabit Ethernet connectivity.

4.6.2 ARCHITECTURE AND CONFIGURATION OF LAYER 4+ SWITCHING

To provide high capacity and high throughput, some Layer 4+ switches are based on an advanced design that features complete physical and logical separation of the application, data, control, and management planes. Typically, this design utilizes modular and specialized hardware to accelerate application processing. It achieves high application processing by using architectures that optimize the distribution and flow of internal traffic to a large number of processor cores.

To maximize flexibility and scalability, the architecture uses a high-speed switch fabric that supports application processing cores, I/O modules, and management modules. The data plane provides high-density multi-gigabit processing with hardware assist for session distribution across multiple application cores. In addition, the management modules could feature field upgradable cards support Secure Sockets Layer (SSL), compression, and any planned future functions.

Like switch/routers, when reconfiguration, scaling, or expansion is required, the chassis of the Layer 4+ switch could be designed to provide a dedicated backplane to support application, data, and management functionality through specialized modules. Typically, the high-end Layer 4+ switches (used in critical high-performing data centers) are based on architectures that support scalability and expansion to meet growing application traffic switching requirements as described below:

- **High-performance, modular design**: This provides a choice of product models starting with compact designs to highly scalable designs with multi-gigabit switching bandwidth. Layer 4+ switches could support expansion slots for management, application switching, switch fabric, line interface, and fan modules to increase performance and port density.
- **Redundant power supplies**: This allows for the support of redundant, hot-swappable power supplies possibly with front-serviceability. In this case, the power supplies typically are hot swappable and load sharing with auto-sensing and auto-switching capabilities, which are critical for power supply redundancy and field deployment flexibility.
- **Hot-swappable modules**: This includes hot-swappable interface modules, fan trays, and power supplies. The hot-swappable power supplies and fan assembly enable the network operator to replace components without service disruption. In addition, several high-availability and fault-detection features help in failover of critical data flows, enhancing overall system availability and reliability.
- **Active/active and active/standby management modules**: This involves using redundant modules for higher availability and performance. Layer 4+ switches can be deployed in multiple high-availability modes with hitless and stateful session synchronization and failover in order to extend availability even through switch failures.

- **Upgradable to hardware-assisted SSL acceleration and compression**: This includes using service modules to add integrated and scalable hardware SSL acceleration and data compression.
- **Reliability**: This includes the support of a resilient switching and routing packets with advanced support for features such as non-stop forwarding, graceful restart, and VRRP.
- **Flexible connectivity options**: This allows for the expansion from, for example, 10 Gigabit Ethernet ports in mixed copper/fiber combinations, to higher Gigabit Ethernet ports.

A management module may have multiprocessor cores, multiple management ports, and external Flash memory slots, along with space for an optional mezzanine daughter card. The application switch module could have several dual-core processors dedicated to processing application traffic, and the switch fabric modules may provide multi-gigabit to terabit switching capacity, providing scalability as I/O modules are added.

4.6.3 CAPABILITIES AND APPLICATIONS OF LAYER 4+ SWITCHING

Layer 4+ switches support a wide range of IP and Web traffic management applications (see Figure 4.11) by providing the following capabilities:

- **Efficient Server Load Balancing (SLB)**: Distributes IP-based application flows and transparently balances traffic among multiple servers while continuously monitoring server, application, and content health in order to increase reliability and availability.
- **System Health Monitoring**: Provides continuous server farm monitoring to detect changes in server health conditions. The Layer 4+ switch monitors and selects server by measuring server, and application responsiveness. Layer 4+ switches provide customizable application-specific health monitoring to help the network operator to determine any degradation or failure of servers and application functions and to redirect clients to alternative resources. The health monitoring messages are typically user-configurable per server and per application port. The switches may send health monitoring messages at a user-configurable periodic interval. Some architectures include a dedicated processor for health monitoring and device management. This design significantly increases server reliability and efficiency and improves overall service availability
- **Disaster recovery and Global Server Load Balancing (GSLB)**: Involves distributing services transparently across multiple sites and server farm locations, and balancing traffic on a global basis while monitoring site, server, and application health. By directing clients to the best site for the fastest content delivery, Layer 4+ switches increase application availability and reduce bandwidth costs. Furthermore, site-level redundancy and fast transparent failover facilitate disaster recovery.

Layer 4+ switches can redirect client traffic geographically among multiple sites based on availability, load, and response time. The switches may also measure client/server proximity as defined by round-trip delay and geographic location. All these features can work in conjunction with the network's existing DNS servers, thereby minimizing network disruption when implementing GSLB. In particular, the Layer 4+ switch may leverage existing DNS servers to minimize disruption to the existing DNS infrastructure. The Layer 4+ acts as a DNS proxy to transparently intercept and modify the DNS responses, thereby directing users to the best site.

The Layer 4+ switches continually monitor multiple sites to detect any changes in servers or services due to varying health and traffic conditions. The use of configurable site load thresholds enables the network operator to align health checking parameters with each site's server and service capabilities.

In addition, Layer 4+ switches may use geographic site selection criteria to keep requests within geographic and continental domains. In this case, continuous application traffic monitoring helps to create a dynamic knowledge base that enables more intelligent GSLB methodologies and site selection criteria.

In addition, Layer 4+ switches could provide a multi-site redundancy solution with Virtual IP (VIP) route health monitoring. This capability matches VIP and server health with intelligent route propagation to the Internet through standards-based routing protocols. This approach provides a knowledge base that enables more intelligent site selection as more clients access a site. This also provides business continuity to IP applications that do not rely on DNS for service name resolution.

- **Intelligent application content inspection and switching**: Avoids replicating application content and functions on all servers, while scaling and optimizing performance. It also helps defeat application-level attacks by using deep Layer 7 content inspection and filtering of application messages. The Layer 4+ switch provides the ability to create rules, policies, and configurations to perform application traffic management operations (from Layer 4 to Layer 7) including server and application load balancing, health monitoring, content inspection, switching, redirection, persistence, and content transformation.
- **Robust application security**: Shields server farms and applications from multi-gigabit rate Denial of Service (DoS), Distributed DoS (DDoS), virus, and worm attacks, while serving legitimate application traffic at peak performance. With built-in intelligence, Layer 4+ switches can detect and discard viruses and worms that spread through application-level messages. The switches load-balance legitimate application traffic while preventing and defeating attacks. Through specialized embedded logic, the switches can reliably protect against many forms of DoS and DDoS attacks.
- **Enterprise applications**: Supports enterprise environments running IP- and Web-based applications, including popular applications such as

Oracle, BEA WebLogic, IBM WebSphere, PeopleSoft, SAP, and Microsoft SharePoint. Layer 4+ switches enable load balancing and persistence to improve availability, security, and performance.

- **SYN-Guard**: Protects server farms against multiple forms of DoS attacks, such as TCP SYN and ACK attacks, by monitoring and tracking session flows. Only valid connection requests are sent to the server.
- **High-availability application switching**: Utilizes active-standby mode, in which the standby Layer 4+ switch assumes control and preserves the state of existing sessions in the event the primary load-balancing device fails. In active-active mode, both Layer 4+ switches work simultaneously and provide a backup for each other while supporting stateful failover.
- **HTTP multiplexing (server connection offload)**: Increases server performance, availability, response time, and security by offloading connection management from the servers. Using persistent HTTP 1.0 and 1.1 connections to the server, Layer 4+ switches stream a large number of client connections to very few server connections. Connection offload enables the servers to dedicate resources for high-performance application content delivery.
- **Application rate limiting**: Protects server farms by controlling the rate of TCP and UDP connections on an application port basis, thereby guarding against malicious attacks from high-bandwidth users.
- **High-performance access control**: Uses extended ACLs to restrict access to specific applications from a given IP address or subnet.
- **Application redirection**: Uses HTTP redirect to send traffic to remote servers if the requested service or content is not available on the local server farm.
- **Hardware SSL acceleration and compression**: Utilizes mezzanine daughter-card service module upgrades to accelerate SSL and compression on Layer 4+ management modules.
- **Advanced firewall and security device load balancing**: Increases firewall and perimeter security performance by distributing Internet traffic loads across multiple firewall and other perimeter security appliances. This approach overcomes scalability limitations, increases throughput, and improves resiliency by eliminating perimeter security devices (such as firewalls, anti-virus gateways, VPN devices, and intrusion appliances) as single points of failure.
- **Transparent Cache Switching (TCS)**: Balances Web traffic across multiple caches, eliminating the need to configure each client browser, thereby improving Internet response time, decreasing WAN access costs, and increasing overall Web caching solution resiliency. Layer 4+ switches improve service availability by implementing cache health checking, redirecting client requests to the next available cache server or directly to the origin server in the event of a cache or server farm failure.
- **IPv6 Gateway**: IPv6 to IPv4 gateway for IPv6 clients provides simultaneous support for both IPv4 and IPv6 real servers behind a single IPv6 VIP, for data center migration.

As discussed above, Layer 4+ switches provide application delivery and traffic management solutions and have been used in the telecommunications industry for decades now. They are helping to mitigate costs and prevent business losses by optimizing the processing of business-critical enterprise and service provider applications. They are effective tools for providing high availability, security, multi-site redundancy, acceleration, and scalability to organizations. The newer generation of these switches are designed to meet growing demand for application connectivity, virtualization, and operating efficiency.

4.7 GENERIC IP HOST ARCHITECTURE

An IP host can be a workstation, laptop, server, portable device such as a smartphone, or any end-system that runs IP-based client and/or server applications. All of these IP hosts support system-level functionalities that allow them to communicate with other devices in a network. The major components of the typical IP host include the following (see Figures 4.12 and 4.15):

- Network interface card (NIC)
- Network interface drivers
- Software interfaces
- Network protocols
- Client and server applications (see Chapter 3)

Any IP host that requires connectivity to a network needs a network interface module. For workstations and servers, for example, the network interface module is usually

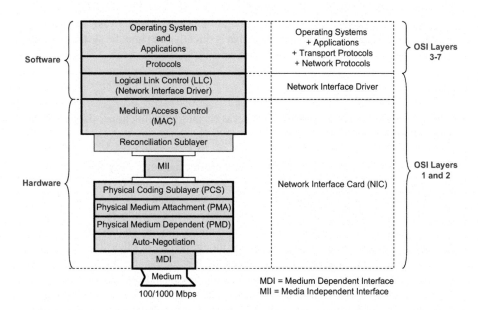

FIGURE 4.12 Protocols in generic host node with Ethernet network interface card.

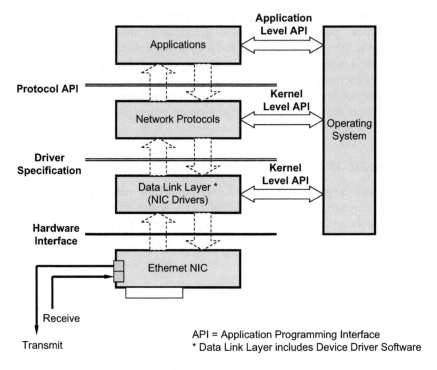

FIGURE 4.13 Generic host node components.

an add-in card called an NIC that is installed in a bus slot (see Figures 4.13 and 4.14). In some cases (e.g., laptops, smartphones and other portable devices), the network interface module is an embedded module that is directly built into the system, usually as part of the motherboard or system baseboard.

4.7.1 TYPES OF NICs

Although NICs have the same logical network interface functions, they come in many different types:

- **Workstation NICs**: These NIC types are inexpensive, easy to install as add-ins on a host system, and are optimized for workstation computers.
- **Embedded NICs**: An embedded NIC is the integration of the MAC and other functions of the NIC directly on the motherboard of the host system and are generally found in laptops and other portable or mobile devices. Embedded NICs can also be implemented on the motherboards of other devices such as workstations and servers. An embedded NIC eliminates the need for an add-in NIC, frees up a bus slot (for other purposes), and ensures that the host is network ready (with the correct drivers installed in the host operating system). The high integration of the NIC functions into the motherboard makes embedded NICs highly cost effective and widely used in many devices.

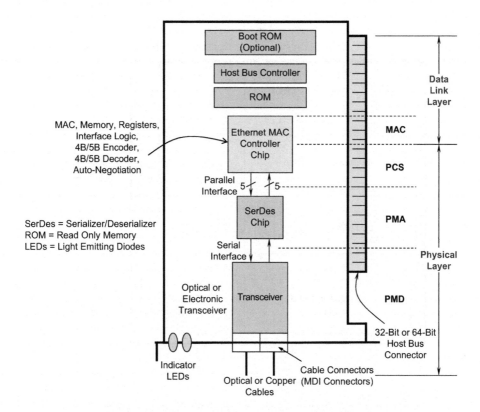

FIGURE 4.14 Architecture of a generic network interface card (NIC) and its relationship to the Ethernet reference model – 100 Mb/s Ethernet example.

- **Server-Based NICs**: These NICs are optimized for high performance (high throughput with very low host CPU utilization) and are designed for servers. Servers place different and tougher requirements on an NIC than the traditional workstation computers. A server typically supports many clients and the requirements for an NIC are different from those of workstation and laptop computers. The high traffic load places an extremely high burden on the server NIC and host bus, a burden that is more pronounced when the server is attached to a high-speed network. The less host CPU time the server NIC consumes when transmitting and receiving frames, the more time the CPU has for processing client requests and local server tasks. The more demand the NIC places on the host CPU leaves less CPU power available for running local tasks, and also limits the number of NICs the server can support to connect to multiple networks. Using multiple NICs with high host CPU utilization can completely offset the benefits of buying a high-end server as most of the CPU power will be used in serving the NICs.
- **Multiport NICs**: These NICs are designed with multiple network interfaces, allowing an NIC to connect the IP host to multiple networks. *Multiport* (or *multiheaded*) NICs have two or more network interfaces (MACs, PHYs,

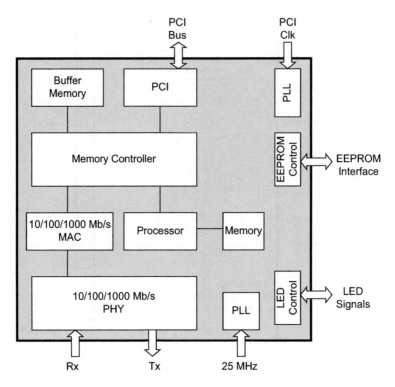

FIGURE 4.15   Example 10/100/1000BASE-T Ethernet controller with integrated transceiver.

and related NIC components), each sharing a single bus interface implemented with a bus-to-bus (e.g., PCI-to-PCI) bridge. This allows a single
NIC to connect to multiple networks, optimizing the available number of
PCI slots on the host. Unlike a workstation which typically connects to one
LAN via a single-port NIC, it is not uncommon to install a multiport NIC
on a single server to enable it connect to multiple LANs. Multiport NICs are
usually designed to have low host CPU utilization to allow the host system
(typically a server) to focus on processing host related tasks and not on
aiding the NIC to transmit and receive data. If the NIC has high host CPU
utilization, this would degrade the performance of the server itself. The traditional single-port NIC consumes a server backplane slot for each network
interface. However, a server will have only a limited number of slots available for network peripherals. Multiport NICs (particularly, those designed
for servers) provide a way for expanding the number of network interfaces without consuming more server backplane slots. The device driver
that comes with the multiport NIC will typically supports either the use of

the NIC ports to connect to multiple networks without link aggregation (as specified in the IEEE 802.1ad standard), or connect to multiple links configured using IEEE 802.1ad Link Aggregation; either case can be used in a server-to-switch configuration or server-to-server configuration for server redundancy or multiprocessing applications. IEEE 802.3ad link aggregation enables multiple Ethernet Physical Layer interfaces to be grouped or aggregated to form a single logical interface, also known as a Link Aggregation Group (LAG) or bundle. Multiport port NICs are not limited to servers but can be used in other devices such as workstations.

NICs are also specific to a networking technology, for example, an NIC that is specific to an Ethernet PHY type (e.g., 1 Gigabit Ethernet copper medium, 10 Gigabit optical fiber medium). or an NIC that is specific to wireless communication (e.g., a WiFi NIC supporting a specific IEEE 802.11 standard).

4.7.2 HIGH-LEVEL NIC ARCHITECTURE

All of today's NIC types are designed using *chipsets* which integrate the multiple functions required for an NIC (see Figure 4.14) into a single chip or device: bus interface, MAC, buffer memory, dedicated I/O controllers, arbitration logic, integrated PHY(s), LEDs, ROM, automatic configuration utilities in ROM, optional boot ROM, and built-in diagnostics. The PHY, usually implemented as a single chip or as a function in larger system-level component, is the interface between the MAC and the external network medium. It is responsible for converting the MAC data into a format that can be transmitted on the network medium. It also convents data signals received from the network medium into MAC data. The MAC is usually implemented in a single chip, and the chip is often responsible for functions such as interfacing to the host system via a host bus. The MAC chip often controls LED indicators and provide an interface for a boot ROM and other NIC features.

The PHY consists of four main parts; Physical Coding Sublayer (PCS), Physical Medium Attachment (PMA), Physical Medium Dependent (PMD), and Medium Dependent Interface (MDI):

- The PCS is responsible for coding outgoing data bytes into symbols and decoding incoming symbols into data bytes (e.g., 4B/5B coding in 100 Mb/s Ethernet, 8B/10B coding in 1 Gb/s, 10 Gb/s).
- The PMA is responsible for serializing outgoing symbols into physical medium bit streams for transmission, and deserializing incoming physical medium bit streams for conversion into symbols.
- The PMD is responsible for actual transmission (reception) of the physical medium bit streams sent (received) over the physical medium (communication channel). The PMD is where physical signal wave shaping and filtering takes place. Conditioning of the signal takes place at this sublayer (appropriate pulse shape, signal power, voltage level, light intensity, etc.).
- The MDI is the electrical or optical connector that allows the rest of the PHY and the overall host system to connects to the physical medium and the external network.

The Medium Independent Interface (MII) is the interface between the MAC and the PHY and allows the MAC to operate completely independent of the type of media to which the network interface is attached. The PHY handles any medium-dependent operations.

The *network interface driver* (generally called a *device driver*) is the software through which the NIC interacts with the host operating system and network protocols (Figure 4.13). The NIC driver abstracts the NIC hardware and provides a standard interface for the operating system and network protocols to interact with the NIC hardware without them having to worry about the hardware specifics of the NIC. The NIC driver allows the network protocols to communicate with the NIC to send and receive data on the network. The interface between the NIC driver and the network protocols is defined by the specific operating system on the IP host (e.g., the Network Driver Interface Specification (NDIS) used in Microsoft Windows).

The optional NIC boot ROM allows the host to boot itself over the network. Other hardware features supported by an NIC are external light emitting diodes (LEDs) which are usually visible from the side of the NIC bracket. The LEDs may indicate the line data rate, link activity, and transmit and receive status on the physical medium. The LEDs are also useful for NIC installation, and troubleshooting of physical network links. The LEDs provide a network administrator, a quick, at-a-glance indication of the status and activity of the NIC without having to remove the host system cover to examine components.

Most NICs are dual-speed (e.g., 10/100 Mb/s Ethernet) or tri-speed (10/100/1000 Mb/s Ethernet) and can automatically detect and configure themselves according to the external device and medium they are connected to. An important requirement for the dual- and tri-speed NIC is the *autosensing driver* and *auto-negotiation*, which makes upgrades to high-speeds easier. The auto-negotiation feature makes it possible for two connecting devices to exchange information about their capabilities (e.g., different media data rates). The NIC and the connecting remote device auto-negotiate to determine the highest data rate each end can support, and then automatically configure to that rate. Using dual- and tri-speed NICs ensures that a network will configure to run at the highest speed auto-negotiated between end-systems. Auto-negotiation is supported by Ethernet interfaces in almost all Ethernet-based network devices (e.g., repeaters, switches (bridges), routers, switch/routers).

4.8 CONFIGURING MAC ADDRESSES IN ETHERNET NICs

The typical NIC used in network devices has a globally unique MAC address that is used as the source MAC address of transmitted frames and as the unicast destination MAC address of received frames. This section explains how the MAC address of an NIC (for example, in a workstation) is configured to be ready for used in the host (see Figure 4.16) [SEIFR2000] [SEIFR2008]. The discussion in this section applies to Ethernet NICs in general although some NIC designs may use their own specific addressing methods (many of which are variants of those discussed here).

When an NIC leaves the factory, the manufacturer programs ("burns") the ROM with a unique MAC address. In some NICs, instead of storing the MAC address in a dedicated NIC ROM, the address is stored in a larger NIC memory that also contains

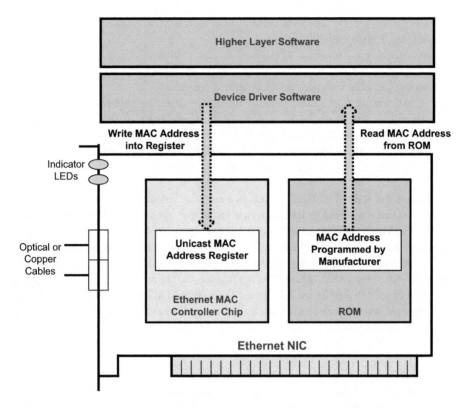

FIGURE 4.16   Configuring MAC address in an Ethernet NIC.

automatic configuration utilities, built-in diagnostic code, and a bootstrap loader. In the typical end station implementation, when the host initializes, its device driver normally reads the programmed MAC address stored in the ROM, and writes this value in a register in the Ethernet MAC controller chip (Figure 4.16). The address value in the MAC controller chip is then used for addressing transmitted frames and comparing against frames with unicast destination MAC addresses.

Usually, the MAC controller chip is not preprogrammed with the unique MAC address. The device driver of the host system loads the MAC address from the ROM into the MAC controller chip's register at host initialization. This initialization feature allows the MAC address used by the MAC controller chip to be the same as the value in the ROM, or to be different, for example, when there is a need to use a locally administered MAC address chosen by the network administrator in the MAC controller chip (see discussion in Section 4.3.5 above).

Also, this initialization feature allows all NICs participating in an IEEE 802.1ad Link Aggregation to have the same source MAC address; the link aggregation driver software loads the same MAC address (from one selected NIC) to all other link NICs in the aggregation. This allows all links/interfaces in the LAG to appear as a single logical network interface; the aggregated links will have the same MAC address instead of having different addresses for each network interface.

Two options are available for creating the source MAC address in transmitted frames [SEIFR2000] [SEIFR2008]:

- Configure the NIC to automatically insert the MAC address in the MAC controller chip's register into transmitted frames.
- Configure the host system to allow the device driver (or higher layer entity) to include the source MAC address in the frame buffer passed to the NIC for transmission. In this case, the NIC will transmit frames without examining or modifying the MAC address passed by the device driver. This is the option used in many implementations.

The reason the second method is used in most implementations is that the device driver is already required to build a frame buffer for transmission that includes the destination MAC address, EtherType field (plus possibly VLAN tag information), and data. Since the source MAC address field is between the destination MAC address field and the EtherType field in Ethernet frames, it is more convenient to just insert also the source MAC address before passing the frame buffer to the NIC. There is no real benefit in having the NIC insert the source MAC address since the device driver has already performed all related works; the device driver already does all the heavy lifting in this case.

REVIEW QUESTIONS

1. Which layer of the OSI model does a hub (also called a repeater) operate and what are its main characteristics?
2. Which layer of the OSI model does an Ethernet switch (also called a bridge) operate and what distinguishes it from a hub?
3. What are the two main paths an arriving frame can take through an Ethernet bridge? How does the bridge decide which path the frame can take?
4. What are the main functions of the MAC Relay Entity in a bridge?
5. What are the main functions of the Forwarding Process in a bridge?
6. What are the main functions of the Learning Process in a bridge?
7. Explain briefly what the bridge port states Disabled, Discarding, Learning and Forwarding represent.
8. What is the role of the Filtering Database (Layer 2 forwarding table) in a bridge?
9. How are static and dynamic filtering entries created and removed from the Filtering Database?
10. What is the main reason for aging dynamic filtering entries in the Filtering Database?
11. What are the Ingress Rules used for in a bridge?
12. What are the Egress Rules used for in a bridge?
13. Explain the main differences between *forwarding*, *filtering*, and *flooding* in an Ethernet switch running the transparent bridging algorithm.
14. What is the different between an Ethernet broadcast MAC address and a multicast MAC address?
15. How does an Ethernet switch handle an arriving packet with destination MAC address of FF.FF.FF.FF.FF.FF?

16. What is the role of the Spanning-Tree Protocol (STP) and its newer variants in an Ethernet LAN?
17. What is a broadcast storm in an Ethernet LAN?
18. What is the purpose of the Organizational Unique Identifier (OUI) in an Ethernet MAC address?
19. What is the purpose of the Individual/Group (I/G) bit in an Ethernet MAC address?
20. What is the purpose of the Universal/Local (U/L) bit in an Ethernet MAC address?
21. What is the function of the network interface driver (device driver) in an NIC?
22. What is the purpose of the auto-negotiation feature in Ethernet NICs?
23. Explain the main functions of the Ethernet PHY sublayers: Physical Coding Sublayer (PCS), Physical Medium Attachment (PMA), Physical Medium Dependent (PMD), and Medium Dependent Interface (MDI).
24. What are the two main methods used by an Ethernet NIC for creating the source MAC address in transmitted frames?
25. Which layer of the OSI model does an IP router operate and what distinguishes it from an Ethernet switch (bridge)?
26. Which layer of the OSI model does a switch/router (also called a multilayer switch) operate and what distinguishes it from an IP router?
27. Explain what distinguishes a Web switch (also called a content switch or Layer 4 switch) from a switch/router.

REFERENCES

[AWEYA1BK18]. James Aweya, *Switch/Router Architectures: Shared-Bus and Shared-Memory Based Systems*, Wiley-IEEE Press, ISBN 9781119486152, 2018.

[CUNNLAN99]. David G. Cunningham and William G. Lane, *Gigabit Ethernet Networking*, Macmillan Technical Publishing, 1999.

[KANDJAY98]. Jayant Kadambi, Ian Crayford, and Mohan Kalkunte, *Gigabit Ethernet: Migrating to High-Bandwidth LANs*, Prentice Hall PTR, 1998.

[IEEE802.1D04]. IEEE Standard for Local and Metropolitan Area Networks: Media Access Control (MAC) Bridges, June 2004.

[IEEE802.1Q05]. IEEE Standard for Local and Metropolitan Area Networks: Virtual Bridged Local Area Networks, IEEE Std 802.1Q-2005, May 2006.

[RFC826]. David C. Plummer, "An Ethernet Address Resolution Protocol", *IETF RFC 826*, November 1982.

[RFC791]. IETF RFC 791, Internet Protocol", September 1981.

[RFC4632]. V. Fuller and T. Li, "Classless Inter-Domain Routing (CIDR): The Internet Address Assignment and Aggregation Plan," *IETF RFC 4632*, August 2006.

[SEIFR1998]. Rich Seifert, *Gigabit Ethernet: Technology and Applications for High Speed LANs*, Addison-Wesley, 1998.

[SEIFR2000]. Rich Seifert, *The Switch Book, The Complete Guide to LAN Switching Technology*, Wiley, 2000.

[SEIFR2008]. Rich Seifert and Jim Edwards, *The All-New Switch Book: The Complete Guide to LAN Switching Technology*, Wiley, 2008.

5 Review of Layer 2 and Layer 3 Forwarding

5.1 INTRODUCTION

This chapter discusses the basics of Layer 2 and Layer 3 forwarding, as well as the methods a switch/router uses to decide which mode of forwarding to use (Layer 2 or Layer 3) when it receives a packet. The discussion covers the forwarding of packets within and between IP subnets, control plane and data plane separation in routing devices, the basics of routing table structure and construction, and the packet forwarding processes in routing devices. The discussion includes the key actions involved in packet forwarding. The IP packet forwarding processes involve parsing the packet's IP destination address, performing a lookup in the IP forwarding table, and sending the packet out the correct outbound interface. This discussion helps in understanding the Layer 2 and Layer 3 processing that takes place in switch/routers. The discussion also helps in understanding the differences between the Layer 2 and Layer 3 processing that takes place in switch/routers.

5.2 DECIDING WHEN TO USE LAYER 2 OR LAYER 3 FORWARDING FOR AN ARRIVING PACKET

The term packet switch is somewhat ambiguous in today's network environment. A packet switch can be a bridge (switch), router, or a hybrid of the two. In its most basic sense, a packet switch is a device that has two or more network interfaces, or ports, and some type of forwarding logic (see Figure 5.1). When a packet is received on a port, the forwarding logic decides which port(s), if any, the packet should be forwarded to. If we think of a packet switch as having two main but separate components, that is, the input/output block and the packet forwarding decision logic block, it is much easier to understand its operation.

The forwarding logic determines the kind of processing the packet switch performs on a packet before passing it to the right output port. For example, the packet processing can be done at Layer 2 (bridging), Layer 3 (routing), or Layer 2 and Layer 3 (as in a switch/router). Figure 5.2 describes how a switch/router decides to forward a received packet at Layer 2 or Layer 3. As discussed in Chapter 4, Layer 4+ switching functions are carried out by so-called Web switches and load-balancers.

When a packet arrives at a switch/router interface and the destination Layer 2 (Ethernet MAC address) of the packet is equal to the Layer 2 address of the receiving interface, the switch/router will attempt to forward the packet at Layer 3 (IP Layer forwarding). Otherwise, the router will attempt to forward the packet at Layer 2 (Ethernet Layer forwarding). We use the word "attempt" here because, even though a packet may require forwarding at a particular layer, other factors may cause the forwarding to fail as discussed in this chapter.

DOI: 10.1201/9781003311249-5   **93**

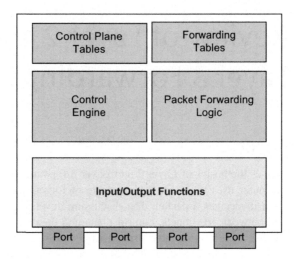

FIGURE 5.1   Components of a packet switch.

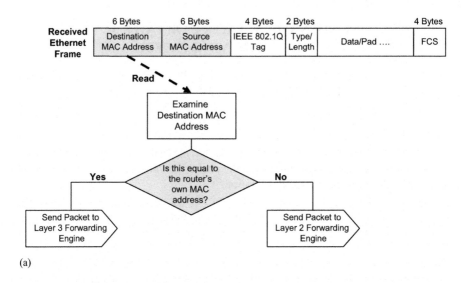

(a)

FIGURE 5.2   Illustrating when a switch/router decides to forward a received packet at Layer 2 or Layer 3. (a) Parsing the Ethernet Destination MAC Address and Deciding How to Forward a Packet Through a Switch/Router,   (*Continued*)

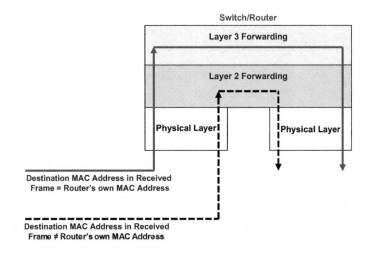

(b)

FIGURE 5.2   **(Continued)**   (b) Paths Taken by Packets Though a Switch/Router.

5.3   LAYER 2 FORWARDING

This section describes the processes involved in Layer 2 forwarding along with the hardware functions that are used. Ethernet switches learn the locations of network nodes (i.e., the switch port to which they are attached or transmitted from) by reading the source MAC address of incoming Ethernet frames.

5.3.1   LAYER 2 FORWARDING BASICS

A transparent bridge is an Ethernet device running the transparent bridging algorithm which connects more than one LAN segment to other bridges and LAN segments to enable Layer 2 forwarding of end-user data across the interconnected LAN segments. As discussed in Chapter 4, a transparent bridge essentially learns the Ethernet MAC addresses of all nodes and the ports to which they are attached, filters incoming frames whose destination MAC addresses are located on the same incoming port, and forwards incoming frames to the destination MAC through their associated port.

The traditional Ethernet switch is basically a multiport transparent bridge on which each switch port either attaches directly to an end-user station, or to another switch connected to an Ethernet LAN segment (broadcast domain) that is isolated from the other LAN segments. A big part of forwarding Ethernet frames involves determining what MAC addresses connect to which switch ports. A switch must either be configured explicitly with where MAC addresses are located (via manual configuration through the switch's Command Line Interface (CLI)), or it must learn this information dynamically (Figure 5.3).

To dynamically learn MAC address locations (i.e., host locations), the Ethernet switch listens to incoming frames on all ports and keeps a table of MAC address and switch port association. When a frame is received on a switch port, the switch examines the source MAC address in the frame to see if it is a new address or an already

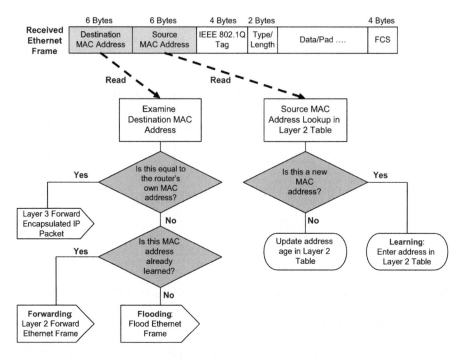

FIGURE 5.3 Layer 2 address learning and forwarding logic.

seen address. If that address has not already been entered into the address table, the
MAC address, switch port, and VLAN on which it arrived are recorded in the address
table. By learning the address locations of the incoming frames, the switch builds the
address (lookup) table used in forwarding frames.

Given that incoming Ethernet frames also contain the destination MAC address,
the switch also looks up this address in the address table, hoping to find the switch
port and VLAN where the address (host) is attached. If the destination MAC address
is found in the table, the frame is forwarded out on the switch port associated with
the address. If the address is not found in the table, the switch floods the frame out all
switch ports that have MAC addresses belonging to the VLAN as the source address
except the inbound port. This process is known as *unknown unicast flooding*, and
happens when the unicast destination location is unknown. In a similar manner,
frames containing a broadcast or multicast destination address are also flooded by the
switch. In real sense, broadcast or multicast destination addresses are not unknown
but are instead addresses used to mark frames destined for multiple locations, meaning they must be flooded by definition.

The basic concept of VLANs can be explained simply as follows. Because the
switch decides on a frame-by-frame basis, which of its ports an incoming frame
should be forwarded on, a useful extension to the broadcast and flooding behavior
of Ethernet switches is to incorporate appropriate logic inside the switch to enable it
select ports (or VLAN tagged end-stations) for special logical groupings. This logical or virtual grouping of ports (or tagged end-stations) results in a broadcast domain

called a VLAN. Each VLAN in a network is identified by a unique VLAN identifier (ID) as discussed in Chapter 6. Each frame sent by a member of a particular VLAN can be tagged with the corresponding VLAN ID. The switch must make sure that traffic from one group of ports (or end-stations), that is, a VLAN is never flooded to other groups of ports (or end-stations). Traffic sent between VLANs can only go through a router, a process referred to as inter-VLAN routing as discussed in Chapters 4 and 6.

Each port or end-station group (VLAN) can each be viewed as an individual LAN segment in its own right; a logical broadcast domain. Routing is used to forward traffic between VLANs as is done between IP subnets. VLANs are also described as logical Layer 2 broadcast domains. This is because the transparent bridging algorithm, which dictates how LAN switches operate, requires that broadcast packets (i.e., packets destined for the "all stations" address (FFFF.FFFF.FFFF)) be forwarded out to all ports that are in the same VLAN. This means that all ports that are in the same VLAN are also in the same broadcast domain. The switch ensures that all Layer 2 broadcasts are restricted to only the end-stations configured to be in a VLAN.

The Spanning Tree Protocol (STP) in the IEEE 802.1D standard [IEEE802.1D04] (or its newer versions Rapid Spanning Tree Protocol (RSTP) in IEEE 802.1w and Multiple Spanning Tree Protocol (MSTP) in IEEE 802.1s) is used by Ethernet switches in a Layer 2 network to maintain a loop-free network, where frames will not be recursively forwarded. Without these protocols, if a loop were to form in the Layer 2 network, a flooded frame could follow the looped path, where it would be flooded again and again, in many cases resulting in a *broadcast storm* and *congestion collapse*.

5.3.2 Lookup Tables Used in Layer 2 Forwarding Operations

Layer 2 switches maintain several data structures and databases that are used in the forwarding process. These are tailored for Layer 2 forwarding but can also be used for Layer 3 forwarding operations (e.g., for Layer 2 address rewrites in outgoing packets). The MAC address table (Figure 5.4) is typically maintained in very fast memory to allow many fields within a frame or packet to be compared in parallel in the address table.

A Layer 2 switch typically uses a Content Addressable Memory (CAM) table [RABCHNIK03] for Layer 2 forwarding (see Figures 5.5, 5.6, and 5.7). Figure 5.5 (first diagram) shows a high-level block diagram of a CAM. A particular entry or record in the CAM is found by matching it with a known pattern. This is done using a *key word*, a *mask word*, and *matching logic* [CHISVDUCK89]. The key (word) contains the pattern which is used for comparison, while the mask (word) enables only those parts of the key word that are relevant in the context of the search. A combination of the key and mask is presented to the *tag memory* and matching logic to perform the actual data comparison.

Figure 5.5 (second diagram) also shows the common data arrangement for a CAM where each work is partitioned into fixed segments. The *tag* bits represent the type of location and indicate whether the location is used or empty. The *label* field is used to match incoming key word search requests, and the *associated data* field stores the search results to be returned or modified.

MAC Address	Port Number	VLAN ID	Age (Timestamp)	Static Bit
MAC Address 1	Port Number 1		Timestamp 1	
MAC Address 2	Port Number 2		Timestamp 2	
MAC Address 3	Port Number 3		Timestamp 3	
MAC Address 4	Port Number 4		Timestamp 4	
⋮	⋮	⋮	⋮	⋮
MAC Address N	Port Number N		Timestamp N	

Static (S) Bit
S = 1 = Do Not Aged Out
S = 0 = Can be Aged Out

FIGURE 5.4   Filtering database (Layer 2 forwarding table).

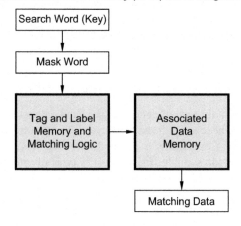

Content-Addressable Memory (CAM) Block Diagram

Search Word (Key)

Mask Word

Tag and Label Memory and Matching Logic → Associated Data Memory

Matching Data

Typical Bit Arrangement for CAM

Tag	Label	Associated Data

FIGURE 5.5   High-level architecture of a CAM.

Figure 5.6 shows a typical CAM architecture that has the following essential elements:

- **Comparand Register**: This contains the data pattern (search or key word) to be compared against the memory array.
- **Mask Register**: This contains the word that is used to mask off portions of the key word that should not participate in the search operations.

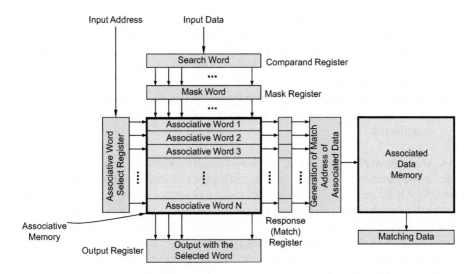

FIGURE 5.6 Typical structure of a CAM.

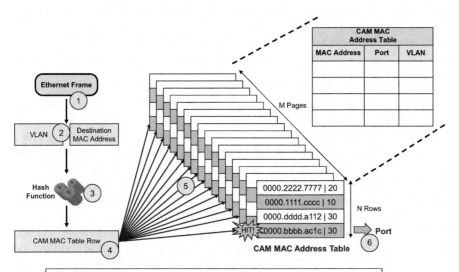

Processing Steps:
1. Ethernet frame received by Layer 2 forwarding engine
2. VLAN ID and destination MAC address extracted from received frame to generate a lookup key
3. VLAN ID and destination MAC address is passed as input to hash function
4. Hash result identifies starting page and row in the MAC address table
5. Lookup key itself (consisting of VLAN and destination MAC address) is compared with entries of the indexed row on all pages simultaneously
6. Lookup type:
    a) Destination MAC address lookup: Matching entry (a "hit") returns the destination port(s)/interface(s). A miss results in flooding of Ethernet frame
    b) Source MAC address lookup: Match results in update of age of matching entry. Miss results in the installation of new entry in the MAC address table

FIGURE 5.7 Generic Layer 2 lookup process using a CAM.

- **Associative Memory Array**: This provides storage and search medium for the labels or associative words used to match incoming masked key words.
- **Associative Word Select Register**: This generates signals based on input address lines to select the associative words that should participate in the search operations.
- **Output Register**: This serves as an interface to output the associated word from the memory array when a search operation has been successfully completed.

A CAM table takes in an index or key value (usually a MAC address from a frame in the simple case where the network does not support VLANs and the CAM does not have VLAN ID entries) as input and performs a lookup that results in a value (usually a switch port). A CAM table allows lookups to be fast; lookups are always based on exact matching of the key. A CAM is most useful for building tables that search on exact matches such as performing a lookup for the MAC address in a MAC address table. In the case of a MAC address table, the switch must find an exact match to a destination MAC address or the switch floods the packet out all ports in the LAN.

In the case where the network supports VLANs and the CAM has VLAN ID entries, the lookup in the CAM is usually done using the destination MAC address and VLAN identifier (ID) fields of the arriving frame as shown in Figure 5.7. These fields are passed through a hash function to generate a search or lookup key which is then used for exact matching lookup in the CAM as explained in Figure 5.7. If no matching entry is found in the CAM, the frame is flooded out all ports in having members in the same VLAN as the flooded frame.

As frames arrive on ports of an Ethernet switch, the source MAC address of the arriving frame is learned and recorded in the CAM table. The port of arrival of the frame and the VLAN to which it belongs are both recorded in the table, along with a timestamp. When a frame arrives at the switch with a destination MAC address that matches an entry in the CAM table, the frame is forwarded out through only the port that is associated with that specific MAC address (entry). The information the switch uses to perform a lookup in the CAM table is called a *key*. The Layer 2 lookup would use the frame's destination MAC address and VLAN ID as a key.

If a MAC address learned on one switch port has moved to a different port, the MAC address and timestamp are recorded for the most recent arrival port (and the previous entry is deleted). To avoid having duplicate CAM table entries, the switch purges/deletes any existing entries for a MAC address that has just been learned on a different switch port. This operation makes sense because MAC addresses are globally unique, and a single host should never be seen on more than one switch port unless problems exist in the network. *Flapping* is said to occur when a MAC address is seen on one switch port, then on another, and then back to the first port, continuously. If a switch notices that a MAC address is being learned on alternating switch ports, it can generate an error message that marks or flags the MAC address as "flapping" between interfaces. If a MAC address is found to have been already entered into the address table for a frame arriving on the correct port, only its timestamp is updated.

Layer 2 switches generally support large CAM tables that can store many MAC addresses and perform fast lookups during frame forwarding. However, these tables generally do not have enough space to hold every possible address on large networks. Thus, to manage the CAM table space to make room for active stations, stale entries (i.e., addresses that have not been active for a configured period of time) are aged out. An Ethernet switch may have a default setting where idle CAM table entries are stored for 300 seconds before they are deleted.

Generally, MAC addresses are learned dynamically from incoming Ethernet frames. However, static CAM table entries can also be configured that contain MAC addresses that are not learned; these addresses are entered and removed manually by the network administrator (see *Static Bit* in Figure 5.4). A *MAC address aging time* is specified for each dynamically learned address (dynamic MAC address) which indicates the time before the entry ages out and is deleted from the MAC address table. For example, the aging time may range from 0 to 1000000 seconds, with a default value of 300 seconds. In many implementations, entering an aging time value equal to 0 for an entry disables MAC address aging for that entry.

The switch deletes a dynamic MAC address entry if the entry is not updated before the *aging timer* expires. The MAC address aging mechanism is used to ensure that a switch can promptly update the MAC address table to accommodate recent network topology changes (i.e., station adds, deletes, and moves). The address aging time may be specified via a timestamp as shown in Figure 5.4. There are several efficient methods for implementing MAC address aging without the use of timestamps as described in [SEIFR2000] [SEIFR2008].

5.4 INTERNETWORKING BASICS

As discussed in Chapter 4, the layer of the OSI model at which a network device operate dictates its fundamental characteristics. The main difference between an Ethernet LAN switch and an IP router is that the LAN switch operates at Layer 2 of the OSI model while the router operates at Layer 3. It is the difference between the protocol make up and the corresponding processing requirements of these layers that affects the way a LAN switch and a router handles network traffic. The switch/router, on the other hand, can operate at both Layers 2 and 3 as discussed in Chapter 4.

5.4.1 Bridging (Switching) in Internetworks

The traditional Ethernet switch by design does not filter broadcast and multicast traffic, or frames arriving with unknown destination MAC address. The lack of filtering for these traffic types might be seen as a bad networking attribute that can lead to severe overload or congestion problems in distributed networks (which many of today's networks are). However, in these networks, broadcast messages can serve a useful purpose by allowing hosts to resolve Data Link Layer and Network Layer addresses and dynamically discover network resources, such as file and print servers.

Broadcast frames originating from each host are received by every other host in that particular broadcast domain (or VLAN) in the network. Even though the net-work devices that are not interested in that particular broadcast would discard the broadcast traffic, large amounts of network bandwidth are consumed by these broadcasts. The propagation of broadcasts within the switched network limits the amount of bandwidth that can be used for real user data.

In some cases, the excessive circulation of broadcasts around the network as well as the generation of control messages by protocols that rely on broadcast mechanisms (e.g., ARP, DHCP, etc.) can saturate the network to the point that no useful bandwidth remains for end-user applications. This situation or phenomenon is commonly known as a *broadcast storm*. A broadcast storm is the excessive transmission or circulation of broadcast traffic within a LAN segment or VLAN.

The problem a broadcast storm creates is, hosts find it difficult to establish new network connections to other hosts, and also existing connections are more likely to be dropped. The more network devices are added to the LAN segment, the more the intensity of broadcast storms increases – the severity increases with each additional device added. Broadcast storms are often caused by loops in the Layer 2 network; loops cause an almost endless circulation of broadcast traffic and can lead to a communication shutdown of an entire network within seconds.

The traditional Layer 2-switched LAN topologies (running the transparent bridging algorithm) are vulnerable to forwarding loops because the network is a flat network. Thus, to prevent looping, the switches in the physical network topologies (which typically contain loops to provide physical path redundancy) need to run the Spanning Tree Protocol (STP) or its newer variants. STP uses the spanning-tree algorithm to build logical topologies on the physical network that do not contain loops. In transparent bridging (switching), switches make topology decisions with the goal of creating loop-free paths by exchanging Bridge Protocol Data Units (BPDUs) [IEEE802.1D04].

5.4.2 ROUTING IN INTERNETWORKS

Switched LAN networks can be designed as physically separate and distributed network segments, but parts or whole of these segments can collectively be mapped to one logical network, such as one IP subnet. Generally, each interface in a network is assigned an address according to at least one of the following addressing structures:

- **Layer 2 Addressing, for example, using Ethernet MAC addresses**: Layer 2 addressing is typically done using a flat address space with each interface in the network assigned a universally unique address. Each interface in the Layer 2 network can be part of one or more broadcast domains (using the concept of VLANs).
- **Layer 3 Addressing, for example, using IP4 and IPv6 addresses**: Layer 3 networks are typically designed using a hierarchical address space with network identifiers within the address space identifying logical network segments, and host identifiers identifying host (or end-systems) or interfaces within those logical networks.

This means a network interface can have both a Layer 2 and a Layer 3 address. A simple address resolution protocol can be used to map the logical Layer 3 address to the physical or hardware Layer 2 addresses in a LAN segment or VLAN. An important example of an address resolution protocol is TCP/IP's Address Resolution Protocol (ARP), which is used to determine the Ethernet MAC address of an interface if its IP address is known.

As discussed in Chapter 4, routers operate at OSI Layer 3, thus providing network managers the means to create networks with hierarchical addressing structures. Routed networks associate a logical addressing structure to a physical infrastructure so that each network segment can be mapped to, for example, an IP subnet. Routed networks also handle traffic flow differently from traditional Ethernet LAN networks. This is because routed networks have complex routing protocols that allow for the determination of optimal paths within the network. They also support more flexible and extensive configuration tools to allow for the management and control of traffic flows.

The traffic flow is more flexibly configured because routers use the hierarchical addressing structures and knowledge of the network topology in determining the optimal path between a source and destination Layer 3 (IP) address. The optimal path is often based on dynamic factors such as link failure, network congestion, path costs, etc. By working at Layer 3 of the OSI model, routers work independent of the underlying Layer 1 and 2 properties of the network. They can work over many physical transmission media (fiber, copper, wireless), and over a wide range of Layer 2 protocols.

Also, routers do not forward broadcast traffic, as a result they do not contribute to broadcast storms. Routers limit broadcast traffic and BPDUs within each LAN segment or VLAN. Routers using the routing protocols can support loop-free paths to network destinations. In a routed environment, routers running routing protocols such as Open Shortest Path First (OSPF) maintain identical topological databases of the network (called Link State Databases (LSDBs)), enabling the network to converge quickly in response to a change in the network topology (resulting from factors such as a link failure, or the addition/removal of a router). Routing protocols are Network Layer functions that are responsible for exchanging information between routers, allowing them to determine the best paths to all known network destinations.

In addition to the standard protocols they run, routers and switches typically support additional features that are used to create more-secure networks as well as control the quality of service (QoS) of end-user traffic flows. Most switches and routers use access control lists (ACLs) and custom filters to provide access control to a network based on a number of packet parameters such as the source or destination address, the Network Layer protocol type (ICMP, IGMP, EIGRP, OSPF), Transport Layer protocol type (UDP or TCP), Transport Layer port number, and other fields within the packet.

Switches and routers may filter or provide access control to the network based on a packet's source and/or destination network address, and on other packet header fields or options within the packet. For example, routers can be configured to permit or deny traffic based on specific TCP/IP information or even on a range of IP addresses (e.g., preventing a group of users from accessing certain websites). Modern-day routers have very extensive QoS and security ACLs, and control tools, that can cater to the needs of a wide range of users.

5.4.3 SWITCHING WITHIN A SUBNET

As discussed above, Ethernet switches (running the transparent bridging algorithm) are essentially self-learning bridges. They discover the host devices attached to their ports to create a simplified picture of the network topology that surrounds them, keeping track of the MAC addresses that identify all attached host devices.

The host devices attached to these Ethernet switches in turn have their own network awareness, provided by mechanisms residing in their communications software. An IP host that wants to send traffic to another host must first know that other host's IP address. This known destination IP address is then used to determine the corresponding destination Ethernet MAC address of the host. ARP is used within the IP host to resolve the destination host's IP address to a corresponding Ethernet MAC address. The resolved MAC address is placed in the destination MAC address field of the Ethernet frame sent by the sending host.

Within the sending host's TCP/IP protocol stack, the host compares its IP addresses with the intended recipient's (known) IP address. This is done by subnet masking (the two IP address fields) to determine whether or not they belong to the same IP subnet. If they do (e.g., a host and server in Subnet B as shown in Figure 5.8), the host broadcasts an ARP request that names the intended IP recipient and asks it to respond with its MAC address (see the process in Figure 5.9).

An Ethernet interface of a host by definition operates within a single broadcast domain which typically maps to one IP subnet. This broadcast domain facilitates the broadcast of ARP requests. When both hosts have acquired each other's MAC address, real data transfer can then begin, and packet transfers between them are forwarded via Layer 2 switching – no routers involved (as illustrated in Subnet B in Figure 5.8).

5.4.4 ROUTING BETWEEN SUBNETS

If any two hosts are not on the same IP subnet (VLAN), then communication between them is achieved through IP routing. The sending IP host initiates this by sending an ARP request to its default gateway which is usually the nearest router (see host in Subnet A and router in Figure 5.10), The default gateway responds by providing the

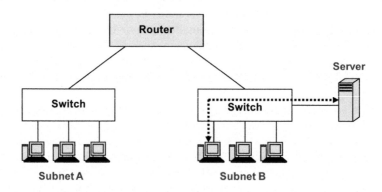

FIGURE 5.8 Switching within a subnet.

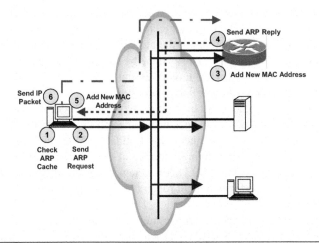

When a host performs a direct or indirect packet delivery, it may need to execute the ARP process using the following steps:
1. Sender consults local ARP cache for an entry for the destination IP address. If an entry is found, Sender skips to step 6.
2. If an entry is not found, sender builds an ARP Request frame containing the MAC address of the sending interface, IP address of the sending interface, and the destination IP address. Sender then broadcasts the ARP Request through the sending interface to the subnet or VLAN.
3. All hosts on the subnet/VLAN receive the broadcast frame and process the ARP Request. If the receiving host's IP address matches the requested IP address (the destination IP address), its ARP cache is updated with the MAC address of the ARP Request sender. If the receiving host's IP address does not match the requested IP address, the ARP Request is silently discarded.
4. The receiving host constructs an ARP Reply containing the requested MAC address and sends it directly to the sender of the ARP Request.
5. When the ARP Reply is received by the ARP Request sender, it updates its ARP cache with the MAC address of the responder. From the ARP Request and ARP Reply frames, both sending and responding hosts have each other's MAC addresses in their ARP caches.
6. The sender transmits the IP packet to the responding host using its newly learned MAC address.

FIGURE 5.9   ARP operations.

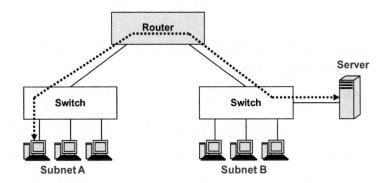

FIGURE 5.10   Switching between subnets.

MAC address of the port that faces the sending host; this port on the default gateway also shares the same IP subnet with the sending host. The host then sends traffic to this router port using the resolved MAC address as the destination address in the transmitted Ethernet frames.

The (default gateway) router in turn may need to broadcast its own ARP request to learn the MAC address of the intended recipient elsewhere in the network. Using the identified destination MAC address, the gateway router performs MAC address rewrite for frames that it transmits. This is done by stripping the source and destination MAC addresses from the frames it receives and replacing them with a new one, a process commonly called *frame or packet rewrites*.

The gateway router replaces the incoming source MAC address with the MAC address of the transmitting port of the gateway router; this serves as the source MAC address of transmitted frames. The incoming destination MAC address is replaced with the MAC address of the receiving interface of the next network device (commonly called the *next-hop*). The MAC address rewrites is an important packet forwarding function that is performed at all routers along the path until the packet gets to its final destination.

Before MAC address rewrites, each router performs IP header Time-to-Live (TTL) and checksum updates. The IP checksum and Ethernet FCS are used to verify data integrity at each router and the destination end-system. The gateway router essentially acts as a middleman by relaying frames on behalf of the sender. The gateway router substitutes its own MAC address so that the node receiving the frame will think that the gateway router is the original sender of the frame. However, the IP destination address in the packet stays unchanged to allow routers on the network to properly route the packet to its final destination.

Meanwhile, every router involved in the data transfer between the two IP hosts has to perform IP forwarding table lookup to determine the next-hop IP address and outgoing interface. The routers may also have to perform next-hop IP address to MAC address resolution to obtain information for destination MAC address rewrites in outgoing Ethernet frames. The forwarding table entries are generated from the routing tables created by the routing protocols (such as RIP, EIGRP, OSPF, IS-IS, BGP). The IP address to MAC address mappings are generated by ARP and stored in ARP caches, or configured statically by the network administrator.

5.5 CONTROL PLANE AND DATA PLANE IN THE ROUTER OR SWITCH/ROUTER

The process of exchanging routing information and forwarding IP packets in a router or switch/router can be divided into two distinct processing planes (Figure 5.11); *control plane* and *data plane* (also referred to as the *forwarding plane*). Processing planes are logical separations used to classify and group together related functions that are performed by the routing device in the network. These processing planes are described below.

5.5.1 CONTROL PLANE

The primary function of a router (or the Layer 3 component in the switch/router) is to use IP routing information created by routing protocols to forward IP packets toward their destination networks (represented by the IP destination address in the IP

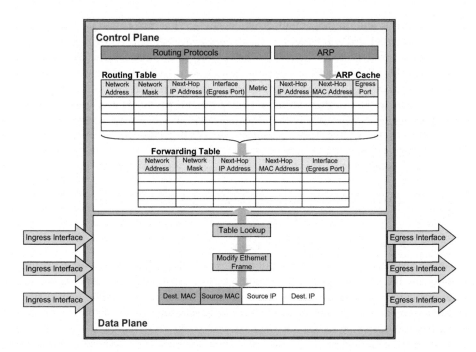

FIGURE 5.11   Control and data planes.

packets). Each IP destination address is associated with a next-hop IP address, which represents the next best router from the local router that leads to the destination. To do this, the router or switch/router needs to search its IP forwarding table (which is generated from the routing information stored in the routing table) for the next-hop and outbound interface information.

The control plane (which relates to the Layer 3 routing operations) is responsible for building and maintaining the IP routing table. The routing table defines where an IP packet should be forwarded based on the IP destination address of the packet. The forwarding information is defined in terms of a next-hop IP address and the outbound interface that the next-hop is reachable from. The module (or collection of processes) that is responsible for building and maintaining the routing table is sometimes referred to as the *control engine* of the router or switch/router (see Chapter 2 and 6).

In this book, we consider the protocols and tools used to access, configure, monitor, and manage a routing device and its resources (i.e., the housekeeping tools) as part of the control plane functions. Some books consider these functions as belonging to a separate plane called the *management plane*. We consider the control plane as encompassing the following:

- **Routing protocols** (e.g., Routing Information Protocol (RIP), Enhanced Interior Gateway Routing Protocol (EIGRP), Open Shortest Path First (OSPF), Intermediate System-Intermediate System (IS-IS), Border Gateway Protocol (BGP))

- **Control and management protocols** (e.g., Internet Control Message Protocol (ICMP), Internet Group Management Protocol (IGMP), Address Resolution Protocol (ARP), Simple Network Management Protocol (SNMP), Dynamic Host Configuration Protocol (DHCP), Domain Name System (DNS), Virtual Router Redundancy Protocol (VRRP))
- **Device access, configuration and management protocols and tools** (e.g., Secure Shell (SSH), File Transfer Protocol (FTP), Trivial File Transfer Protocol (TFTP), Telnet, Ping, Remote Authentication Dial-In User Service (RADIUS), Terminal Access Controller Access-Control System Plus (TACACS+), Command-Line Interface (CLI), performance monitoring tools, log files).

Reference [AWEYA1BK18] discusses in detail the different planes in a routing device (the control, data, and management planes).

5.5.1.1   Basic Control Plane Operations

Layer 2 switches also have a notion of control plane (e.g., STP for loop-free operations), but the control plane operations in routers or switch/routers relate mainly to the operations of routing protocols and the maintenance of the routing tables. The routing protocols contain network topology discovery procedures that allow for the construction of routing tables.

Hence the control plane must provide the processing resources capable of supporting the complex algorithms and data structures associated with protocols such as RIP, EIGRP, OSPF, IS-IS, and BGP. Depending on the routing protocol(s) configured on a router (for intra-autonomous or inter-autonomous routing), the control plane operations and the resources required can vary significantly between the different routing protocols.

Typically, router manufacturers use a general-purpose CPU to run the control plane operations because they have relatively more processing complexity compared to data plane operations. The general-purpose CPU is capable of supporting high-level programming languages that allow vendors to easily develop, maintain, and support the complex software code associated with the various routing protocols. The flexibility provided by a processing platform supporting a high-level programming language allows code to be developed/evolved to accommodate new protocol changes/versions and other control plane operation requirements.

For these reasons, the control plane (implemented in software) is sometimes called the *software processing plane*, *process path*, or *slow-path*. Routers and switch/routers adopt this approach when implementing the control plane operations (associated with IP routing) that is, by using software that runs on a general-purpose CPU. This arrangement allows a general-purpose CPU to perform routing protocol-related functions, along with other tasks such as system maintenance and providing CLI access.

5.5.1.2   Routing Table or Routing Information Base

The routing table (sometimes called the routing information base (RIB)) in a router or switch/router contains information about the topology of the network or internetwork the device is operating in (Figure 5.11). The routing table stores information about networks and how they can be reached (either directly or indirectly). The routing table may have a series of static entries according to the operational requirements of the network. Generally, routing table entries are entered either manually through node management utilities or dynamically through interaction with routers (i.e., via the routing protocols).

The routing table provides the following key information when an IP packet is to be forwarded:

- **The next-hop IP address**: For a direct delivery (i.e., to a directly connected network), the next-hop IP address is the destination IP address in the IP packet. A directly connected network is a network that is directly attached to one of the router interfaces. When a router forwards a packet directly to a host, such as a web server, that host is on the same network as the router's forwarding interface.

  For an indirect delivery (i.e., to a remote network), the next-hop IP address is the IP address of a router (i.e., the next best router leading to the remote network). A remote network is a network that can only be reached by forwarding the packet to another router. Remote networks are added to the routing table using either a dynamic routing protocol or by configuring static routes (i.e., manually configured).

- **The interface to be used for the forwarding**: The interface identifies the physical or logical interface that is used to forward the packet to either its final destination or the next-hop router.

The typical IP routing table consists of a number of information fields some of which include the following (Figure 5.11):

- **Network Prefix (or Network Identifier (ID))**: This is the IP destination network prefix which can be a class-based IP address, an IP subnet, an IP supernet (or aggregate) address, or the IP address of a host route. An IP supernet is formed from the combination/aggregation of two or more networks (or subnets) and has a common Classless Inter-Domain Routing (CIDR) prefix [RFC1517] [RFC1518] [RFC1519] [RFC4632]. The new routing or network prefix for the combined network essentially aggregates the prefixes of the constituent networks. The network prefix can also be represented by a combination of an IP destination address and a network mask:
  - *Destination IP Address*: This is the IP address of a particular network or host
  - *Network Mask (Netmask)*: This is the mask that is used to determine which portion of an IP destination address constitutes the network prefix. The 32-bit IPv4 network mask is used to divide an IPv4 address into the network-specific portion and the host- or subnet-specific portion. Bits set to 1 in the network mask correspond to the network prefix portion of the IP address. For example, destination address 192.168.0.0 and network mask 255.255.255.0 can be written as network prefix 192.168.0.0/24. The destination network and network mask together describe the network prefix.

- **Next-Hop**: This is the IP address of the next-hop node. The next-hop IP address is the IP address of the next network node (i.e., router, switch/

router, gateway, etc.) to which the packet is to be sent on its way to the final destination. The network prefix/next-hop association indicates to a router that a particular destination can be optimally reached by sending the packet to a specific router that represents the best node leading to the final destination. The next-hop information may also include the outgoing interface to the final destination as discussed below.

- **Interface**: This gives an indication of the network interfaces (or ports) on the router to be used to forward an IP packet. The interface can lead to a directly connected network (direct delivery) or a remote network (indirect delivery).

  o *Interface leads to a directly connected network*: This outbound interface represents the interface that leads directly to the IP destination address as carried in the IP packet. A router interface configured with an IP address and subnet mask and attached to a directly connected network becomes a host on that attached network. The network address and subnet mask of the interface, along with the interface type and number, are entered into the routing table as a directly connected network.

  o *Interface leads to a remote network*: This outbound interface leads to one or more routers (next-hops) and finally to the destination network (the final remote network).

- **Metric**: This is the cost associated with the path (route) through which the packet is to be forwarded and is routing protocol-dependent. A routing metric is a number used to indicate the cost of the route so that the best route among possible multiple routes to the same destination can be selected. One example of a routing metric is the metric used in RIP, which is the number of hops (routers to be crossed) to the final destination network.

The above discussion shows that a router interface can connect either to a directly connected host (e.g., email server, web server), directly connected network, or remote network via the next-hop. This means the routing table entries (each associated with a network prefix field) can be used to store in greater detail additional routing information such as the specific types of routes:

- **Directly Attached Network Prefixes**. For directly attached networks, the next-hop field can be blank or contain the IP address of the destination interface on that network. IP packets destined for the directly attached network are not forwarded to a router but sent directly to the destination.
- **Remote Network Prefixes**. For remote networks, the next-hop field is the IP address of another router in between the local forwarding node and the remote network.
- **Host Routes**. This is a route to a specific IP address. Host routes allow routing to occur on a per-IP address basis. For host routes, the network prefix is the IP address of the specified host and the network mask is 255.255.255.255.

- **Default Route**. A default route, when configured in the routing table, is used when no specific routing information about a network destination is found. The default route is used when a more specific network prefix or host route is not found in the routing table. The default route network prefix is represented as a pseudo-network with all 0s in the IP address; the network part is 0.0.0.0 and the network mask is 0.0.0.0. If the default route is chosen because no better routes are found, the IP packet is forwarded to the IP address in the next-hop column of the routing table using the interface corresponding to that IP address in the interface column.

Depending on design and application, some routers and switch/routers will even go further to include the following information in the routing table to facilitate path selection and also specify how a packet should be treated:

- **Quality of Service (QoS)**: This is a metric, number, or label that represents the QoS associated with the route.
- **Filtering and Access Control**: This represents the filtering criteria/access lists associated with the route. This can be used for security, access control, QoS, and traffic control purposes.

5.5.1.2.1 Routing Table Structure and Representation of Information

As discussed above, the routing table stores all the routing information and related parameters needed by a routing device. Each entry of the table contains information such as a network prefix (i.e., the address range covered by a route), outbound interface, next-hop IP address, and so on. In Cisco routers, for example, the information is split into the following parts [ZININALEX02]:

- **Prefix Descriptor**: This part contains parameters such as the network prefix, the routing information source (i.e., the source supplying the route), and the administrative distance of the route.
- **Path Descriptor**: This part contains the outbound interface, the intermediate network address (e.g., the next-hop IP address), and the routing metric.
- **Interface Descriptor Block (IDB)**: This part is more routing device-specific and contains information about the interfaces on the device. Each physical and logical (or virtual) interface on the routing device has an interface descriptor instance. An interface descriptor may contain information such as Layer 2 encapsulation type, interface output buffer pool address, reference to the interface output queue structure, pointers to the functions (or software modules) supported by the interface drivers (i.e., the software modules that communicate with the interface controllers), and so on. Internal system entities not necessarily connected with interfaces may use the interface descriptors. The parameters in the interface descriptors are routing device and interface specific.

5.5.1.2.2   Routing Information Sources and the Routing Table

Routing devices populate their routing table using routing information provided from the following sources (see Figure 2 in Chapter 2):

- **Directly attached (or connected) networks**: A *directly attached or connected network* is a major network or subnet that is attached to a particular interface of a routing device and is known to the routing device from the IP addresses and network masks assigned to the attached network. The route to a directly connected network is also known as an *interface route* or a *connected route* [ZININALEX02]. A route to a directly connected network is derived from the IP address and network mask configured on the interface by the network administrator. Routing tables in Cisco routing devices display interface routes (or connected routes) with the "C (Connected)" code [ZININALEX02]. A connected route is calculated by applying the network mask to the IP address assigned to the interface, and this route is installed in and removed from the routing table whenever that IP-enabled interface goes up or down, respectively.
- **Manually configured routes (static routes)**: A *static route* is one that has been entered into the routing table manually and does not adapt to network changes. Although most of today's networks run dynamic routing protocols, network administrators still use static routes because they allow the trajectories of traffic flows to be easily changed and arbitrary routing policies to be applied. A network administrator can use static routing as complementary to dynamic routing when the forwarding decisions of routing devices need to be influenced without significant changes to the configurations of the dynamic routing protocols running. A network administrator can also use static routing when dynamic routing is not necessary in a particular network scenario or not desirable because of certain reasons such as bandwidth constraints, security, or network topology reasons. However, it should be recognized that improper use of static routing can lead to improper use of network bandwidth and routing loops.

    To configure static routes, the network administrator generally has knowledge about the network topology and the policy to be applied to traffic flowing to the destinations of the static routes. To implement static routing, the network administrator manually configures the routing tables of all the relevant routers leading to a destination with the routing information needed to correctly forward packets to that destination. Note that a route that references only an interface without a next-hop IP address is known as a *directly connected route* which is not the same as an *interface route* or *connected route* as described above [ZININALEX02]. The term *directly connected route* refers to the amount of next-hop information specified for the route rather than the routing information source. *Directly attached networks* (i.e., *interface routes* or *connected routes*) and *static routes* are the only two routing information sources that supply *directly connected routes* [ZININALEX02].

- **Routes learned by the dynamic routing protocols (dynamic routes)**:
  Dynamic routes are those that are learned by the dynamic routing protocols
  (such as RIP, EIGRP, OSPF, IS-IS, and BGP) and are capable of adapting
  to network changes dynamically. Routing protocols allow routing devices
  to continuously exchange information about the network, providing routing
  information that can be used to compute new routes to network destinations
  when network changes occur.

The different types of routing information sources are discussed in greater detail in
[AWEYA2BK21V1] [AWEYA2BK21V2].

5.5.1.2.3 Routing Information Sources and Route Selection

It is possible that different routing information sources may supply routes to the
same network destination for the routing table at a given time (e.g., directly connected networks, static routes, or dynamic routes from dynamic routing protocols).
For example, a routing device may obtain routing information about a particular destination network from RIP and OSPF. However, because RIP and OSPF use different
routing metrics (hop count and cost, respectively) which also have different significance in each protocol, the routing device cannot choose between these two routing
information sources by simply comparing their routing metrics (see Chapter 1 of
[AWEYA2BK21V1]).

Thus, because different routing protocols use different metrics that are not com-
parable or incompatible, the *administrative distance* (also called the *route preference*) is used as a tiebreaker mechanism when different routing information sources
provide routes to the same network destination (see Figure 2 in Chapter 2). The
administrative distance which is expressed as an integer value represents the trustworthiness or believability of the routing information source. A routing information
source with a lower administrative distance is more trustable or believable that one
with a higher administrative value.

Typically, before a routing device participates in routing in a network, default
administrative distance values (which are configurable) are assigned to the routing
information sources in the device. For example, a router inserts routes to *directly
connected networks* into the routing table with an administrative distance of 0, and
installs *static routes* with a default administrative distance of 1. Figure 2 in Chapter 2
shows the Cisco default administrative distance values of different routing information sources. Compared to directly connected networks and static routes, routes provided by dynamic routing protocols have higher administrative distance values. A
route that is assigned the administrative distance value of 255 is never installed in the
routing table; it is a nontrusted route.

Note that for some routing information sources such as static routes and BGP, the
router will first validate a route before installing it in the routing table. For these
sources, the router will check if the referenced local router interface is up or the referenced intermediate IP address is resolvable (see discussion below). For other routing information sources such as IGPs (RIP, EIGRP, OSPF, and IS-IS), the router does
not need to perform this check because IGPs always reference both an outbound

interface and a next-hop IP address (which is an address that does not require resolution). OSPF, for example, will never install a route in the routing table that references a local interface that does not have an active OSPF adjacency.

The route selection process in Figure 2 of Chapter 2 installs a new route in the routing table using the following logic:

1. The routing table is checked to see if it already contains a route to the same network destination.
2. If no route to the same destination is found, the new route is installed.
3. If a route to the same destination is found, the administrative distance and routing metric values of the old and new routes are compared.
4. If the administrative distance values of the old and new routes are equal, and both routes are supplied by the same routing information source, the route with the best routing metric is preferred.
   a. Note that RIP and OSPF may install multiple equal-cost routes to the same destination when using equal-cost multipath (ECMP) routing (see RIP in Chapter 5 of [AWEYA2BK21V1] and OSPF in Chapter 1 of [AWEYA2BK21V2]).
   b. EIGRP may install multiple unequal-cost routes to the same destination because it is a protocol capable of supporting unequal-cost multipath routing (see EIGRP in Chapter 6 of [AWEYA2BK21V1]).

5. If the administrative distance values of the old and new routes are different, the route with the lower value is preferred.
   a. If the new route is selected, the old route is removed from the routing table and the new route is installed along with its administrative distance and routing metric values.
   b. If the old route has the lower administrative distance value, the new route is not installed but the routing device may save information about this route in case a backup route is needed in future.

When a route with a better administrative distance is removed from the routing table because the referenced interface is down or because the referenced intermediate IP address is unresolvable (see discussion below), other available routes to the same network destination (but with worse administrative distance values) can be installed in the routing table.

5.5.1.2.4 Understanding Route Resolvability

Before a router installs a route in its routing table, it will first check whether the route is valid, that is, whether the route is resolvable. Reference [RFC4271] defines the *Route Resolvability Condition* as follows:

1. A route *Route_X* that references only an intermediate IP network address is considered resolvable if there exists in the IP routing table at least one resolvable route *Route_Y* that matches the intermediate IP network address of *Route_X* and is not recursively resolved (directly or indirectly) through

Route_X. If there exist multiple matching routes, only the longest matching prefix route (i.e., the more specific route) is considered.

2. A route that references a router interface (with or without an intermediate IP address) is considered resolvable if the state of the interface being referenced is up (i.e., operational) and if IP processing is enabled on that interface.

Condition 1 refers to a non-recursive route and implies that a route that goes through (or references) an intermediate address can be resolved via another route in the routing table. This condition also implies that the route being checked, *Route_X*, cannot be used to resolved its own intermediate address, or the intermediate address of any other route that is (implicitly or explicitly) used to resolve the intermediate address of *Route_X*.

Condition 2 defines the condition for exiting a recursive lookup; that is, a route that specifies an intermediate address and not an interface, must finally be resolved by a route that references an interface. Figure 5.12 shows an example of recursive routes that do not satisfy the Route Resolvability Condition. The Route Resolvability Condition is meant to ensure that all recursive lookups end with an interface for a route that specifies only an intermediate address.

A router excludes unresolvable routes from the routing table. A route that is unresolvable prevents the IP forwarding process in the router from correctly forwarding packets and other competing valid routes from being installed in the routing table. As is discussed next, the routes in the IP routing table consist of routes that specify (outbound) interfaces and those that may not. Routes to directly connected networks and IGP routes specify their associated outbound interfaces. However, BGP routes specify only intermediate addresses, while static routes can specify their outbound interfaces, intermediate IP addresses, or both.

Note: BGP routes specify only intermediate addresses only because the IP address that is used as the next-hop for advertised network prefixes in BGP is almost never directly connected to the router. For example, most BGP routers use loopback interfaces for BGP peering [AWEYA2BK21V2]. When this happens, the next-hop IP address of received network prefixes is the loopback address of the BGP peer which is not connected to the local router. This means a router will have to perform recursive lookups to determine the next-hop IP address and corresponding outbound interface for a BGP route (i.e., an advertised prefix). The next-hop IP address is also used

```
Route_X: ip route 10.10.0.0  255.255.0.0  20.20.2.2
Route_Y: ip route 20.20.0.0  255.255.0.0  30.30.3.3
Route_Z: ip route 30.30.0.0  255.255.0.0  10.10.1.1
```

- Route_X's intermediate address is resolved via Route_Y, Route_Y's intermediate address is resolved via Route_Z, and Route_Z's intermediate address is resolved via Route_X.
- The router cannot install or keep Route_X in the routing table because its presence will lead to endless recursive lookup operation when a packet is to be forwarded over Route_X.

FIGURE 5.12 Example of recursive routes that do not satisfy the route resolvability condition.

to determine the corresponding Layer 2 address of the next-hop when the router performs Layer 2 packet rewrites during packet forwarding.

5.5.1.2.5   *Routes in the Routing Table, and Their Interfaces and Intermediate IP Addresses*

Different routing information sources provide different types of routes and information for the routing table of a routing device. The routes in the routing table can be categorized according to the amount of information provided for the outbound interface and the IP addresses of other intermediate routing devices along the route, for example, the next-hop IP address [ZININALEX02]:

1. **Routes that specify only local router interfaces**:
   - *Directly connected networks*: Directly connected networks (i.e., connected routes or interface routes) reference only the local router interface on which they are connected. The only requirement here is, the interface referenced must be up and IP-enabled. This type of route references only interfaces and not a next-hop IP address or other intermediate addresses and do not rely on other routes in the routing table. If for a received packet, a lookup in the routing table produces a route that specifies only an outbound interface (meaning a directly connected network on the interface), the packet is sent out that interface *but the destination address in the packet is used as the next-hop IP address*. The IP destination address is usually the final destination of the packet – it is the "final next-hop node" and is the destination end-system of the packet. This IP destination address is used to determine the Layer 2 (Ethernet MAC) address of the receiving interface of the next-hop node (using, for example, ARP as described in Figure 5.9). The Layer 2 address is used in Layer 2 frame rewrites before the frame is sent out the outbound interface (see discussion in the "Basic Data Plane Operations" section below).
   - *Static routes*: A static route can be configured to specify only the outbound interface explicitly, which means additional step has to be taken to determine the next-hop IP address. When a packet is to be forwarded over a static route that references only an interface, the IP destination address carried in the packet is used as the next-hop IP address just as discussed above for directly connected networks or hosts [ZININALEX02]. For a static route in this case, the forwarding engine takes the packet's IP destination address and uses ARP or an ARP cache to determine the Layer 2 address (Ethernet MAC) address corresponding to the receiving interface of the next-hop node.

     The next-hop IP address in some situations is not necessary, for example, on static routes configured on serial point-to-point links, where specifying only the outbound interface is sufficient. In such a case, all packets destined for a specific subnet through an interface over a static route can reference only the corresponding interface; the forwarding engine does not need the next-hop address for packet delivery since the packet's destination IP address is sufficient. However, as discussed in [ZININALEX02], static routes referencing interfaces only, if used at all, should be used only if the corresponding interfaces are point-to-point. Otherwise, it is best practice to configure static routes specifying both next-hop IP addresses and outbound interfaces. Also, it is best practice to avoid configuring static routes referencing intermediate addresses only (as discussed below) since such routes need to be resolved (into corresponding interfaces), which also consumes more processing time during IP address lookups, slows down convergence, and can create unexpected routing loops.

2. **Routes that specify only intermediate IP addresses**:
   - *Static routes*: A static route can be configured to specify only the next-hop IP address explicitly, which means the outbound interface is determined by performing a recursive lookup in the routing table. Note that the next-hop IP address is used to determine the corresponding Layer 2 (Ethernet MAC) address of the next-hop node's receiving interface. If a route does not specify an outbound interface but instead only an intermediate IP address (i.e., the address of the next-hop or any other intermediate node), the route is considered a *recursive route* (also applicable to BGP routes as discussed below).

     In Cisco routers, for example, a static route can be configured to a remote network where the intermediate IP address specified for the static route does not belong to a directly connected network as discussed above [ZININALEX02]. Here, the intermediate address does not reference a local interface, and therefore, must be resolved via another route (connected route, static route, or dynamic route). In this case, the route to the remote network and intermediate address can be installed in the routing table of the local router by any routing information source, for example, a connected route, a static route, or a dynamic route supplied by a dynamic routing protocol. Before the local router will install the static route to the intermediate address in its routing table, it will check if the intermediate address can be resolved. If the router determines that the intermediate address is currently not resolvable, it will not install the route in its routing table. However, the router does not discard the route but retains it in its memory until such time that the routing table has information allowing the route (i.e., the intermediate address) to be resolved and installed in the routing table.

     Regardless of the routing information source providing a route to the intermediate address, when the router determines that a route that can be used to resolve the intermediate address of the static route has appeared in its routing table, it will install the corresponding static route, and vice versa. Also, whenever a route disappears from the routing table, all routes in the routing table that are resolvable by the disappeared route are removed from the routing table, provided no other routes can be used for resolving the intermediate address [ZININALEX02].
   - *BGP learned routes*: Routes learned by BGP (i.e., BGP routes) reference only intermediate addresses, meaning the outbound interface is determined by performing a recursive lookup in the routing table. The intermediate address is the address used in the next iteration of the recursive

lookup operation in the routing table in search of the outbound interface. If the result of a lookup is another address, this address is again used to perform another search in the routing table until the outbound interface is found. The recursive lookup process is described in Chapter 3 of [AWEYA2BK21V2] and Chapter 3 of [ZININALEX02].

3. **Routes that specify both local router interfaces and intermediate IP addresses**:
   - *Static routes*: A static route can be configured to specify both the outbound interface and the next-hop IP address to be used to reach a particular network destination. It is important to note that when an interface is also specified as in this case, there is no need to perform a recursive lookup to determine the outbound interface (also applicable to IGP routes). The outbound interface and next-hop IP address are used only to find the Layer 2 (Ethernet MAC) address of the next-hop node's receiving interface.
   - *IGP learned routes*: Routes learned by IGPs (i.e., IGP routes) reference both the outbound interface and the next-hop IP address to be used to reach a particular network destination. IGP routes are always installed in the routing table and specify both the outbound interface and next-hop IP address of each route. IGP routes do not rely on the presence of other routes in the routing table. A packet is always sent out its outbound interface to the address corresponding to the route's next-hop IP address. As noted above, if a route specifies an outbound interface and a next-hop IP address (also applicable to static routes as discussed above), the next-hop address is the address used to determine the corresponding Layer 2 (Ethernet MAC) address of the receiving interface of the next-hop node.

5.5.1.2.6   Routing Table Maintenance

The following are events that can cause changes in the routing table of a routing device:

- **An interface goes up**: An interface on a routing device goes up when the Layer 2 (i.e., Data Link Layer) protocol associated with the interface declares it as up and it is not disabled by a shutdown command [ZININALEX02]. If IP processing is enabled on the interface, the routing device proceeds to install the routes derived from the interface's (primary and secondary) addresses, the static routes that directly reference the interface, as well as, other routes that are resolvable over routes that reference the interface. The IP processing function also notifies all dynamic routing protocols enabled on the interface about the interface being up.

  If some static routes reference the newly activated interface, or their intermediate addresses are resolvable over the connected routes accessible over the interface, these static routes will be installed in the routing table. Dynamic routing protocols enabled on the interface will perform their own protocol-dependent checks to determine if the interface should participate in the routing process. OSPF, for example, when enabled on the interface, will discover OSPF neighbors, establish adjacencies on the interface, and send routing updates (see Chapter 1 of [AWEYA2BK21V2]). RIP, on the other hand, will just send routing updates and listen for incoming updates (see Chapter 5 of [AWEYA2BK21V1]). The routes learned by a dynamic routing protocol from the newly added interface will be added to the routing table.

- **An interface goes down**: An interface goes down when the Layer 2 protocol goes down because it has experienced a communication fault, it has gone through a physical state transition, or an administrative action such as a shutdown command has been issued [ZININALEX02]. If IP processing is enabled on the interface, this event causes the routing device to remove routes that are derived from the interface's (primary and secondary) addresses, routes that directly reference the interface, and routes that are resolved over the deleted routes. The IP processing function also notifies the dynamic routing protocols enabled on the interface about the interface going down. This even renders the interface inoperational from the IP routing perspective.

  When an interface goes down, the IP routing process has to check all routes installed in the routing table. The exact actions taken by the routing protocols enabled on the interface are protocol-specific. Most importantly, all routes that were known over the interface will be considered inaccessible, and all neighbors will be notified about this. If some static routes to a specific network destination are deleted after the interface goes down, other static routes to the same destination but with higher administrative distance values and referencing other interfaces or intermediate addresses can be installed in the routing table. The routing device deletes all routes derived from the IP addresses of the inoperational interface, and requests backup routes that may come up.

- **A routing information source requests for the installation of a new route**: When the network administrator has configured a new static route or a dynamic routing protocol has learned a new route, the new route is passed to the IP processing function for further processing. The IP processing function performs a general administrative distance and routing metric value checks before it installs the route in the routing table [ZININALEX02]. The IP processing function must ensure that static routes that are resolvable through the newly installed route are also installed in the routing table. Events such as this can be recursive, because the installation of a route can cause other routes to be installed.

  Note that as discussed above, the installation of IGP routes does not lead to the installation of other dynamic routes. Also, routes derived from interface addresses do not rely on other routes. The situation is different for BGP which installs routes referencing only intermediate addresses. Other than BGP, static routes that reference intermediate addresses rely on the presence of other routes, which can be routes derived from interface addresses, static routes, or dynamic routes. This means the routing device has to periodically reexamine static routes to check if changes made to the routing table by a dynamic routing protocol affect the resolvability of the intermediate addresses referenced by the static route.

- **A routing information source requests for the deletion of an existing route**: When the network administrator has deleted an existing route or a dynamic routing protocol request for a route to be invalidated, the IP processing function consults the routing table to see if it contains the route. If the routing table contains the route, the IP processing function proceeds to delete the route. The IP processing function must ensure that static routes that become unresolvable after the route is deleted are also removed from the routing table [ZININALEX02]. Then the IP processing function must request all routing information sources in the routing device for a backup route to the same network destination, if any exists. Note that when a route is deleted, some competing routes to the same destination may get installed in the routing table. If no competing routes exist, some routes that reference the deleted one can become unresolvable and must be removed from the routing table.

The discussion above shows that when an interface on a routing device goes up or down, the device must update its routing table to reflect the event. Also, all routing protocols must be notified about the event to be able to update their internal databases, or to send routing updates notifying neighbor routers about the interface state change.

5.5.1.2.7 Recursive Route Lookup in an IP Routing Table

Traditional centralized CPU-based router architectures use the suboptimally structured software-based IP routing table for route lookups which may contain recursive routes and require recursive lookups. In these architectures, the router's CPU uses software-based lookups and is involved with every IP packet forwarding decision. For each IP packet, the CPU searches the IP routing table again and again (i.e., recursively) until it finds the outbound interface for the packet, a process that can slow down the router considerably.

Figure 5.13. explains the process of recursive lookup in an IP routing table containing recursive routes. The routes (other than the fourth route) are recursive, specifying only the intermediate addresses. The lookup engine takes an immediate address as the current next-hop address and performs a lookup in the routing table again. At the end of the search, the outbound interface and the next-hop IP address are found and used for forwarding the packet.

The recursive routing table lookup described in Figure 5.13. is mainly used in forwarding architectures where a CPU is involved with every forwarding decision. In this case, searching the routing table again and again until the outbound interface is found slows down the packet forwarding process; a considerable amount of processing time can be spent on destination address lookups. Internet core routers typically have large routing tables that mostly contain BGP routes that are recursive, requiring the router to perform many time-consuming recursive lookups (on these BGP routes) if the routing table is used directly for packet forwarding.

Modern forwarding architectures such as those using IP forwarding tables contain pre-resolved or pre-computed entries for each route (IP address prefix, Next-hop IP address, Next-hop MAC address, Outbound interface), allowing them to perform

Network IP Address	Next-Hop IP Address	Outgoing interface
172.16.1.0/24	172.16.2.126	-
172.16.2.0/24	172.16.3.126	-
172.16.3.0/24	10.1.0.2	-
10.1.0.0/30	-	Gi0/1

Network 10.1.0.0/30 is directly connected through interface Gi0/1

- The router performs a lookup of the IP Routing Table in order to forward a packet to the destination IP address of 172.16.1.222.
- The route 172.16.1.0/24 is the best-match with the next-hop IP address 172.16.2.126.
- The router performs another look up in the IP Routing Table for 172.16.2.126 and the route 172.16.2.0/24 is the best-match with the next-hop IP address 172.16.3.126.
- Again, the IP Routing Table is searched to find the best match for the next-hop IP address172.16.3.126. The route 172.16.3.0/24 is the best-match with the next-hop IP 10.1.0.2.
- Finally, the next-hop IP address 10.1.0.2 matches the route 10.1.0.0/30 in the IP Routing Table and the packet is forwarded over the outgoing interface Gi0/ towards the destination 172.16.1.222.

FIGURE 5.13   Recursive route lookup in an IP routing table.

lookups in the forwarding table non-recursively even if the underlying routing table contains routes (i.e., recursively chained entries). In these architectures, recursive routes are flagged as recursive in the routing table, and the router searches the recursive chain of routing table entries up to the entry pointing to the outbound interface. The router resolves a recursive route as soon as it is created in the routing table, and enters the corresponding next-hop IP address and outbound interface in the forwarding table.

5.5.1.2.8   Resolving Recursive Routes before Installation in the IP Forwarding Table

To enable faster IP destination address lookups in an IP forwarding table, recursive routes (which do not directly point to a router interface) are first completely resolved into their next-hop IP addresses and corresponding interfaces before being installed in the forwarding table. Typically, routing devices that use IP forwarding, which is a database extracted from the routing table to allow faster address lookups as discussed below, have a mechanism for resolving recursive routes before they are installed in the forwarding table.

In the absence of such a route resolution mechanism, if the routing table containing recursive routes is used for destination address lookups, several lookups may have to be carried out in the routing table if a given IP destination address involves recursive routes. Such lookups result in extra forwarding delays which can be avoided if a forwarding table is used with pre-resolved recursive routes. The forwarding table in this case does not contain recursive routes allowing the forwarding engine to avoid the extra delays associated with recursive lookups.

Figure 5.14 summarizes the key steps involved in installing routes in the routing and forwarding tables (Route Resolvability Condition check and resolving recursive routes). The algorithm for routing and forwarding table maintenance must be sophisticated enough to handle complicated situations as described in Chapter 4 of [ZININALEX02] and Chapter 3 of [STRINGNAK07].

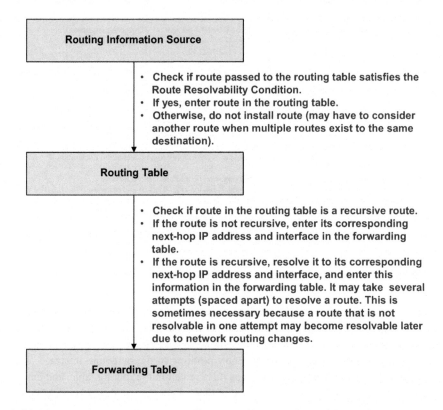

FIGURE 5.14   Steps involved in installing routes in the routing and forwarding tables.

5.5.1.3   Forwarding Table or Forwarding Information Base

As discussed above, the routing table contains all the routing information learned by all routing protocols including those configured manually. The forwarding table (sometimes referred to as the FIB), which is a much smaller table, contains the routes actually used to forward packets (Figure 5.11). The reason behind this is the routing table contains far more information beyond what is actually needed to forward a packet.

For this reason, the information in the routing table is distilled to generate the information needed for the smaller forwarding table. The forwarding table can be seen as a more compact form of the routing table containing only the best routes for packet forwarding. The forwarding table is often generated in a condensed/compressed or pre-compiled format that is optimized for hardware storage and lookup.

It is not uncommon to see software routers also having a forwarding table generated from the routing table. This is to allow the use of much faster software lookup algorithms and storage structures even though all processing is done in software. Also, in router architectures that optimally separate the control plane functions from the forwarding plane functions, lookups and forwarding can be done in an uninterrupted fashion using optimized and specialized custom-built forwarding software and/or hardware architectures.

5.5.2 DATA PLANE

The data (or forwarding) plane is responsible for forwarding IP packets toward their destinations using the optimal routing information learned by the control plane. Whereas the control plane defines where an IP packet should be forwarded to (i.e., maps out the best path it should take), the data plane defines exactly how an IP packet should be handled on a node-by-node basis as it goes through the best path (Figure 5.15).

The forwarding information includes the underlying Layer 2 addressing information required for the IP packet to enable it to reach the next-hop node, as well as other operations required for IP forwarding, such as decrementing the IP header TTL field and recomputing the IP header checksum [RFC1812]. Figure 5.15 describes the IPv4 forwarding process when the underlying network is based on Ethernet. Note that,

---

Receive Layer 2 Function:
- Layer 2 packet (e.g., Ethernet frame) received on an interface
- Basic Layer 2 packet length and Frame Check Sequence (FCS) are verified, as well as, other Layer 2 packet sanity checks are performed
- Network Layer protocol demultiplexing is performed to determine the upper layer protocol to receive the encapsulated data

---

Data Passed by Receive Layer 2 Function:
For each valid received Layer 2 packet, the Layer 2 receive function passes the following information to the IP Layer forwarding function:
- Encapsulated IP packet
- Length of the data portion excluding the Layer 2 framing data in the received Ethernet frame
- Interface (ID) on which the Ethernet/IP packet was received
- Type of destination MAC address: Layer 2 unicast, broadcast, or multicast
- Source MAC address of Ethernet frame
- Incoming VLAN ID, if applicable
- Layer 2 packet priority field value and related packet priority marking/tagging information

---

IP Layer Forwarding Function:
IP forwarding function receives the IP packet and performs the following:
- Validate IP packet header including IP header checksum (also, source and destination addresses should be valid addresses)
- Process any IP Options, if applicable
- Examine destination IP address to determine type of forwarding: local delivery; forward to next hop; both local delivery plus forward to next hop
- Perform longest matching prefix lookup in IP forwarding table using destination IP address
- Decrement IP TTL (Time-to-Live)
- Update any IP header packet priority fields, if required
- Continue with IP Options processing, if applicable
- Update IP header checksum
- Perform any necessary IP fragmentation (using MTU of outbound interface), if required and supported
- Determine Layer 2 address of the packet's next hop (using ARP, for example)

Transmit Layer 2 Function:
- The IP Layer forwarding function passes the following information to the Layer 2 transmit function for each transmitted packet:
  - IP packet
  - Length of the IP packet
  - Outbound interface
  - Next hop IP address
  - Next hop Layer 2 address, if immediately available
  - Outbound VLAN ID, if applicable
  - Layer 2 packet priority value and related packet priority marking/tagging information
- Update Layer 2 packet priority value and related packet priority marking/tagging information
- Encapsulate the IP packet (or each of the packet's fragments) in an appropriate Layer 2 packet
- Update Layer 2 (Ethernet) FCS
- Transmit Layer 2 packet on outbound interface toward the next hop

---

FIGURE 5.15 Key items in the IPv4 unicast forwarding process.

according to [RFC1812], the TTL value of an IP packet *MUST NOT* be checked by a router except when the packet is being forwarded. We present below a discussion about TTL checks and local delivery of packets. The module responsible for performing the data plane operations and forwarding the packet is sometimes called the *forwarding engine*. Readers interested in IP multicast forwarding can refer to the discussion in [AWEYA2BK19].

It should be noted that some important value-added functions such as packet encryption/decryption, inspection, filtering, marking, policing, shaping, tagging, and field rewrites/translation can also be performed as part of the data plane operations of a routing device. For improved performance, most of these functions are implemented using specialized hardware (ASICs), an approach commonly referred to *hardware assist*. Typically, in the event a packet cannot be processed by the specialized hardware or processing unit, it is passed to the route processor for further processing.

5.5.2.1 Basic Data Plane Operations

Data plane operations are performed for every packet that a router forwards. The basic data plane operations are simple and fixed in their implementation – essentially, the same forwarding operations are executed on the transiting packet, regardless of the routing protocols involved in the construction of the routing table (i.e., regardless of where a transit packet should be forwarded).

Data plane operations are performed much more frequently than control plane operations, while control plane operations must be performed only for routing topology changes during which the routing table is updated. This means the performance of the data plane implementation ultimately dictates how fast a routing device can forward packets.

When a Layer 2 packet (e.g., Ethernet frame) is received by a router interface, the interface controller first performs a validity check on the Ethernet frame by examining its length, calculating and verifying the FCS, and checking if it has well-formed and valid MAC addresses. If the Ethernet frame is valid, the interface controller examines the destination MAC address to determine if the frame is destined for the router itself. If the destination MAC address of the frame is not equal to the active MAC address of the receiving interface, is not a broadcast MAC address (all consisting of all "F"s or "1"s), or is not the MAC address of a multicast group in which the router is currently participating in, the frame is ignored and dropped by the interface.

After the interface controller completes the Ethernet frame examination, it examines the Ethertype field of the arriving Ethernet frame to determine the upper layer or Network Layer protocol (e.g., IPv4, ARP) that is to receive the encapsulated data in the frame (0x0800 for IPv4, 0x86DD for IPv6, and 0x0806 for ARP, all in hexadecimal notation). For each arriving packet, the interface controller passes a number of parameters to the IP forwarding function including, at a minimum, the following (Figure 5.15):

- Length of the data encapsulated in the Ethernet frame (i.e., data in the Data field of the Ethernet frame)
- Identifier of the router interface on which the Ethernet frame is received
- Type of Ethernet destination MAC address: unicast, multicast, or broadcast

Upon receiving an arriving packet from the network interface, the IP forwarding function performs a number of IP packet verification checks including the following:

- The total length of the data passed by the interface must not be less than the minimum legal length of an IP packet (i.e., it must be equal to or greater than 20 bytes).
- The checksum of the IP packet is calculated and compared with the IP header Checksum field value.
- For IPv4 routers, the IP Version field in the IP header is checked and must be equal to 4.
- The IP header Length field value is checked and must be at least 5 (i.e., the number of 32-bit or 4-byte words in the IP header). Note that the minimum IP header length is 20 bytes, which is equal to five 4-byte words.
- The IP header Total Length field value is checked and must not be less than the IP header length indicated in the Length field. That is, the Total Length field value must be greater than 4 multiplied by the value in the IP header Length field. The multiplier 4 is used because the total length of an IP packet is measured in bytes, while the IP header length is measured in 32-bit or 4-byte words.

If an arriving packet does not satisfy any of these verification checks, it is silently discarded, or ignored without any notification being sent to the packet originator. A packet that satisfies these conditions, may further be subjected to filtering via an inbound access control list (ACL) if one is configured for the inbound interface. For a packet that is filtered by the ACL, the router may send an ICMP "Destination Unreachable" message with the code value set to 9 (which represents "Administratively Prohibited") to the originator of the packet.

After IP packet verification, the router may perform unicast Reverse Path Forwarding (uRPF) checks (see Section 5.5.2.2 below), if the router is configured to perform this function. With uRPF, the router performs a forwarding table lookup for the source IP address of an arriving packet and checks if this address is resolvable through the router interface on which the packet is received. If the source address is not resolvable through the receive interface, the packet is silently discarded by the router (see uRPF details below).

For an IP packet that is being forwarded (by the router or switch/router to the next hop routing device through an Ethernet interface), the following fields in the outgoing packets must be modified:

- **IP TTL**: This value must be decremented by one as required by the normal rules of IP routing.
- **IP Header Checksum**: This value must be recalculated, as the TTL field has changed.
- **Source MAC Address**: The MAC address of the outgoing interface of the router or switch/router must be written to this field in the Ethernet frame.
- **Destination MAC Address**: The MAC address of the receiving interface of the next-hop node must be written to this field in the Ethernet frame.

The MAC address information is built from the ARP process or is configured manually.
- **Ethernet Frame Checksum**: This value must be recomputed as the Source
  and Destination MAC addresses have changed.

5.5.2.1.1 Adjacency Information
To complete the forwarding operations of a packet, knowledge of the outbound interface along with rewrite of the destination and source MAC addresses is required. The
adjacency information (i.e., Layer 2 address of the next-hop's receiving interface),
which is typically obtained through ARP or configured manually, specifies the destination MAC address needed for the frame's MAC address rewrites. Two nodes are
considered to be adjacent if they can reach each other over a Layer 2 network (point-
to-point or broadcast). A router that is directly connected to a host or another router,
or shares a common IP subnet or VLAN with a host or another router, is considered
adjacent. Figure 5.16 explains the packet processing at the different protocol layers
in a routing device supporting Ethernet and IPv4.

5.5.2.1.1.1 Creating Adjacency Information
The adjacency information in a router can be integrated with the Layer 3 forwarding
table or implemented as a separate table to be used by the Layer 3 forwarding engine.
Either way, the Layer 2 address entries of the next-hop can be populated using any
one of the following methods:

- **Created by Sending ARP Requests**: These entries are obtained from ARP
  requests sent by the local routing device to the next-hop node and neighbor
  devices on directly attached networks (i.e., routers and hosts on the same IP
  subnet or VLAN).
- **Gleaned from ARP Request Received**: These entries are gleaned from
  ARP request sent by neighbor devices to devices on the same IP subnet or
  VLAN including the local routing device (as explained in Figure 5.9).
- **Gleaned during Packet Forwarding to Directly Attached Networks**:
  These entries are gleaned when the local routing device sends packets to
  directly attached networks – the entry is gleaned for a specific host-route
  adjacency.
- **Manual Configuration**: These entries are configured manually by the network administrator by considering the devices that are directly connected
  to the local routing device, i.e., connected by a Layer 2 network (point-to-
  point, VLAN/IP subnet).

In the first method, upon receiving an ARP reply, the router stores the information
in an ARP cache so that it can use this information the next time a packet is to be
forwarded to the same node. Each entry of the ARP cache contains the IP address (of
the next-hop), the learned next-hop MAC address, the local interface through which
the MAC address was learned, a timer indicating the age (i.e., elapsed time) of the
entry from the moment of MAC address insertion, and flags indicating whether the

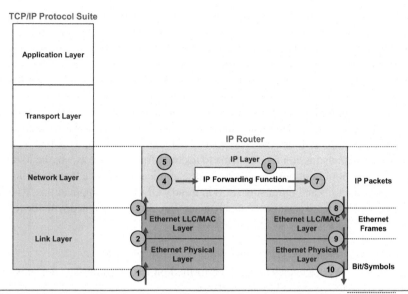

TCP/IP Protocol Suite

| Application Layer |
| Transport Layer |

IP Router

| Network Layer | IP Layer ⑤ ⑥ |
| | ④ → IP Forwarding Function → ⑦ | IP Packets |

③ | ⑧
Ethernet LLC/MAC Layer | Ethernet LLC/MAC Layer | Ethernet Frames

| Link Layer | ② | ⑨
Ethernet Physical Layer | Ethernet Physical Layer

① | ⑩ Bit/Symbols

The following steps summarized the processing at the different protocol layers in the routing device:
1. **Bit/Symbol Reception:** Interface receives bits and Ethernet symbols from the transmission medium and constructs Ethernet frame
2. **Data Link Frame Verification**: Interface performs verification of Ethernet frame length, Ethernet checksum (or Frame Check Sequence (FCS)), destination MAC address, etc.
3. **Encapsulated Protocol Demultiplexing**: Interface demultiplexes the encapsulated packet according to its Ethertypeor protocol number (IPv4 (= 0x0800), IPv6 (= 0x86DD), ARP (= 0x0806), etc.)
4. **IP Packet Validation**: IP Layer validates the IP(v4) packet by verifying the total data length passed by the Data Link Layer, IP checksum, IP version, IP header length, IP packet total length, etc.
5. **Local or Remote Packet Delivery Decision**: IP Layer decides if received IP packet is for local delivery or is to be forwarded to another external node (a next-hop node).
6. **IP Forwarding Table Lookup and Packet Forwarding Decision**: IP forwarding function performs a longest prefix matching (LPM) search in its IP forwarding table to determine the next-hop node and outbound interface for the IP packet. IP Layer also decrements the IP TTL and updates the IP header checksum.
7. **Data Link Layer Parameter Mapping**: IP Layer determines the Data Link Layer parameters to be used in encapsulating the IP packet (e.g., source and destination Link Layer addresses, VLAN mappings, Class-of-Service (CoS) mappings, etc.).
8. **Data Link Layer Frame Construction and Frame Rewrites**: Data Link Layer encapsulates the IP packet in a Data Link Frame with appropriate source and destination Data Link addresses, and updates all relevant fields in the frame such as VLAN and CoS fields, and then updates the Ethernet checksum.
9. **Mapping of Data Link Layer Frame into Symbols**: Physical Layer receives the Ethernet frame and maps it into corresponding Ethernet symbols
10.**Transmission of Symbols/Bits**: Interface transmits the Ethernet symbols and bits on the transmission medium.

FIGURE 5.16 Packet processing at the different protocol layers in a routing device (Ethernet and IPv4 example).

state of the entry is "*complete*", "*incomplete*", "*expired*", etc. The interface through which the MAC address is learned is important because when routing changes occur, the IP address of the next-hop may be reachable via another interface, and the MAC address of the next-hop may be different. This makes the old MAC address learned via the previous interface ineligible for use on the new interface.

Some architectures implement ARP such that when the first packet for a destination IP address arrives and there is no ARP entry for the next-hop, the packet is dropped to save the packet forwarding engine from initiating the ARP process and waiting for an ARP reply [ZININALEX02]. The forwarding engine does not have to wait for an ARP reply to determine the next-hop's MAC address because the ARP reply may potentially never be received, and could result in the forwarding engine holding up the processing and forwarding of other packets through the interface.

In these architectures, the forwarding engine discards the first packet but still initiates the ARP process to determine the next-hop's MAC address to be used for other packets going to that node. An entry with state "*incomplete*" is created in the ARP cache while an ARP request is sent out. Once an ARP reply is received, the remaining fields of the entry are filled in, and the next packet that arrives and is to be forwarded to this destination will use the "*complete*" ARP entry. Note that TCP-based applications have retransmission mechanisms to account for lost packets, so, the loss of the first packet should not pose problems for the applications (since they were designed with packet losses and retransmissions in mind). UDP-based applications, which are designed without retransmission mechanisms, are always designed with the assumption that packets may be lost in the network.

5.5.2.1.1.2   Network Types and Adjacency Information

The number of adjacencies that can be formed on a router interface depends on the type of network connected to the interface:

- **Broadcast Multiaccess Network**: Networks based on Ethernet technologies have inherent broadcast capabilities and are examples of a broadcast multiaccess network. A broadcast interface usually attaches to an IP subnet or VLAN and can have multiple destinations attached to it (i.e., next hops that can be hosts and/or other routers). The adjacencies on the broadcast interface can be manually or dynamically discovered using ARP.
- **Point-to-Point (P2P) Link**: Only a single adjacency can be formed on a P2P link. A P2P interface implies only one destination can be formed through the interface (i.e., a single next-hop that is a host or another router).
- **Point-to-Multipoint (P2MP) Network**: A P2MP interface can have multiple destinations (next-hops) attached to it each with a unique Layer 2 receive connection address. For example, a P2MP interface can be formed on connection-oriented protocols such as ATM. Although ATM is now a legacy protocol, it is a suitable protocol for illustrating how P2MP interfaces can be created. In this case, the P2MP interface consists of multiple logical or virtual ATM connections or circuits each going to a different next-hop node.

ATM uses (manually configured) static mapping or ATM ARP [RFC2225] to map IP addresses to Layer 2 (ATM) addresses. A connection ID can be an ATM Virtual Path Identifier (VPI) or Virtual Channel Identifier (VCI). A P2MP interface may have multiple connections with non-distinct next-hop IP addresses, that is, all connections have the same next-hop (receive interface) IP address. Thus, in the case where a lookup produces a combination of a next-hop IP address, connection ID, and outbound interface configured as a P2MP interface with a non-distinct next-hop IP address, the connection ID is used to decipher which ATM connection to forward a packet on; the connection ID distinguishes which connection to use [STRINGNAK07]. Using ATM, next-hop IP addresses need to be first resolved into (next-hop) ATM addresses.

The local router then signals to establish an ATM connection to the next-hop node (with destination address being the resolved ATM addresses). An ATM connection is represented by a VPI/VCI. The local router must use this VPI/VCI to send packets to the next-hop (destination); the VPI/VCI represents the ATM connection. RFC 2225 assumes the existence of an ATM ARP server on the P2MP network (which is configured as an IP subnet and support interfaces that have both IP and ATM addresses). Every client on the IP subnet communicates with the ATM ARP server to resolve the destination's IP address to an ATM address. The ATM ARP server holds the IP-to-ATM address information for all hosts in the subnet. P2MP interfaces using VPIs/VCIs can also be configured manually on the local router.

5.5.2.1.1.3 Controlling and Maintaining Adjacency Information

To control the amount of information in the ARP cache, "*incomplete*" entries are purged after a maximum time period (e.g., 1 minute). "*Expired*" ARP entries are also purged and have to be refreshed when new packets arrive and the ARP process is triggered once again. An *ARP aging process* periodically processes the entries in the ARP cache. An implementation may use one of the following methods for resetting the *ARP entry age timer*:

- Reset the ARP entry age timer every time a packet is forwarded to the corresponding destination, and let the timer timeout after a maximum inactivity period.
- Do not reset the ARP entry age timer but let it time out after a maximum time period (e.g., 4 hours) from the moment a MAC address is inserted in the ARP cache [ZININALEX02].

The second method is less complex to implement and maintain. In the second method, the ARP aging process calculates the remaining lifetime for each ARP entry and, if the lifetime is less than, say, 1 minute, the ARP aging process refreshes the entry by sending an ARP request out the interface associated with the IP address listed in the entry. If the remaining lifetime is zero, the ARP entry is purged. Note that the ARP aging process does not send an ARP request when an "*incomplete*" ARP entry is removed from the ARP cache.

5.5.2.1.2 Implications of Classless Inter-Domain Routing on IP Address Lookups

CIDR also referred to as supernetting [RFC1518] [RFC1519] [RFC4632] is a method of assigning IP addresses with the goal of improving the allocation of IPv4 address. CIDR replaces the older classful addressing system based on Class A, Class B, and Class C addresses. CIDR provides efficient IPv4 allocation and was initially proposed to slow down the growing routing table sizes on routers in the Internet as well as to decrease the rapid exhaustion of IPv4 addresses. The now obsolete classful IP addressing scheme was highly inefficient and drained the pool of IPv4 addresses available and unassigned faster than was necessary.

CIDR is based on the concept of Variable-Length Subnet Masking (VLSM) [RFC950] [RFC1878] which is a method that allows IPv4 addresses to be specified with arbitrary-length prefixes (and not on classful address boundaries). CIDR introduced the CIDR notation which is a new method of representing IP addresses. In the CIDR notation, an IPv4 address or address prefix is written with a suffix that indicates the number of bits of the prefix, for example, the IPv4 address 192.168.100.0/22 has a prefix length of 22 bits (addresses from 192.168.100.0 to 192.168.103.255). With the introduction of VLSM and CIDR, the number of IPv4 addresses available for use has greatly increased.

Although CIDR did help reduce the size of Internet routing tables, it also made the address lookups in IPv4 forwarding tables more complex when compared to lookups in forwarding tables composed of classful IPv4 addresses. The prefix lengths in classful address Class A, B, and C are 9, 16, and 24 bits, respectively. The fixed classful address prefix lengths allow forwarding table lookups to be performed using exact prefix matching algorithms, for example, using search techniques such as the standard binary search technique.

With CIDR, the routing and forwarding tables contain addresses with arbitrary prefix lengths, thereby requiring lookups to be performed using longest prefix matching (LPM) algorithms rather than exact prefix matching algorithms. LPM (or best match) algorithms involve taking the IP destination address (the search key) from an arriving packet, and searching in the forwarding table for the address entry that has the longest prefix that matches the search key (the longest matching prefix). As discussed above, the forwarding table is a database generated from the routing table that contains at a minimum, IP destination addresses (prefixes) along with their corresponding next-hops and outbound interfaces. The objective of an IP destination address lookup is to find the table entry that best matches the search key and to determine the next-hop node and outbound interface to which the packet should be forwarded.

With the continuous growth of enterprise networks, service provider networks, and the Internet, routing table sizes, link rates, and packet forwarding requirements also continue to increase. To support wire-speed forwarding rates, especially at network aggregation points and at the core, routers and switch/routers must have high-speed, high-performance forwarding table lookup mechanisms. The design of the lookup mechanism is crucial to the packet forwarding performance at aggregation and core routing devices. In general, the performance of a lookup scheme can be characterized by the time it takes to perform a lookup (the lookup time), the time it takes to update the forwarding table when routing changes occur (the update time), and the memory required to implement the lookup scheme.

5.5.2.1.3 Handling Local Packet Delivery during IP Packet Forwarding

Just like end-users can send data to other end-users, routers and other network devices, in general, can be senders and receivers of information too. A bulk of the traffic sent to a router is transit traffic but a smaller portion is destined to the router itself. Routing protocol packets (e.g., RIP, EIGRP, OSPF, IS-IS, BGP), control and management protocol packets (e.g., ICMP, IGMP, SNMP), and packets destined to applications running in the route processor (e.g., SSH, TFTP, FTP, Telnet) are examples of packets are delivered locally in a routing device. Figures 5.15 and 5.16 describe the steps involved in processing and forwarding IP packets in a router.

5.5.2.1.3.1 Performing Lookups for the Local Router IP Addresses

IP packets to be delivered locally in a router include packets that are addressed to the router itself (i.e., directly to addresses of every IP-enabled interface) and addresses of IP multicast groups that the router has joined. One way a router can determine if a received IP packet is destined for the router itself is to use a small list or table of local router IP addresses as a front-end lookup table before the main lookup process (Figure 5.17). If this small lookup list does not contain the destination IP address in the received packet, then the router proceeds to use the main FIB (or forwarding table) to determine the next-hop IP address and outbound interface for the packet.

It is possible to conceive many other different ways (other than the one proposed above) to determine if a packet is due for local delivery. It is possible to integrate the local router IP addresses into the main forwarding table and then use the main lookup algorithm in the system to determine if the packet's destination address matches any local IP address. In this case, any local IP address is treated as if it is that of a directly attached host/network, except in this case, the destination is a local client process in the router itself. There are endless ways of handling local delivery and the particular design approach depends on the complexity of the router architecture. The discussion

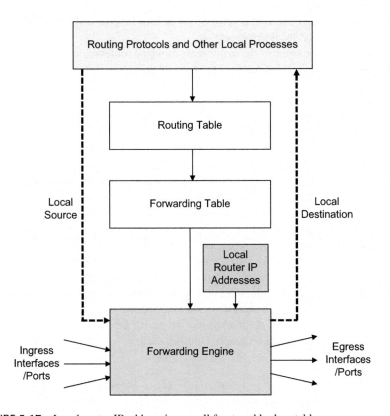

FIGURE 5.17   Local router IP address in a small front-end lookup table.

here is only to highlight the key steps involved in IP packet forwarding rather than present an optimal way of forwarding packets.

5.5.2.1.3.2 IP Time-to-Live and Local Delivery

Every IP (IPv4 and IPv6) packet has a TTL field that carries a value to indicate the lifetime of a packet in the network. The IPv4 TTL is a 9-bit field in the IPv4 packet header. In IPv6, it is called the Hop Limit field (also a 9-bit field resulting in a maximum value of 255 hops in a network). The IPv4 TTL is typically implemented as a hop count that is set by the sender and decremented by one at every IP node (router) on the route a packet takes. When the TTL value reaches zero at an intermediate node (before the packet reaches at its destination), then the packet is discarded at that node and an ICMP error message (Type 11 or Time Exceeded message) is sent back to the sender. The TTL is used to place an upper bound on the lifetime of an IP packet in the network, and prevent packets from circulating endlessly, for example, when routing loops occur in the network.

According to [RFC1812], *a router must not check and decrement the TTL value of a received IP packet before determining if the packet is destined for the router itself.* A received IP packet must not be discarded by the router just because it arrived with TTL value equal to one or zero. If the packet is destined for the router itself and is valid in other aspects, the router must do its best to receive it (barring other issues such as overwhelmed local resources).

The following are some of the major reasons why a router must not check the TTL of an arriving IP packet if is destined for the router itself:

- Since the packet has arrived and is already destined for the router itself, it is much better for the router itself to decide when it is appropriate to discard the packet.
- Furthermore, if a packet is destined for the router and has already arrived, examining or decrementing the TTL and then deciding to discard the packet could most likely deprive the router of much-needed control and management information (and most often critical, information) for its operations. There is no point and it makes no sense in discarding a packet (with 0 TTL value) if the packet has already arrived at its destination, that is, the local router. It is much better for the router to decide if it does not need the arriving packet.
- Not checking the TTL also avoids discarding critical/important network control packets carrying information such as routing updates needed by the router to participate and maintain proper operation and stability of the overall network. Critical packets destined to the router include routing updates, network control, and error messages (e.g., ICMP messages, IGMP messages, IP packets with IP header Options, etc.).

5.5.2.1.4 Benefits of Packet Forwarding via the IP Forwarding Table (or FIB) Rather than the Routing Table (or RIB)

The growth in the size of enterprise and service provider networks as well as internetworks such as the global public Internet, has created the demand for core routing devices that support large routing tables and higher packet forwarding speeds. As discussed in Chapter 2 (section "Second Generation Routing Devices") and Chapter 6 (section "Architectures Using Route Cache Forwarding"), architectures that use route caches for packet forwarding have cache entries that are created on demand when the first packet of a stream of packets heading to the same destination (i.e., a flow), is forwarded via the more processing and memory-intensive IP routing table.

Because of the use of VLSM and CIDR which results in variable-length address prefixes, lookups in the RIB and FIB are based on longest prefix matching (LPM). As discussed in Chapter 2, after a successful RIB (or FIB) lookup using LPM, the /32 destination IP address (as seen in the first packet of a flow), the next-hop IP address, the outbound interface, and other Layer 2 information required for Layer 2 packet rewrites are written in the route cache to be used for forwarding subsequent packets of the same flow.

Lookups in the route cache are faster and more efficient because they are based on the exact matching of the (fixed) /32 destination IP addresses of arriving packets (see the "Exact Matching in IP Route Caches" section in Chapter 6). However, route cache-based forwarding is seen as unsuitable, especially, for core networks where the traffic mix is high and many flows are short-lived.

5.5.2.1.4.1 *Limitations of Route Cache-Based Forwarding*

The limitations of route cache-based forwarding can be summarized as follows:

- Route cache entries are generated on demand and in core networks, in particular, continuous cache updates can easily overwhelm the control or route processor which is responsible for updating the route cache. This forwarding method is not scalable in large enterprise and service provider networks and the Internet, as core routers have to process and forward a considerably higher amount of first packets (of flows) for which no route cache entries are available. Such packets have to be forwarded via the routing table which may contain recursive routes, causing the route processor to spend more processing time performing recursive route lookups. Recall from the discussion above that all BGP routes specify only intermediate network addresses and not interfaces, a situation that calls for recursive lookups.
- The entries of a route cache are destination-based, which means, core routers using such a method have to process and forward packets going to a large number of network destination addresses, which in turn means a large memory has to be used to hold the route cache entries. So, given that the cache memory has to be limited, cache overflows can occur, resulting in continuous cache invalidation and creation.
- Route caches are not designed to support features such as per-packet load balancing on parallel routes to a common destination. This means when route cache-based forwarding is used and per-packet load balancing is required, such a feature has to be delegated to the route processor which has more flexibility for advanced feature implementation but results in performance degradation.

5.5.2.1.4.2   Characteristics of FIB-Based Forwarding

An architecture that forwards packets using the network topology-based IP forwarding table (or FIB), as described above, remedies the limitations of route cache-based forwarding. A FIB-based forwarding architecture has the following characteristics:

- An FIB is a forwarding database that is generated from the information in the routing table; it is not cache-based.
- The data structure in the FIB is optimized for high-speed IP destination address lookups, and also provides efficient storage of routing information. Being a scheme that is based on classless addressing as noted above, lookups in an FIB are based on LPM.
- Recursive routes in the routing table are completely resolved before being placed in the FIB; no recursive lookups required in the FIB. A router can forward packets along a route that does not reference an interface in the routing table, but it has to perform a recursive routing table lookup to determine the associated outbound interface.
- The FIB is always synchronized with the routing table; all changes in the routing table are reflected in the FIB. This means FIB entries are totally network event-driven and not packet-driven as in the route cache.
- Each entry in the FIB may contain adjacency information about the next-hop node (see Section 5.5.2.1.1 above). The adjacency information may include Layer 2 (Ethernet) packet headers plus all Layer 2 parameters needed for rewriting and encapsulating Layer 2 packets.
- An FIB can be used for per-packet and per-destination load balancing on parallel routes unlike the route cache which can support only per-destination load balancing. Note that multiple parallel routes (i.e., next-hops) to a given network destination must exist and be installed in the routing table for a router to be able to perform equal path load balancing (as in RIP, OSPF, EIGRP) or unequal path load balancing (as in EIGRP).

Each entry of the FIB contains a network address, network mask, routing information source/protocol, next-hop address (or multiple next-hop addresses in the case of equal/unequal cost multi-path routing), next-hop Layer 2 parameters or a pointer to an adjacency entry that holds this information, and possibly, load balancing/distribution parameters. The adjacency table contains next-hop information that is used for rewriting and encapsulating Layer 2 packets heading to the next-hop. Each entry of the adjacency table may also contain pre-computed Layer 2 headers for the Layer 2 packet to be forwarded to the next-hop.

When an FIB-based routing device receives a packet, it makes all attempts to forward it using the FIB, and if this fails, the packet may be dropped or forwarded to the route processor for further attention. If the FIB-based routing device does not support special encapsulation or any other feature for a received packet, the forwarding engine typically forwards the packet to the route processor for further processing. Note that in a network undergoing routing transition, some routes may not yet be resolved or some Layer 2 parameters may yet to be known.

5.5.2.1.4.3   Generating the FIB from the Routing Table

The typical routing device using an FIB has an *FIB maintenance process* that is responsible for creating and maintaining the FIB, and an *FIB route resolution process* for resolving routes that specify intermediate addresses only. The FIB is generated as follows [ZININALEX02]:

- All routes in the routing table (structured based on network address prefixes) are passed to the *FIB maintenance process* which creates the entries of the FIB. Each network prefix in the routing table has a corresponding FIB entry.
- Routes that specify intermediate addresses only such as BGP routes, are processed by the *FIB route resolution process* which walks through each route and tries to resolve any unresolved route.
- Whenever network changes occur and the contents of the routing table change, the *FIB maintenance process* is notified, which then uses the new routing information to change the affected FIB entries.

The FIB may contain special entries that contain the IP addresses of the local router itself (e.g., IP addresses of the local interfaces of the router). Packets destined to these addresses are delivered to the router itself and are not transit packets.

5.5.2.1.4.4   Special Adjacencies

Most routing devices such as Cisco routers use special types of adjacencies to instruct the FIB forwarding process on how to handle certain special or excerption packets [STRINGNAK07] [ZININALEX02]:

- **Punt Adjacency**: This type of adjacency is used when a received packet has features that are not supported by the FIB forwarding process and has to be punted to the route processor for further processing.
- **Drop Adjacency**: This is used for routes that reference the Null Interface. Packets forwarded to the Null Interface are dropped by the routing device. The Null Interface is some sort of a "black hole" interface; all packets sent to this interface are discarded. It is mostly used for filtering unwanted packets that arrive at the routing device.
- **Incomplete Adjacency**: This is used to indicate that an adjacency is not operational such as when an interface to a next-hop has gone down.

5.5.2.1.5   Reasons for Dropping Packet in Routing Devices

In this section, we describe a number of reasons why a routing device will have to drop (discard) a packet rather than forwarding it:

- **Lack of routing information**: A packet may be dropped by a routing device due to lack of routing information about the destination of the packet.
- **Lack of processing and memory resources**: A packet may be dropped due to lack of processing and memory resources to process the packet which

can occur when the routing device is overloaded or overwhelmed by too much traffic than is expected. Queue overflows can result in packets being dropped.

- **Expired TTL**: A packet may be dropped when the value of its TTL reaches 0 (see TTL discussion above). The TTL is a mechanism used to prevent packets from looping around endlessly when routing loops occur in the network.
- **Inability to fragment a packet**: A routing device may drop a packet with size greater than the Maximum Transmission Unit (MTU) of an interface. This happens when the routing device is unable to fragment the packet and transmit it on that interface. The interface's MTU is the maximum packet size (i.e., data block size) that can be sent on the interface. Typically, every interface has a known MTU which allows the routing device to determine if a packet cannot be transmitted on that interface and if it requires fragmentation.

When a packet is about to be sent over an interface, the routing device will check if the packet fits within the interface's MTU. If the router determines that the packet is bigger than the interface's MTU and it supports fragmentation, it will split the packet into smaller pieces (fragments) that fit into the interface and transmit each as a separate IP packet. The process of creating the IP fragments is called *IP fragmentation*. The destination host of the fragmented packet is responsible for reassembling all fragments into the original unfragmented IP packet. Intermediate network devices treat each fragment as a separate independent IP packet. Note that, depending on the MTUs of the interfaces along a route, it is possible for a fragment to be again fragmented by other routing devices.

An IP packet that has the "do not fragment (DF)" bit set in its IP packet header must not be fragmented. When a routing device capable of fragmentation receives an IP packet with a size greater than an interface's MTU and the DF bit is set, the router will simply drop the packet and send an ICMP "Destination Unreachable" message to the packet's originator with code field set to "Fragmentation needed and DF set". A routing device that is not capable of fragmentation will just drop such a packet and send the same message type.

Most routers do not support fragmentation since this process takes up further processing and memory resources. This means such routing devices will simply drop a packet with size greater than the interface's MTU rather than attempt to process it further. The default MTU for Ethernet interfaces is 1500 bytes.

5.5.2.1.6   Why Data Plane Operations Are Amenable to Hardware Processing that Control Plane Operations

Traditional software-based routers use the general-purpose CPU used to perform control plane operations for data plane operations – data plane operations are also handled in software. Present-day high-end routers and switch/routers, on the other hand, use an application-specific integrated circuit (ASIC) to handle data plane operations. This is mainly because it is very easy to program/implement the very simple operations required for the data plane into an ASIC.

The reasons for partitioning the control plane and data plane into software processing and hardware processing, respectively, are straightforward. This is because this presents the best way to optimize packet forwarding speed and still support the complex processing required by the routing protocols. A general-purpose CPU is structured to support the computation of many (complex) different functions (like those involved in routing protocols). An ASIC, on the other hand, is structured to support the processing of a smaller number of specific and simple functions such as those required for the data plane operations and packet forwarding. The data plane operations tend to be very simplistic and repetitive in nature, making them more amenable to ASIC implementations.

An ASIC is able to operate much faster (than a general-purpose CPU) because the internal architecture of the ASIC can be optimized just to perform the operations required for data plane operations. A general-purpose CPU can handle much better a series of complex functions that do not relate to data plane operations. In addition to handling control plane operations, the CPU must support other applications such as those related to system configuration and management.

In the traditional software-based router, a high-level programming language is combined with the generic functions of the general-purpose CPU to provide the specific functions required to perform both the complex control plane operations and the data plane operations. This integrated approach provides flexibility in the implementation of complex operations but comes at the price of decreased forwarding performance and scalability. For these reasons, a router or switch/router that performs data plane operations using ASICs tends to forward packets much faster than a traditional router that performs data plane operations using a general-purpose CPU.

5.5.2.2 Unicast Reverse Path Forwarding

The data or forwarding plane also plays a key role in the implementation of a number of network security mechanisms. It is now common knowledge that IP address spoofing may occur during a denial-of-service (DoS) attack. IP spoofing allows an intruder or malicious user to send IP packets to a destination with the intent of disguising it as genuine traffic, when in fact the packets are malicious and not actually genuine and should not be forwarded to the destination. This type of spoofing is harmful because it consumes network and destination host resources, and sometimes can bring down the operations of the network and the destination. *Unicast Reverse Path Forwarding (uRPF) check* [RFC3704] is a tool used to reduce the forwarding of IP packets that may be carrying spoofed IP addresses.

Network administrators can use uRPF to limit malicious traffic on a network. uRPF helps to prevent problems that are caused by the introduction of malformed or forged (spoofed) IP source addresses into a network, by discarding IP packets that lack a verifiable (i.e., resolvable) IP source address. uRPF limits such attacks by forwarding only packets that have source addresses that are valid and are resolvable through the receive interface in the IP forwarding table.

With uRPF, the router examines each packet received on an interface to make sure that the packet's source IP address and an associated interface are in the forwarding table and match the interface on which the packet was received. uRPF is implemented as an input function in the packet forwarding process and is applied only on the inbound interface of a router and toward the upstream end of a flow or connection. uRPF checks to see if any packet received on a router interface has arrived on one of the best return paths to the source of the packet.

uRPF does this by doing a *reverse lookup* in the forwarding table using the source IP address of packets. If the packet was received from an interface that has one of the best reverse paths (i.e., one of the best routes that leads back to the packet's source), the packet is forwarded as normal. If there is no reverse path on the same interface from which the packet was received, this might mean that the source address was modified or forged. If uRPF does not find a reverse path for the packet, the packet is silently dropped (without any notification sent to the source).

One major disadvantage of uRPF checks is that they may cause valid and genuine packets to be discarded in a network with asymmetric routing, that is, if the forward path and reverse path between two points in the network are not topologically identical. In such a case, asymmetric routes will cause the uRPF checks to fail and valid packets to be discarded. This means the network administrator must ensure that asymmetric routing is not present before enabling uRPF checks at a router.

5.5.2.3 Multicast Reverse Path Forwarding

The RPF check is also essential for multicast implementation in routers or switch/routers. RPF is a fundamental concept in multicast routing that enables routers to correctly forward multicast traffic down the multicast distribution tree. As discussed above, unicast forwarding decisions are typically based on the IP destination address of the packet arriving at a router. As a result, the unicast routing table is organized by destination network prefix (subnet, supernet, or host) and mainly set up to forward the packet toward the destination.

In multicast, the router forwards the packet away from the multicast source, and the packet then travels along the multicast distribution tree, and with each forwarding stage designed to prevent routing loops. The router's multicast forwarding state runs more logically by organizing multicast forwarding tables based on the reverse path, from the receiver back to the root of the multicast distribution tree. The process of RPF is used when organizing the multicast forwarding tables based on the reverse path. The router adds a branch to a multicast distribution tree depending on whether the request for traffic from a multicast group passes the RPF check.

When handling multicast traffic, RPF uses the existing unicast routing table to determine the upstream and downstream neighbors. A router will only forward a multicast packet if it is received on the upstream interface leading back to the source (Figure 5.18). This RPF check helps to guarantee that the distribution tree will be loop free.

Every multicast packet received must pass an RPF check before it is eligible to be replicated or forwarded on any interface. When a multicast packet arrives at a router, the router will perform an RPF check on the packet, and if the RPF check is successful, the packet will be forwarded, otherwise, it will be dropped. The router looks up the IP source address of the packet in the unicast routing table to determine if it has arrived on the interface that is on the reverse path back to the source. If the packet has arrived on the interface leading back to the source, the RPF check is successful, and

RPF Check Fails:

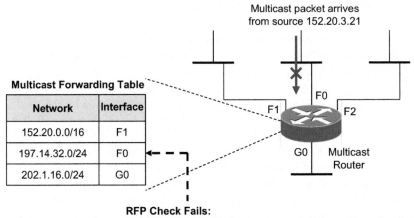

RFP Check Fails:
Multicast packet arrived on wrong interface. Discard packet.

RPF Check Succeeds:

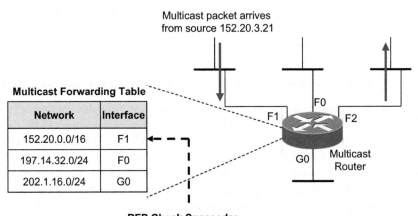

RFP Check Succeeds:
Multicast packet arrived on correct interface. Accept packet.

FIGURE 5.18 Multicast reverse path forwarding.

the packet will be forwarded. If the RPF check fails, the packet is dropped. The following are additional features of multicast RPF:

- RPF checks are performed only on unicast IP addresses to find the upstream interface for the multicast source or Rendezvous Point (RP). The routing table used for RPF checks can be the same routing table used to forward unicast IP packets, or it can be a separate routing table used only for multicast RPF checks. In either case, the RPF table contains only unicast routes, because the RPF check is performed on the IP source address of the multi-cast packet, not the multicast group destination address.

- Note that a multicast address is forbidden from being used in the source address field of an IP packet. The unicast address is used for RPF checks because there is only one source host (IP address) for any given stream of IP multicast traffic sent to a multicast group address, although the same content could be available from multiple sources.
- If the routing table used to forward unicast packets is also used for the RPF checks, the routing table is populated and maintained by the traditional unicast routing protocols such as BGP, IS-IS, OSPF, and RIP. If a dedicated multicast RPF table is used, this table must be populated by some other method. Some multicast routing protocols (such as the now obsolete Distance Vector Multicast Routing Protocol (DVMRP)) essentially duplicate the operation of a unicast routing protocol and populate a dedicated RPF table. Others, such as Protocol Independent Multicast (PIM), do not duplicate routing protocol functions and must rely on some other routing protocol to set up this table (PIM is protocol independent).
- Using the main unicast routing table for RPF checks provides simplicity. However, a dedicated routing table for RPF checks allows a network administrator to set up separate paths and routing policies for unicast and multicast traffic, allowing the multicast network to function more independently of the unicast network.

5.5.3 EXAMINING THE BENEFITS OF CONTROL PLANE AND DATA PLANE SEPARATION

We discussed above that the process of routing IP packets through a routing device can be decomposed into two interrelated components, the control plane and the data (or forwarding) plane. These components can be implemented independently in the routing device and even do not have to be co-located. Through the exchange of routing information with other routing devices, the control plane supplies the information needed for constructing and maintaining the routing tables from which the forwarding tables are generated. The forwarding plane is responsible for performing time-critical tasks such as receiving a packet, reading the IP destination address, looking up the address in the forwarding table for the next-hop node and outbound port, updating certain packet fields, and transferring the packet to the outbound port.

In this section, we examine the key motivations for control plane and data plane separation. Certain design attributes for routing devices are dependent on the ability to decouple the control plane from the data plane.

5.5.3.1 Scalability and Distributed Forwarding Architectures

The primary motivation for decoupling the control plane from the data plane is the ability to scale the packet forwarding performance of a routing device (see Figures 5.19 and 5.20). This can be done through the use of multiple forwarding engines (i.e., a pool of parallel forwarding engine), and distributing the forwarding engines to the line cards (see "Architectures Using Topology-Based Forwarding

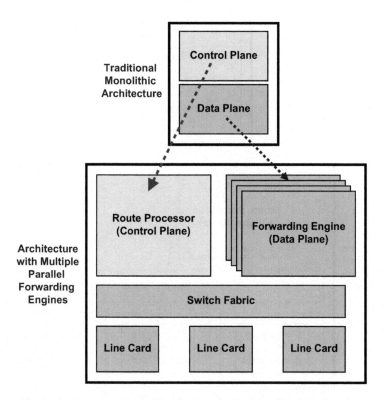

FIGURE 5.19 Scaling the forwarding capacity using a pool of parallel forwarding engines.

Table Forwarding" section in Chapter 6 and "Distributed Forwarding – Express Fast Path Forwarding" section in Chapter 7).

5.5.3.1.1 *Facilitates the Design of Distributed Forwarding Architectures*

One of the key motivations of using a distributed architecture is the ability to group together the non-time-critical control plane tasks and implement them on a centralized processor (i.e., the control engine or route processor), and to implement all the time-critical data plane tasks on multiple parallel forwarding engines or on distributed forwarding engines in line cards. The control plane tasks of exchanging routing information with neighbor devices, building and maintaining the routing tables can easily be implemented on a centralized processor as discussed earlier.

An architecture with a single processor running every task is not scalable because the processor is shared among the multiple tasks running on it, some of which can be processing intensive and time consuming such as the processing of BGP routing updates [AWEYA2BK21V2]. This centralized single processor architecture can create situations where a flood of routing information updates can lead to network instability. This happens because the processor cannot adequately and timely process the flood of routing updates which in turn can degrade the packet forwarding performance of the overall system.

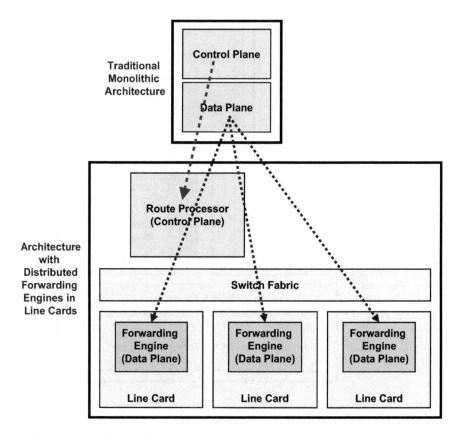

FIGURE 5.20 Scaling the forwarding capacity using distributed forwarding engines in line cards.

Particularly, in core routers handling hundreds of simultaneous BGP sessions, network instability can be exacerbated when the routers are not able to process routing updates timely. Thus, decoupling the routing protocols tasks (the exchange of routing information and routing table maintenance) from the actual process of forwarding packets, provides a more scalable solution. A distributed architecture allows these two processing planes to be implemented in their own separate processing modules.

The time-critical tasks such as destination address parsing and lookups plus packet field rewrites tend to be simplistic and repetitive in nature and can be implemented on the distributed forwarding engines and line cards. This architecture allows the time-critical tasks to be optimized on each line card according to the link speeds and line card requirements.

5.5.3.1.2   Scaling the Packet Forwarding Capacity of a Routing Device

As the traffic in enterprise networks, service provider networks, and the Internet continue to grow, link rates, aggregate bandwidth requirements of core routers, and the size of routing tables also continue to increase. Although the use of optical fiber transmission, for example, has allowed link rates to keep pace with traffic growth, these factors call for the packet forwarding rates of routers to increase to match the traffic they receive. However, one of the major factors limiting the ability to increase the forwarding rates of routing devices is the bottleneck created by IP address lookup operations.

The packet forwarding performance of a routing device can be scaled by adding more forwarding engines when the device uses a pool of forwarding engines, or the time-critical forwarding tasks can be optimized and implemented on multiple distributed ASIC or specialized processors. The data plane performs time-critical tasks such as parsing IP destination address from packets and forwarding table lookups. Generally, forwarding table lookups constitute the biggest processing bottleneck in routing devices. This means providing more processing resources for packet forwarding is an effective way of scaling up the packet forwarding performance of a routing device. The forwarding capacity of a routing device can be scaled up as the aggregate arriving traffic and link rates increase.

Also, given that the forwarding table is a critical component of the data plane (i.e., on the time-critical forwarding path), an efficient implementation of the forwarding table is one way of scaling up the packet forwarding performance of a routing device. Designers are always looking for ways to optimize the forwarding table to achieve the smallest address lookup times. Designers are interested in efficient lookup algorithms, and ways to achieve the lowest forwarding table update times, as well as the smallest memory required for address information storage and lookup operations (including efficient address information data structures).

Routing tables on the other hand are typically optimized to reduce routing information storage and routing update times (insert/modify/delete operations) to allow the routing device to react quickly to routing changes. Typically, high-performance routing devices use hardware or a combination of hardware and software architectures for faster forwarding table lookups and lower memory consumption.

5.5.3.2 Control Plane Redundancy and Fault Tolerance

Another important motivation for separating the control plane from the data plane is the ability to build routing devices with control plane redundancy through the use of multiple route processors (or control engines) as shown in Figure 5.21. Separation of the control plane and the data plane is a prerequisite for implementing control plane redundancy. If the control plane and the data plane are not decoupled and made independent, it would be very hard to localize and separate control plane failures.

Adding control plane redundancy improves the fault tolerance, reliability, and availability of a routing device (see "Node Redundancy and Resiliency" section in Chapter 1 of Volume 2 of this two-part book). Control plane hardware and software failures are major contributors to routing device outages [HUSSFAULT04]. Thus, the ability to separate the control plane from the data plane allows techniques such as Stateful Switchover (SSO) to a standby route processor and Non-stop Forwarding (NSF) to be used. Such techniques, when used, lead to a high-availability routing device design. Without the ability to decouple the control plane from the data plane, it would be difficult to perform control plane switchover and recover control plane state information without disrupting the packet forwarding operations.

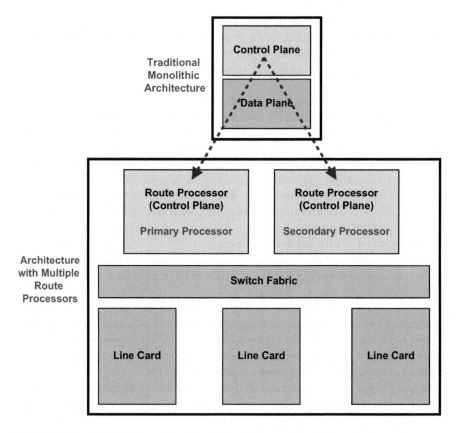

FIGURE 5.21   Control plane redundancy through the use of multiple route processors.

5.5.4   Lookup Tables Used in Layer 3/4 Forwarding Operations, QoS, and Security ACLs

Improvements in optical communication technologies such as Dense Wavelength Division Multiplexing (DWDM) and Coarse Wavelength Division Multiplexing (CWDM) have enabled the design of communication networks with higher link speeds. Link speeds continue to increase with new advances in optical technologies. However, the forwarding speeds in devices such as routers have not increased correspondingly to keep up with the higher link speeds. The main reason for this lag is the relatively complex packet processing and forwarding operations required at a router when compared to the simpler Ethernet switch. As a result, the problems of speeding up routing/forwarding table lookup and packet classification have received considerable attention by network device manufacturers.

When an IP packet is being routed from its source to its destination, a destination address lookup is performed at every router that is encountered on the path of the packet. The address lookup operation at each router involves performing a comparison of the packet's IP destination address against every entry in the forwarding table of the router. The correct forwarding table entry for the incoming packet points to the

IP next-hop address and outbound interface on the router for the packet. The packet is forwarded out this outbound interface (to the next-hop address) on its way to the final destination.

However, since it would be prohibitively expensive or impossible to record the individual destination address of every end-system in a network, routers store entries in the routing or forwarding tables in a compact form. This is achieved by storing destination addresses as network prefixes (referred to simply as prefixes). Each entry has a network mask that records the bits of the network address that need to be considered when the router performs the forwarding table lookup.

As illustrated in Figure 5.11, every entry in the forwarding table contains a network prefix, a mask, the next-hop address, and an outgoing interface. Each time the router receives an IP packet, it extracts its destination IP address and performs a bitwise-AND of the IP destination address with the network mask of each forwarding table entry. The resulting data is compared with the network prefix of the corresponding forwarding table entry. If a match is found, the IP packet is forwarded to the outbound interface and next-hop pointed to by the matching forwarding table entry.

However, in some cases, an IP destination address may match two or more entries in the forwarding table. In this case, the router forwards the packet to the interface corresponding to the forwarding table with the longest network prefix. This is referred to as *Longest Prefix Matching* (LPM). The router compares the prefix lengths of each entry, finds the longest matching prefix, and forwards the packet to the corresponding interface. A longer prefix indicates that more specific forwarding information is available for all the matching prefixes, and therefore the router should forward the packet to the next-hop associated with the longest prefix.

IP addresses were originally partitioned using a class-based addressing scheme. Class A, B, and C addresses utilized 9, 16, and 24 bits for the network part, respectively. Because the number of class-based IP addresses was rapidly being depleted, there was a strong need to utilize IP addresses more efficiently. This motivated the introduction of Classless Inter-Domain Routing (CIDR) [RFC1519] in 1993. In this addressing scheme, networks were permitted to have an arbitrary number of network bits (i.e., arbitrary length network prefixes), allowing more flexible IP address allocation.

The downside of this addressing scheme was that it resulted in an increase in the size of IP routing tables, due to the fine granularity of addressing that results from the use of CIDR addressing. In the CIDR scheme, routing table entries could have arbitrary length prefixes, allowing for more efficient assignment of IP addresses and route aggregation (also called route summarization or supernetting.

In order to provide enhanced services, such as packet filtering, traffic shaping, policy-based routing, routers, switch/routers, and switches also need to support the ability to identify and classify flows. A flow is a set of packets that can be identified based on some rule (also called a policy), which is done by looking at some or all of the header fields of the packet. These fields can include IP source and destination addresses, source and destination port numbers, protocol, and other patterns in the packet. For example, packets with a specified IP destination source address may be identified as a single flow by the router using a rule defined by these special packet identifiers. A collection of rules is often referred to as a *policy database* which in turn is used by a classification engine or classifier. The identification of the flow to which an incoming packet belongs is called *packet classification* and is viewed as a generalization of the routing table lookup operation.

Packet classification requires the network device to find the "best-matching rule" among the set of rules in a given policy database that match the relevant patterns in an incoming packet. A rule may be specified by defining a network address prefix, a range of addresses, or an identifiable pattern that can be compared against each of several fields of the packet header. It may happen that the patterns in the header of an arriving packet may satisfy the conditions of more than one rule in the policy database. One approach typically used to resolve this situation is to use the rule with the highest priority to classify the arriving packet.

The address lookup operation is preferably done in hardware instead of software because the former provides extremely fast lookup speeds. One common hardware implementation for IP address lookups is the Ternary Content Addressable Memory (TCAM) [MCAFRA93] [PEIZUKO91] [WADSOD89]. TCAMs are designed to store an entire forwarding table and allow the simultaneous comparison of all entries against the IP destination address of an arriving packet. A good overview of existing TCAM approaches is found in [MCAFRA93].

The TCAM is a popular hardware device for performing fast IP address lookups and packet classification. The TCAM is typically used in Layers 2 to Layer 4 packet forwarding decisions. Current commercial offerings of TCAMs focus on flexible lookup table configuration, and higher forwarding table capacity and lookup frequency. Current TCAMs support CIDR with variable prefix lengths and extensions for QoS and security processing.

A TCAM is similar to a CAM [RABCHNIK03] with the additional ability to disregard a subset of address bits while performing the lookup. The address bits that are considered during an address lookup operation correspond to the non-zero mask bits of the forwarding table entry. TCAMs used in routers must be able to perform LPM operations to resolve situations where multiple entries match a packet's IP destination address.

Each entry in a router's TCAM has a network prefix and a mask, along with the associated next-hop information. Network prefixes and masks can be of arbitrary length and together specify the network portion of the network's addressing space. A mask value is used to decide which bits of the key are actually relevant. When a mask bit is a "1", the corresponding prefix bit ("0" or "1") is considered in determining a match. Conversely, when a mask bit is a "0", the corresponding prefix bit is disregarded.

A TCAM cell (which consists of a prefix and mask bit pair) can logically take on one of three states: 0, 1, or X (i.e., don't-care) bit values; this results in a three-way or *ternary* outcome for a bit position (Figure 5.22). Note that in a typical implementation, all "don't-cares" (X's) are limited to successive right-hand bits for any entry. The last entry in Figure 5.22 consists of all X's to represent a default route.

Generally, for IP address lookups and packet classification applications, TCAM entries consist of *Value*, *Mask*, and *Result* (VMR) combinations. Fields from Ethernet frame or IP packet headers are passed to the TCAM, where they are matched against

Prefix/Mask	TCAM Format
101/3	101X
111/3	111X
10/2	10XX
0/0	XXXX

FIGURE 5.22   Simplified example of TCAM format with each TCAM entry storing a 4-bit word.

the value and mask pairs to yield a result. In some Cisco routing platforms, for example, Catalyst 6500, these can be described as follows:

- **Values:** These are always 134-bit quantities and consist of IP source and destination addresses and other relevant protocol information. The information that is concatenated to form the *Value* is dependent upon the type of database (e.g., access control list (ACL)) to be configured. *Values* in the TCAM come directly from any IP address, UDP/TCP port, or other protocol information.
- **Masks:** These are also 134-bit quantities, in exactly the same format, or bit order, as the *Values*. A *Value* consists of a number of bits and the *Mask* selects only the *Value* bits of interest. A *Mask* bit when set (i.e., equal to 1) exactly matches a *Value* bit, and when not set (i.e., equal to 0) means a *Value* bit should be ignored.
- **Results:** These are numerical values that represent the action to be taken after the TCAM lookup is performed. Where traditional access lists support only a permit or deny result, TCAM lookups can support a number of possible results or actions. For example, the *Result* can be a permit or deny decision, an index value to a QoS policer, a pointer to a next-hop routing table, and so on.

The LPM computation is typically done in hardware, either using dedicated hardware [KOBAYA00], or by arranging the routing table entries in a specific order as described in [SHAGUP01]. Typical TCAM-based hardware IP address lookup approaches store the entries in groups in increasing order of their mask lengths [SHAGUP01] as illustrated in Figure 5.23a. Typical TCAM implementations store forwarding table entries in clusters [SHAGUP01] [WADSOD89], where each cluster contains IP address entries of a particular mask length. This allows for fast lookups but results in a worst-case insertion penalty.

In case of multiple matches, the LPM computation simply requires that we find the match from the group with the largest prefix length. The major drawback in these approaches is that insertion of a new entry may require O(n) entries to be re-arranged (i.e., to create the space required to add the new entry and ensure that the groups are

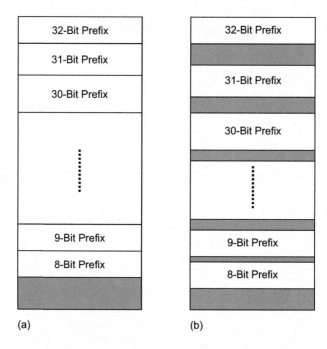

FIGURE 5.23 TCAM memory pool organization: a) This simple solution keeps the free space pool at the bottom of memory, b) This solution improves the average case update time by keeping empty spaces interspersed with prefixes in the TCAM.

maintained in increasing order of prefix lengths), where n is the length of the network address. For IPv6, this could result in worst case insertion delays of 128 clock cycles, which is undesirable in large backbone routers.

An alternate LPM implementation keeps some entries unused for each group, for possible use at a later time [SHAGUP01] as illustrated in Figure 5.23b. When a new entry is inserted, it is placed in the free space of the group corresponding to its prefix length. The drawback of this scheme is that portions of the TCAM memory remain unused, and the worst-case insertion still requires n clock cycles.

Often, the free space pool shown at the bottom of Figures 5.23a and 5.23b is located in the center of the TCAM, resulting in a halving of the worst-case insertion delay. Even with such an implementation, the worst-case insertion cost of such schemes is linear in n [SHAGUP01]. In the scheme of [SHAGUP01], memory management was performed external to the TCAM. Although this resulted in a reduction of the worst-case insertion delay, the proposed memory management was performed in software, reducing the effectiveness of the technique.

Some commercial TCAM solutions allow the insertion of new routing entries in arbitrary locations within the TCAM. In such approaches, the LPM operation requires the use of a priority encoder [CILETT02]. The drawback of this technique is that the priority encoder circuit required for the LPM task has an implementation whose complexity grows linearly with n.

An alternative approach [KOBAYA00] also allows entries to be stored in arbitrary locations. In this scheme, IP address lookup is a non-pipelined, two-stage operation.

The TCAM performs the lookup in the first phase and performs a bitwise OR of the matching entries' masks to produce the longest mask. This "longest mask" is fed back to the TCAM to further constrain the original matching entries to produce the entry with the longest prefix. The main drawback of this approach is that in lowering the cost of insertion, the cost of each lookup is doubled.

In the implementation in [GAPFKH03], routing table entries can be stored in any order, thus eliminating the large worst-case insertion cost of typical TCAM implementations, as described in [SHAGUP01]. In addition, the method utilizes a Wired-NOR-based LPM circuit, whose delay scales logarithmically with n, thus improving over the linear complexity (in the size of the TCAM) of priority encoder-based circuits. The goal of the design is to simultaneously achieve both fast updates to the IP forwarding table by allowing arbitrary insertion of the entries and high-speed search throughput as well. The architecture of the TCAM is pipelined and provides 1 lookup per clock cycle with a latency of 3 clock cycles.

Figure 5.24 illustrates Layer 3 forwarding using a typical TCAM. In the Cisco Catalyst 6500, the TCAM is always organized by masks, where each unique mask

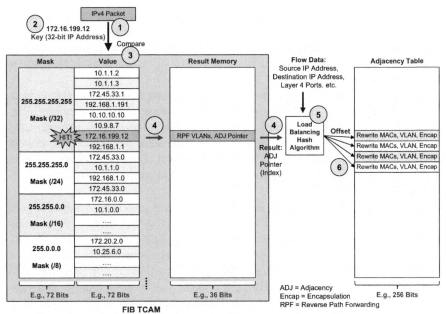

Processing Steps:
1. IP packet is received and destination IP address is read from the packet
2. Lookup key created based on destination IP address in packet
3. Lookup key is compared to TCAM entries while applying associated mask
4. Longest prefix match entry returns an index to an adjacency table and the adjacency or number of adjacencies involved in load-sharing, if applicable
5. The adjacency index and packet field data applicable to the load-sharing scheme are fed to a load-sharing hash function
6. Load-sharing hash result returns an adjacency offset value that is used to select an adjacency entry in the indexed adjacency table (containing the appropriate next-hop information)

FIGURE 5.24   Generic Layer 3 forwarding using a TCAM.

has eight value patterns associated with it. The Catalyst 6500 TCAM (one for security ACLs and one for QoS ACLs) holds up to 4096 masks and 32,768 value patterns. Each of the mask-value pairs is evaluated simultaneously, or in parallel, yielding the best or longest match in a single table lookup. The Catalyst IOS Software has two components that are part of the TCAM operation:

- **Feature Manager (FM):** After a security or QoS ACL has been created or configured via the Catalyst IOS Software, the Feature Manager software compiles, or merges, the access control entries (ACEs) into entries in the TCAM table. The TCAM can then be consulted by the forwarding engine for packet forwarding.
- **Switching Database Manager (SDM):** The TCAM can be partitioned on Catalyst switches into areas for different functions. The SDM software configures or tunes (reorganizes) the TCAM partitions, if needed.

5.6   SPECIAL FOCUS: CONTROL PLANE MANAGEMENT SUBSYSTEMS

Network devices such as switches, switch/routers, and routers generally support network management capabilities that allow a network administrator to perform device configuration, monitor system and network performance, and troubleshoot and isolate system faults. This section describes the control plane functions that are related to network device management.

Recall that in this book, we consider the control plane to include all system management functions (i.e., functions related to device monitoring, information collection and exchange, device access, and device configuration and management). Figure 5.25 shows a high-level view of the management subsystem in a network device such as a switch/router. It should be noted that registers and counters (both hardware and software-based) play an important role in information/statistics collection and device monitoring.

Most network devices such as switches, routers, and switch/routers are designed to be managed over the network of which they are part of. This mode of device management is called *in-band management*. Most managed devices also have serial (console) port connections that allow the device to be configured and managed via a terminal, or a workstation or laptop (running a terminal emulation program). This mode of device management is called *out-of-band management*. Out-of-band management can also be done via a dedicated Ethernet management port on the network device (see Chapter 3 of Volume 2).

Most switches, routers, and switch/routers support SNMP and RMON which are extremely useful for troubleshooting and diagnosing network problems. Such devices also support full-screen (some web-based) menu-based interfaces via SSH or Telnet. These interfaces are typically graphical user interfaces (GUIs). These interfaces are usually very easy to use (unlike a CLI) and can often display device and network status and statistics. Often the full-screen interfaces can update statistics automatically, giving the network administrator a live view of the network operations.

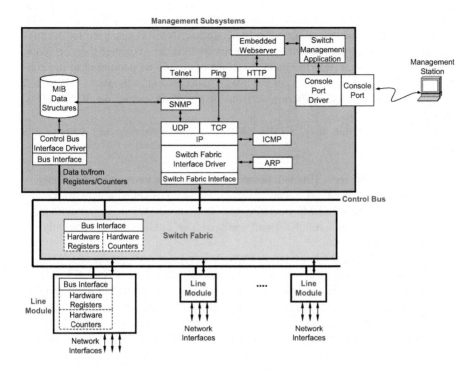

FIGURE 5.25   Generic switch/router architecture showing management subsystems.

Most devices support both CLIs and full-screen interfaces (see Chapter 3 of Volume 2 for more discussion on CLIs). The CLI can be extremely handy for performing quick checks and configuration changes on the device.

Unlike Telnet, SSH (SSH-2 as discussed in Volume 2) provides secure communication between two network nodes. SSH provides authentication and confidentiality for information transfer and can be used for remote management access to a network device. Although SSH offers the same benefits as Telnet, it provides additional features such as end-to-end security, broad compatibility with many SSH clients and servers in use today, and the ability to access and manage multiple sessions over a single SSH connection. Some of the security features in SSH-2 include Diffie-Hellman key exchange, data integrity and authenticity checking using Message Authentication Codes, and multiple sessions over a single SSH connection.

In addition to access protocols such as Telnet and SSH, and path tracing and troubleshooting protocols such as Ping, SNMP and RMON play an important role in both device and network monitoring and management. We discuss below these two important network management protocols.

5.6.1   Simple Network Management Protocol

As discussed in Chapter 3, SNMP [RFC1157] is an IP-based protocol that runs on UDP and is used for monitoring, collecting and organizing information, and managing devices (e.g., switches, routers, switch/routers, servers, IP hosts, IP telephones, IP video cameras, modems) in IP networks. SNMP is also used for modifying parameters in managed devices in the network in order to change their behavior. This section discusses only the main functions of SNMP as pertaining to the control plane of a routing device; control plane management subsystems. Chapter 2 of Volume 2 discusses the different versions of SNMP in greater detail (SNMPv1, SNMPv2, SNMPv3).

An IP network that uses SNMP consists of the following key components (Figure 5.26):

- **Managed Devices**: This is a device in the network that is monitored and managed via SNMP. A managed device (sometimes called a network element) supports an SNMP interface that allows node-specific information to be accessed – bidirectional (read and write) or unidirectional (read-only) information access.
- **Network Management System (NMS)**: In typical deployment of SNMP, one or more computing stations (under administrative control) called SNMP managers, are delegated the task of monitoring and managing a group of devices in a network. The software application that runs on the SNMP manager is generally called an NMS. The SNMP manager runs applications that monitor and control the managed devices. Many times, the SNMP and

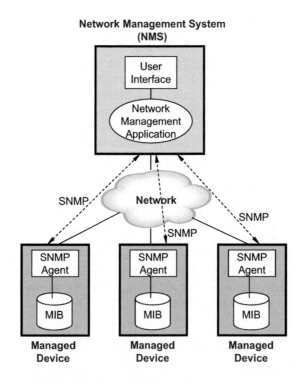

FIGURE 5.26 Network management model.

NMS are used interchangeably. A managed device exchanges node-specific information with the SNMP manager. The NMS can gather and correlate information about the network to provide more human-readable presentation that includes the following:

o Network maps (indicating device status (down, up, initializing, etc.), network connectivity, and topology)
o Network load displays (that may include current and historical link utilization levels)
o Performance reports
o Error logs (indicating link and device failures and time of occurrence)

- **SNMP Agent**: Each managed device supports a network management software component called an SNMP agent which reports information to the SNMP manager via SNMP. The SNMP agent runs on managed devices while the NMS runs on the SNMP manager. The agent has local knowledge of the managed device's environment and translates management information to SNMP-specific form to be sent the SNMP management and vice versa (see Figures 5.27 to 5.30).
- **Management Information Base (MIB)**: SNMP exposes information about managed devices in the form of variables that are organized in an MIB (MIB-II [RFC1213]). The information in an MIB describes the system configuration and status. The MIB variables can be queried (and, in some cases, manipulated) by remote management applications (e.g., NMS). SNMP allows active management tasks, such as device configuration changes to be carried, through remote modification of MIB variables. The MIB is an ASCI database that describes data objects and structures in a network device running an SNMP agent. The descriptions of the data objects (i.e., *managed objects*) in the MIB are in the Abstract Syntax Notation 1 (ASN 1) format. Data objects and structures can be retrieved and/or modified using SNMP. A device that supports an MIB and SNMP is referred to as a managed device. The managed objects in the MIB of a given device may include the following:
  o Configuration parameters (e.g., addresses, timer settings/values, disabled/enabled features)
  o Information about system operational state (e.g., interface status, mode of operation)
  o Performance statistics (e.g., packet counters, byte counters, error logs)

An SNMP agent receives SNMP messages on UDP port 161. The SNMP manager may use any available UDP source port to send messages to UDP port 161 on the SNMP agent. The SNMP agent sends back a response to the UDP source port on the SNMP manager. The SNMP manager receives notifications (via SNMP *Trap* and *InformRequest* messages) on UDP port 162. Note that the SNMP agent may send notification messages from any available UDP port. Chapter 2 of Volume 2 gives a detail description of the different SNMP message types (*GetRequest*, *SetRequest*, *GetNextRequest*, *GetBulkRequest*, *Response*, *Trap*, and *InformRequest*).

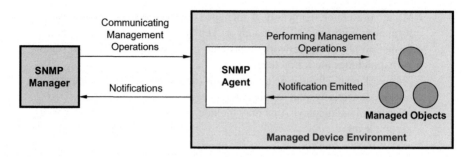

FIGURE 5.27   Interaction between SNMP manager, agent, and objects.

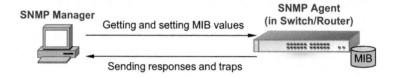

FIGURE 5.28   Communication between an SNMP manager and agent.

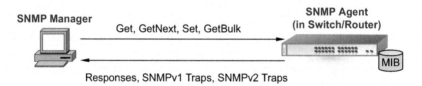

FIGURE 5.29   Flow of management operations requests, responses, and traps between the SNMP manager and the agent.

In normal operation, the SNMP manager automatically polls each SNMP agent in the network at regular intervals, retrieving the contents of the local device's MIB and combining this with those of other SNMP agents into a global NMS data store. Each SNMP agent provides a myriad of raw information about the local device's internal state and performance.

SNMP Traps are asynchronous notifications sent by an SNMP agent to a manager. Traps are sent by the SNMP agent without being explicitly requested by the manager. SNMP Traps are unsolicited SNMP messages that allow an agent to notify the manager of significant local events, possibly, triggered by alarms (see Figure 5.31).

5.6.2   Remote Network Monitoring

RMON is a standard for monitoring and protocol analysis of traffic flows on network segments to facilitate troubleshooting, and traffic pattern analysis and profiling [RFC2819] [RFC3577]. RMON enables a network administrator to detect a number of network problems and to implement the actions necessary to remedy those

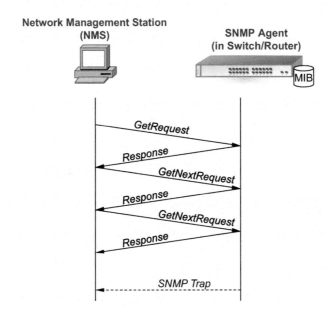

FIGURE 5.30   SNMP event interaction and timing.

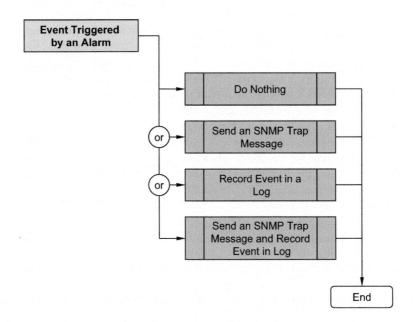

FIGURE 5.31   Alarm events and device action options.

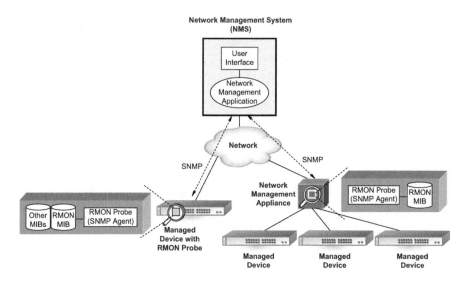

FIGURE 5.32   Management using RMON probes.

problems. Chapter 2 of Volume 2 discusses RMON in greater detail. RMON supports a network-specific MIB and collects statistics about an entire network segment.

Figure 5.32 shows the use of RMON in a network management context. RMON allows various network agents (called *RMON probes*, *RMON monitors*, or *RMON agents*) and a centralized *management console system* (the NMS) to exchange network monitoring data. An RMON probe is hardware/software or software components that is embedded in a network device, such as a switch or routing device (Figure 5.32). Through the use of remote RMON probes, RMON can monitor various network operational activities. An RMON probe in a network device collects traffic flow information to be further analyzed by the management console.

A network administrator can use the RMON alarms and events to monitor managed devices in a network. The probe can monitor traffic passing through the managed device and may set an alarm when certain conditions occur. Performance thresholds are set and the RMON probe sends automatic alerts when the thresholds are crossed. RMON probes are now commonly built into many switches and routing devices, even low-cost devices.

The main difference between SNMP and RMON is that SNMP is mainly used for "device-based" management (i.e., configuration and status monitoring of individual devices in a network), while RMON is designed with "flow-based" traffic monitoring in mind. The data collected by RMON deals mainly with network traffic patterns (and profiling) similar to flow-based monitoring technologies such as NetFlow [RFC3954] and sFlow [RFC3176] (see NetFlow and sFlow in Chapter 2 of Volume 2). Using SNMP, a network administrator can perform the necessary configuration changes on network devices and track their basic health metrics. RMON allows the network administrator (from a central management console) to collect traffic flow data for troubleshooting, traffic profiling, and implement the controls necessary for better network performance.

The communication between an RMON probe and the management console follows the client and server model. The RMON probe supports *RMON software agents* that analyze packets and collect information. A probe acts as a server and the network management applications on the management console act as clients. The RMON probe's data collection and communication with the management console is through SNMP-based systems; communication is via SNMP. However, unlike the traditional SNMP agent, RMON probes take control of data collection and processing, which helps reduce SNMP traffic in the managed network and the processing load on the clients.

To further reduce network traffic, the RMON probe only transmits information when required instead of continuous information monitoring and polling by the management console. One disadvantage of the periodic monitoring behavior of RMON is that the remote RMON probe shoulders more of the management burden, and requires relatively more processing resources to operate in this manner. For this reason, some RMON implementations try to reduce the RMON probe burden by implementing only a subset of RMON capabilities. This results in a reduced RMON probe that supports only a few management features. The probe periodically analyzes/audits packets as well as collects statistics to be sent to the management console.

While SNMP and its MIBs are extremely useful and play an important role in network management, the MIBs must be polled by the SNMP manager (NMS) to gather data. This polling can be problematic because it can waste network bandwidth and does not scale well. It is challenging for a single SNMP manager to actively poll many devices in a network, a situation that can lead to the SNMP manager running out of processing power to poll the many devices. The RMON probe solves the polling-related problems by performing the polling and data collection in the network device itself. An RMON probe performs periodic sampling of statistics and records this information in an RMON MIB. This process takes place independently of the SNMP manager. The SNMP manager first configures an RMON probe to record data and only communicates with the RMON probe when it needs statistics information. This significantly reduces the amount of traffic needed to gather network-level statistics.

An RMON implementation has MIBs on the managed devices as shown in Figure 5.32. The data in the RMON MIB is gathered by an RMON probe as stated above. The SNMP agent within RMON probe collects information and communicates this via SNMP to an SNMP management application on the management console. The contents of the RMON MIBs feature the objects that need management. A *network management appliance* is a hardware module with processing power and memory to host RMON probes as add-on for monitoring a number of managed devices (see Figure 5.32). The network management appliance has the necessary hardware and software to support RMON functionality and operate as a probe.

An RMON probe may be implemented on only one managed device or on a device interface (per IP subnet). The RMON agent software runs on the device's port, monitors, and collects network statistics for the attached IP subnet. The (SNMP) management console contacts the RMON probe only when it needs to collect statistics to help the network administrator analyze trends in network traffic. With RMON, a network administrator has more flexibility in selecting RMON probe types and locations to meet the particular needs of the network.

Most practical implementations of RMON agents support only a subset of RMON MIB groups (namely, Statistics, History, Alarm, and Event) out of the total RMON1 and RMON2 MIB groups. This is done mainly to minimize the complexity of the RMON agent and its processing load. RMON1 focuses on OSI Layer 1 and Layer 2 monitoring while RMON2 focuses on upper layers up to the Application Layer (Layer 7).

The following are the most commonly used RMON groups (in a minimal RMON agent implementation):

- **Ethernet Statistics Group**: This contains statistics for each RMON monitored Ethernet interface on the managed device (e.g., frame length, Cyclic Redundancy Check (CRC) errors, packets dropped, etc.). This group consists of the *etherStatsTable*.
- **Ethernet History Group**: This contains periodic statistical samples from an Ethernet network which are stored for later retrieval. This group consists of the *etherHistoryTable*.
- **Alarm Group**: This contains statistical samples that are periodically taken from variables in the RMON probe and compared to previously configured thresholds. If a monitored variable crosses a configured threshold, an *event* is generated. This group also holds definitions for RMON SNMP Traps to be sent by the RMON agent when variables exceed defined thresholds. Typically, an RMON implementation includes a hysteresis mechanism to limit the generation of alarms. This group requires the implementation of the Event group. This group consists of the *alarmTable*.
- **Event Group**: This group controls the generation and notification of events from the managed device. The RMON agent may send alerts (i.e., SNMP Traps) for the events as shown in Figure 5.31. This group consists of the *eventTable* and the *logTable*.

The basic RMON groups (Statistics, History, Alarm, and Event) are much easy to implement than the other RMON groups which are sometimes called the *Advanced* groups. The basic RMON groups deal only with statistics, and in modern devices, these statistics are typically gathered in hardware and can be tracked with little host CPU processing power, even in low-cost network devices. On the other hand, for the Advanced groups, information is gathered by physically examining each frame that is transmitted on the network segment, a situation that calls for a more complex RMON probe design; require more processing and memory resources to implement.

RMON allows a network administrator to specifically define the information that any RMON probe in network should provide. Figures 5.33, 5.34, and 5.35 show the general architecture of an Ethernet interface with components for statistics collection. This architecture may be used for statistics collection in an RMON probe implementation.

Using the architecture in Figures 5.33, 5.34, and 5.35, an RMON probe may capture Ethernet statistics from an Ethernet interface such as the following: bytes received, packet drop events, packets received, broadcast packets received, multicast packets received, CRC and alignment errors, undersized packets (less than 64 bytes)

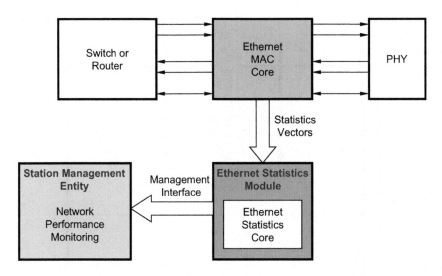

FIGURE 5.33   Statistics system interfaces.

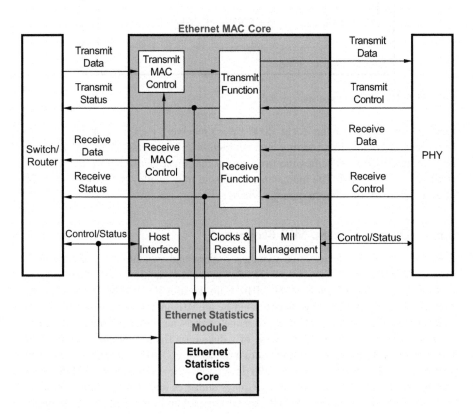

FIGURE 5.34   Ethernet statistics collection application.

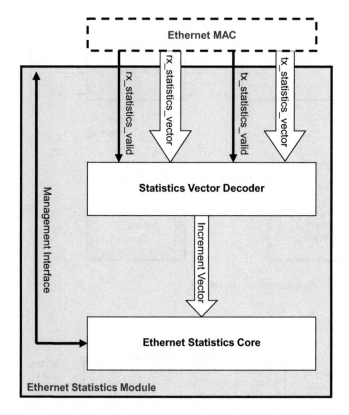

FIGURE 5.35   Ethernet statistics module.

received, oversized packets (over 2000 bytes) received, packets with less than 64 bytes received (excluding framing bits, but including FCS bytes). A *refresh rate* may be configured which specifies the time period that must elapse before the interface statistics are refreshed. The management console may display the RMON statistics of all the ports of the chosen managed device.

5.6.3 DEVICE GUIs AND MIB BROWSERS

Today, most network devices like switches, routers, and switch/routers provide GUIs for accessing MIB data. These interfaces often provide a graphical view of a device and network, hiding actual details of the MIBs from the user. An interface may allow the user to access all the data objects described by an MIB in a device using SNMP. The user can view the MIB object IDs, object names, values, and description for each object.

Some interfaces can present a view of a device on the screen and display status information that mirror those on the actual device (e.g., LEDs that blink), and click-able hotspots. For example, to control a device port, the user can simply click on the port to bring up a GUI window that displays statistics and other information about the port. These kinds of device GUIs are very useful and easy for the user to understand and manipulate. A good device GUI may display information in different formats using, for example, graphs, tables, and pie charts. The GUI can often be configured to generate reports automatically.

With the widespread use of Web-based applications and Web browsers, many network devices are equipped with internal Web servers (as an alternative to traditional SNMP-based management). A managed device that supports a Web server is able to communicate internal state and management information to a computing platform with an appropriate Web browser. Such a system can implement security through the use of passwords and encrypted message exchanges.

The information exchanged between the device's internal Web server and the browser is the same as that provided by SNMP. The Web-based approach allows better security (than using most older SNMP versions) and a more user-friendly interface at the SNMP manager. The underlying transport mechanism between the Web server in the managed device and the Web browser in the management station is still SNMP, and the MIB data structures are still kept consistent in both end-systems. As in the traditional SNMP-based approach, the Web-based approach operates in-band.

REVIEW QUESTIONS

How does a switch/router decide when to forward a received packet at Layer 2 or Layer 3?

What are the three main sources of routing information for populating the routing table of routing devices?

What information does the adjacency table of a routing device contain and what is this information used for?

Describe the three main methods for populating the adjacency table of a routing device.

How many adjacencies can be formed on a router interface attached to a broadcast multiaccess network, point-to-point link, and point-to-multipoint network?

Why do ARP cache entries have to be aged and purged from the cache?

What is the purpose of the Route Resolvability Condition?

What are the basic IP packet (Layer 3) rewrite operations a routing device must perform when Layer 3 forwarding an IP packet to the next hop?

Why does a routing device need to recompute (update) the IP header checksum of an IP packet when it is being forwarded at Layer 3?

Why does a routing device need the Layer 2 (i.e., Ethernet MAC address) of the next-hop device when forwarding an IP packet?

How does a routing device obtain the Ethernet MAC address of the next-hop IP device?

What are the basic Ethernet frame (Layer 2) rewrite operations a routing device must perform when forwarding an IP packet over Ethernet to the next hop?

What are the four main reasons why a routing device will have to drop (discard) an IP packet instead of forwarding it?

Explain briefly why data plane operations are easier to implement in ASIC and not control plane operations.

When does a router have to perform a recursive lookup when forwarding a packet?

How does the use of VLSM and CIDR impact the size of IP routing tables?

How does the use of VLSM and CIDR affect IP address lookups in IP routing or
    forwarding tables?
Explain the main benefits of using the IP forwarding table (FIB) rather than the
    IP routing table (RIB).
Explain briefly the main benefits of control plane and data (forwarding) plane
    separation in routing devices.
What is unicast Reverse Path Forwarding (uRPF) and what is its purpose?
What is the main difference between in-band management and out-of-band man-
    agement of a network device?
What is the main difference between SNMP and RMON in network management?

REFERENCES

[AWEYA1BK18]. James Aweya, *Switch/Router Architectures: Shared-Bus and Shared-
    Memory Based Systems*, Wiley-IEEE Press, ISBN 9781119486152, 2018.
[AWEYA2BK19]. James Aweya, *Switch/Router Architectures: Systems with Crossbar Switch
    Fabrics*, CRC Press, Taylor & Francis Group, ISBN 9780367407858, 2019.
[AWEYA2BK21V1]. James Aweya, *IP Routing Protocols: Fundamentals and Distance
    Vector Routing Protocols*, CRC Press, Taylor & Francis Group, ISBN 9780367710415,
    2021.
[AWEYA2BK21V2]. James Aweya, *IP Routing Protocols: Link-State and Path-Vector
    Routing Protocols*, CRC Press, Taylor & Francis Group, ISBN 9780367710361, 2021.
[CHISVDUCK89]. L. Chisvin and R. J. Duckworth, "Content-Addressable and Associative
    Memory", IEEE Computer, July 1989, pp. 51–64.
[CILETT02]. M. Ciletti, *Advanced Digital Design with the Verilog HDL*. Prentice-Hall, 2002.
[HUSSFAULT04]. I. Hussain, *Fault-Tolerant IP and MPLS Networks*, Cisco Press, 2004.
[GAPFKH03]. B. Gamache, Z. Pfeffer, and S. P. Khatri, "A Fast Ternary CAM Design
    for IP Networking Applications" *The 12th International Conference on Computer
    Communications and Networks*, 2003 (ICCCN 2003), 20-22 Oct. 2003, pp. 434–439.
[IEEE802.1D04]. IEEE Standard for Local and Metropolitan Area Networks: Media Access
    Control (MAC) Bridges, June 2004.
[KOBAYA00]. M. Kobayashi, T. Murase, and A. Kuriyama, "A Longest Prefix Match Search
    Engine for Multi-Gigabit IP Processing," *Proceedings of IEEE International Conference
    on Communications*, Vol. 3, 2000, pp. 1360–1364.
[MCAFRA93]. A. J. McAuley and P. Francis, "Fast Routing Table Lookup Using CAMs,"
    Proceedings of IEEE INFOCOM, March-April 1993, pp. 1382–1391.
[PEIZUKO91]. T. B. Pei and C. Zukowski, "VLSI Implementation of Routing Tables: Tries
    and CAMs," *Proceedings of IEEE INFOCOM*, Vol. 2, 1991, pp. 515–524.
[RABCHNIK03]. J. M. Rabaey, A. Chandrakasan, and B. Nikolic, *Digital Integrated Circuits*.
    Prentice Hall, 2nd ed., 2003.
[RFC950]. J. Mogul and J. Postel, "Internet Standard Subnetting Procedure", *IETF RFC 950*,
    August 1985.
[RFC1157]. J. Case, M. Fedor, M. Schoffstall, and J. Davin, "A Simple Network Management
    Protocol (SNMP)", *IETF RFC 1157*, May 1990.
[RFC1213]. K. McCloghrie and M. Rose, "Management Information Base for Network
    Management of TCP/IP-based internets: MIB-II", *IETF RFC 1213*, March 1991.
[RFC1517]. R. Hinden, Ed., "Applicability Statement for the Implementation of Classless
    Inter-Domain Routing (CIDR)", *IETF RFC 1517*, September 1993.

[RFC1518]. Y. Rekhter, T. Li, "An Architecture for IP Address Allocation with CIDR", *IETF RFC 1518*, September 1993.

[RFC1519]. V. Fuller et al., "Classless Inter-Domain Routing (CIDR): An Address Assignment and Aggregation Strategy," *IETF RFC 1519*, 1993.

[RFC1812]. F. Baker, Ed., "Requirements for IP Version 4 Routers", *IETF RFC 1812*, June 1995.

[RFC1878]. T. Pummill and B. Manning, "Variable Length Subnet Table For IPv4", *IETF RFC 1878*, December 1995.

[RFC2225]. M. Laubach and J. Halpern, "Classical IP and ARP over ATM", *IETF RFC 2225*, April 1998.

[RFC2819]. S. Waldbusser, "Remote Network Monitoring Management Information Base", *IETF RFC 2819*, May 2000.

[RFC3176]. InMon Corporation's Flow, "A Method for Monitoring Traffic in Switched and Routed Networks", *IETF RFC 3176*, September 2001.

[RFC3577]. S. Waldbusser, R. Cole, C. Kalbfleisch, and D. Romascanu, "Introduction to the Remote Monitoring (RMON) Family of MIB Modules", *IETF RFC 3577*, August 2003.

[RFC3704]. F. Baker and P. Savola, "Ingress Filtering for Multihomed Networks", *IETF RFC 3704*, 2004.

[RFC3954]. B. Claise, Ed., "Cisco Systems NetFlow Services Export Version 9", *IETF RFC 3954*, October 2004.

[RFC4271]. Y. Rekhter, T. Li, and S. Hares, Ed., "A Border Gateway Protocol 4 (BGP-4)", *IETF RFC 4271*, January 2006.

[RFC4632]. V. Fuller, T. Li, "Classless Inter-domain Routing (CIDR): The Internet Address Assignment and Aggregation Plan", *IETF RFC 4632*, August 2006.

[SEIFR2000]. Rich Seifert, *The Switch Book, The Complete Guide to LAN Switching Technology*, Wiley, 2000.

[SEIFR2008]. Rich Seifert and Jim Edwards, *The All-New Switch Book: The Complete Guide to LAN Switching Technology*, Wiley, 2008.

[SHAGUP01]. D. Shah and P. Gupta, "Fast Updating Algorithms for TCAMs," *IEEE Micro*, Vol. 21, Jan/Feb 2001, pp. 36–47.

[STRINGNAK07]. N. Stringfield, R. White, and S. McKee, *Cisco Express Forwarding, Understanding and Troubleshooting CEF in Cisco Routers and Switches*, Cisco Press, 2007.

[WADSOD89]. J. Wade and C. Sodini, "A Ternary Content Addressable Search Engine," *IEEE Journal of Solid- State Circuits*, vol. 24, Aug 1989, pp. 1003–1013.

[ZININALEX02]. Alex Zinin, *Cisco IP Routing: Packet Forwarding and Intra-Domain Routing Protocols*, Addison-Wesley, 2002.

6 Packet Forwarding in the Switch/Router
Layer 3 Forwarding Architectures

6.1 INTRODUCTION

Multilayer switching in this book refers to the capability of a network device to forward packets based on information in the Layer 2 and Layer 3 packet headers. The device learns how to forward packets at Layer 3 by communicating with other routers in the network. The distinction between a router and a switch/router (also called a multilayer switch) has become increasingly vague because of the evolution of highly intelligent Layer 3-aware ASICs used in packet forwarding. In current switch/router designs, the capability of the routing (Layer 3) component to interact efficiently with the Layer 2 forwarding component has led to a dramatic increase in device compactness and versatility (i.e., tightly integrated Layer 2 and 3 forwarding) and packet forwarding performance.

Switch/routers have become a primary component in today's enterprise and service provider networking environments. In such a critical role, the switch/router must provide a reliable switching platform that offers in addition high performance and intelligent network services like security and QoS processing. This chapter discusses the basic packet forwarding functions in switch/routers and the different design methods and architectures used in switch/routers. The discussion includes the basic packet forwarding functions in the typical switch/router as well as some of the well-known switch/router architectures used in the industry.

In particular, the discussion describes details about the control plane and data (or forwarding) plane functions in each architecture; the traditional centralized CPU-based forwarding architectures, the centralized and distributed route cache-based forwarding architectures, and the distributed forwarding architectures using network topology-based forwarding tables (or FIBs). The methods and architectures discussed here lay out the fundamental ideas for the discussions in subsequent chapters of the book.

6.2 PACKET FORWARDING IN THE ROUTER OR SWITCH/ROUTER

Figure 6.1 illustrates the main control plane and data plane operations in a typical router as an IP packet is forwarded through it [AWEYA2001] [AWEYA2000] [MENJUS2003]. In this figure, Host A is sending an IP packet to Host B over a

DOI: 10.1201/9781003311249-6

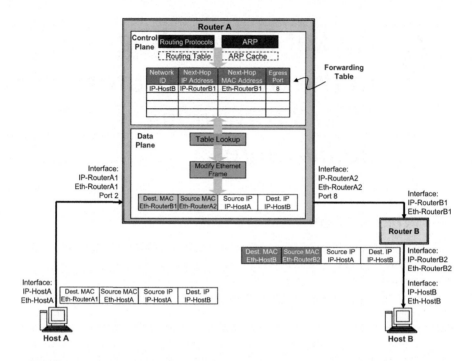

FIGURE 6.1   Router control plane and data plane operations.

network that includes a number of routers. The main events are described by the following steps:

Step 1: Packet to Be Sent to Default Gateway (Router A) and Host A Sends ARP Request

- Host A has an IP packet that is destined to Host B on a different IP subnet or VLAN. By examining the IP address and subnet mask assigned to its network interface and the IP address of Host B, Host A determines that Host B is on a different subnet or VLAN and, therefore, requiring Host A to send the IP packet to its configured default gateway, Router A. Let us assume Host A, Host B, and the routers are connected to an Ethernet network. With this, Host A must deliver the IP packet in an Ethernet frame to Router A. Host A is configured with the IP address of the default gateway, Router A.

- To properly address the Ethernet frame that is to be delivered to Router A, Host A needs to know the Ethernet MAC address of Router A's receiving Ethernet interface. Host A examines its local ARP cache to see whether there is an entry for the MAC address of Router A's receiving interface. If one exists, then this means Host A has recently communicated with Router A. If the ARP cache of Host A does not contain the MAC address, Host A broadcasts an ARP request, which is forwarded to all devices on its IP subnet or VLAN to requests for the

MAC address associated with the IP address of Router A's receiving interface (i.e., Port 2). Host A and Router A's receiving interface both share the same IP subnet or VLAN.

Step 2: Default Gateway (Router A) Sends ARP Reply to Host A
- Router A which is configured with the IP address on the Ethernet interface facing or leading to Host A (i.e., Port 2), responds to the Host A's ARP request by sending a unicast ARP reply, which provides the MAC address of that router interface (Port 2).

Step 3: Hosts A Sends Packet to Default Gateway (Router A)
- Knowing the MAC address of Router A's receiving interface (Port 2), Host A encapsulates the IP packet in an Ethernet frame and sends it to Router A. The destination MAC address field of the Ethernet frame contains the MAC address of the receiving Ethernet interface of Router A, which indicates to Router A that the received IP packet requires routing (i.e., Layer 3 forwarding) as discussed in Chapter 5.
- The IP destination address in the IP packet originated by Host A, however, remains that of Host B, the true destination of the IP packet and not that of Router A. The routing processing from Host A to Host B does not modify the IP source and destination addresses in the original IP packet transmitted by Host A.

Step 4: IP Forwarding Table Lookup in Router A
- Router A receives the Ethernet frame sent by Host A and performs the forwarding (data plane) operations. In order for Router A to forward the packet on to the appropriate next hop, it must know the outgoing interface, next-hop and the MAC address of the next-hop's receiving interface.
- To determine the next-hop, Router A extracts the IP destination address of the IP packet and performs an address lookup (using longest prefix matching) in its forwarding table (also called Forwarding Information Base (FIB)) for an IP network prefix entry that best matches the IP destination address (i.e., IP address of Host B). The forwarding table lookup indicates to Router A that Host B is reachable through Port 8 to a next-hop having the IP address of Router B.

Step 5: Router A Sends ARP Request to Router B, Modifies Packet, and Sends It to Router B
- As stated above, we assume that Router A is connected to Router B via Ethernet, thus requiring Router A to send the IP packet encapsulated in an Ethernet frame addressed to Router B. Router A examines its local ARP cache to see if an entry exists for the MAC address associated with the IP address of the next-hop, Router B. If no ARP cache entry exists, then Router A must generate an ARP request for the MAC address associated with the next-hop IP address.
- Once Router A determines the MAC address of the next-hop's receiving interface, the destination MAC address of the outgoing Ethernet frame can be rewritten in the appropriate field. The source MAC

address in the frame is also rewritten to be the MAC address of the outgoing Ethernet interface on Router A (i.e., Port 8) to indicate to Router B that the originator of the received frame is Router A. As stated above, a router does not modify the IP source or destination addresses of IP packets that are being forwarded but must rewrite the destination and source MAC address so that the IP packet can be delivered over the Ethernet network to the next-hop.

- As part of the forwarding operations, Router A modifies some important fields in the IP header. Router A decrements the IP time-to-live (TTL) field and also recomputes the IP header checksum, since the TTL field has been modified. In some networking environments, IP addressing might also have to be modified, for example, if Network Address Translation (NAT) is configured. However, this operation is performed by a separate function and considered outside the scope of this discussion. After rewriting the destination and source MAC addresses of the outgoing Ethernet frame, Router A also recomputes the Frame Check Sequence (FCS) of the frame. Router A then sends the rewritten Ethernet frame containing the IP packet through Port 8 to Router B.

Step 6: Router B Receives Packet from Router A and Forwards It to Host B

- Router B receives the Ethernet frame from Router A and performs a forwarding table look again using the IP destination address of the packet which is that of Host B. The forwarding table lookup shows that the IP destination address is that of a host that is locally connected (i.e., directly attached). Router B completes the delivery of the IP packet originating from Host A by sending the packet to Host B.
- Host B is connected via Ethernet to Router B, requiring Router B to send the IP packet encapsulated in an Ethernet frame addressed to Host B. The same rewrite of the destination (and source) MAC address described in Step 5 takes place here, and the Ethernet frame is delivered to Host B, its final destination.

In the forwarding steps described above, it is important to highlight that the MAC addresses written in the Ethernet frames are specific only to each local LAN and need to be known only within each LAN. Host A is not required to know Host B's MAC address or even Router B's MAC address. Host A needs to know only the MAC address of Router A receiving interface (its default gateway) so that it can deliver IP packets in Ethernet frames locally to Router A to be routed. Router A then forwards the packet to the next-hop and this process is repeated on a hop-by-hop basis until the IP packet reaches its final destination.

6.3   PACKET FORWARDING ARCHITECTURES

A number of factors influence how fast the data plane can forward packets through a routing device, which is usually referred to as the packet *forwarding speed* of the device (expressed in packets per second (pps)). One factor is the extent to which the control plane and data plane are decoupled as discussed in Chapters 2 and 5. Another factor is the type of processing device on which the data plane is implemented, for example, implementing the basic forwarding operations on custom-built application-specific integrated circuits (ASICs) versus on a general-purpose processor. It is discussed in Chapter 5 that the basic data plane operations required for forwarding IP packets are simple enough to allow implementation on ASIC.

A major factor that affects the packet forwarding speed, which is a fundamental requirement for data plane operation, is the speed of forwarding table lookups to determine the outbound interface, and the next-hop IP address and its associated MAC address for transiting IP packets. The MAC address of the next-hop has to be written on the outgoing frame. The process of looking up the next-hop parameters may also involve retrieving information for access control lists (ACL) for security processing, and for quality-of-service (QoS) control.

The forwarding table lookup process (which takes place during the data plane operations) can become a bottleneck if not properly implemented. Most often, the way the lookup process is implemented (e.g., custom ASIC, special-purpose processor, general-purpose processor) determines the forwarding speed of the data plane, or equivalently, the routing device as a whole. To ensure that the lookup process does not significantly slow down packet forwarding and delay the rewrite processes of the data plane operations, high-end routers and switch/routers in particular, use custom-built ASICs or special-purpose processors with specialized routing information data structures that allow fast network address lookups. These data structures can be categorized as those based on route caches (also called loosely flow/route caches or flow caches), and those based on optimized network topology-based forwarding tables (or FIBs). These routing information data structures are described in this chapter.

6.3.1 TRADITIONAL CENTRALIZED CPU-BASED FORWARDING ARCHITECTURES: SOFTWARE-BASED FORWARDING USING THE IP ROUTING TABLE

The earlier generation of routers was typically based on centralized forwarding architectures where both the control plane and data (or forwarding) plane are implemented on a general-purpose processor (CPU). Even most present-day low-end routers such as residential gateways and other low-end devices still used this architecture. Figure 6.2 shows a high-level architecture of the traditional CPU-based routing device with the main transmit and receive line module subsystems. Figure 6.3 shows an example Ethernet network interface controller (NIC) design in the line module, while Figure 6.4 shows a block diagram of a Gigabit Ethernet Media Access Control (MAC) and Physical Coding Sublayer (PCS) of the Ethernet Physical Layer (PHY) [IEEE802.3].

The traditional CPU-based forwarding is still the basic mode of forwarding (i.e., the basic forwarding path) for control and management traffic in all routing devices (low-end, mid-end, and high-end routing devices). The terms *route processor* or *control engine* are used to describe the module in the routing device that contains the processor (CPU), memory components, and management interfaces, and is responsible for running the routing protocols, management and control protocols, and system configuration and management software tools.

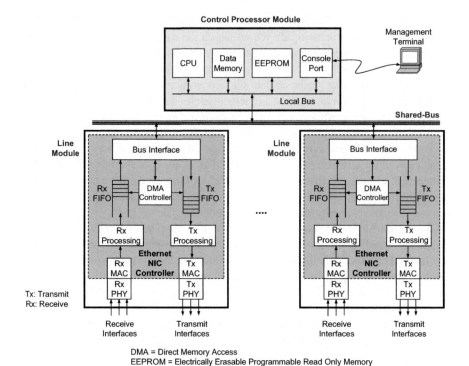

DMA = Direct Memory Access
EEPROM = Electrically Erasable Programmable Read Only Memory
NIC = Network Interface Controller
PHY = Physical Layer

FIGURE 6.2 Centralized architecture showing line module subsystems.

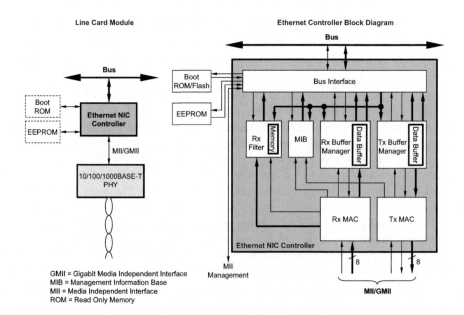

GMII = Gigabit Media Independent Interface
MIB = Management Information Base
MII = Media Independent Interface
ROM = Read Only Memory

FIGURE 6.3 Example Ethernet network interface controller design in a line module.

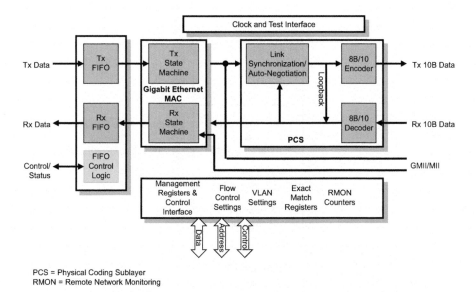

PCS = Physical Coding Sublayer
RMON = Remote Network Monitoring

FIGURE 6.4 Block diagram of gigabit Ethernet MAC/PCS.

The traditional CPU-based centralized forwarding architectures use the mostly sub-optimally structured database, that is, the software-based IP routing table, for all destination address lookups. The routing table in these architectures is involved in all packet forwarding operations and may contain recursive routes. Recursive routes require the router to perform recursive lookups to find the outbound interface of packets sent on such routes as discussed in the "Recursive Route Lookup in an IP Routing Table" section in Chapter 5.

6.3.1.1 Packet Forwarding in the Traditional CPU-Based Forwarding Architectures Using the IP Routing Table

A key feature of the forwarding process in the traditional CPU-based architecture (also called *process switching* (in Cisco terminology), *process forwarding*, or *slow-path forwarding* (see Figure 6.5)) is that the CPU handles all system tasks in addition to normal packet forwarding: all routing protocol tasks (e.g., RIP, OSPF, etc.); control and management protocol tasks (ICMP, IGMP, SNMP, etc.); and device/system management and configuration tasks (Command-Line Interface (CLI), secure logins and system access, file copying, etc.). The CPU handles all forwarding tasks such as verification of arriving packets, IP address lookups, IP packet field updates, and Layer 2 packet rewrites. All these processes have to compete for system resources with all other non-routing-related processes.

In the traditional CPU-based forwarding architecture, all routing and control packets, and all packets requiring normal forwarding are received, queued, and processed along with other application traffic to/from software components running on the CPU. In addition to processing routing and control packets, transit packets are queued for the CPU which then makes all forwarding decisions at the process level.

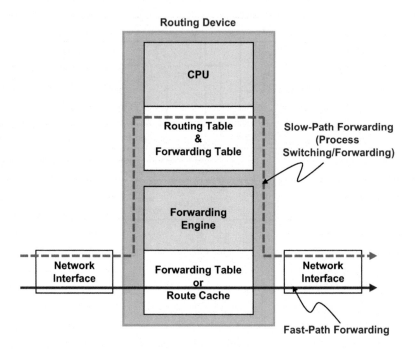

FIGURE 6.5 Slow-path versus fast-path forwarding.

6.3.1.1.1 Understanding Process Switching/Forwarding

The term *process switching* (Cisco terminology) [ZININALEX02] refers to packet
forwarding by a software process that receives CPU control from the Operating
System (OS) process scheduler as shown in Figure 6.6. In this chapter, we sometimes
use the term *process forwarding* instead of process switching. The OS process scheduler running in the CPU switches to the IP forwarding process as soon as it detects
packets in the IP input processing queue.

A multitasking OS is one that can execute multiple tasks (also called processes)
concurrently over a certain period by switching between the tasks very rapidly. The
tasks may pertain to multiple different users or all relate to a single user. Usually,
multitasking OSs (e.g., UNIX, Linux, newer versions of Microsoft Windows from
Windows 2000) have several levels of execution; the *interrupt level*, the OS *kernel
level*, and the *process level*. The interrupt level is where the hardware and software
interrupts are processed, while, the OS kernel level is where the time critical OS tasks
are processed.

The process level is where tasks (or processes) not related to the OS itself such as
user applications, and network service daemons are run. Special features such as priority queuing, Weighted Round-Robin (WRR), Weighted Fair Queuing (WFQ), and
Weighted Random Early Detection (WRED) can be implemented at the process level.
The *scheduler process* (which is part of the OS), gives a time slice to each process.
Segments of multiple tasks are executed by the CPU in an interleaved manner, while
they share the common processing resources and main memory. In the event several

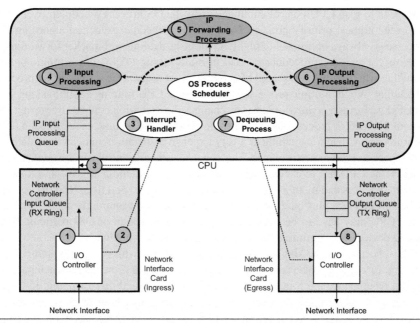

The following steps summarized the processing in the software based router using a centralized processor:

1. The ingress input/output (I/O) controller (in the ingress side of a network interface card (NIC)) receives a packet from the network, performs a number of required Layer 2 packet verification tasks, and if the packet is valid, stores it in its receive (RX) ring to be free to start receiving the next packet.
2. The I/O controller at the same time sends an interrupt request (IRQ) to the CPU.
3. The interrupt handling routine in the CPU (interrupt handler) then takes the received packet from the RX ring, performs some required basic packet verification (e.g., determining the protocol type in the packet (IP, ICMP, IGMP)), and places it in the IP input processing queue if it has some free space. The interrupt handler then returns control to the CPU.
4. The packet is dequeued from the IP input processing queue and passed to the IP input processing function which performs a number of IP packet verification tasks.
   a. If the packet is valid, the IP input processing function determines if the packet is for the router itself (i.e., local delivery) or is a transit packet to be forwarded to a next-hop node in the network.
   b. If the packet is a transit packet to be sent to another node, the IP input processing function passes the packet to the IP forwarding process.
5. The IP forwarding process reads the IP destination address in the packet and performs a longest prefix matching (LPM) lookup in its IP forwarding table to determine the next-hop IP address and the outbound (or egress) interface (or port) of the packet. The IP forwarding process also performs a number of IP packet processing tasks (e.g., decrement IP TTL, update IP header checksum, process IP options, etc.) as well as determine the appropriate Layer 2 parameters to be used in the Layer 2 encapsulation of the packet.
6. After completing all of its tasks, the IP forwarding process passes the packet to the IP output processing function which is responsible for encapsulating the IP packet in the appropriate Layer 2 packet, updating some relevant fields in the Layer 2 packet (e.g., rewriting the Ethernet frame source and destination MAC addresses, calculating the Ethernet frame checksum, etc.), and enqueueing it for transmission on the correct outbound (or egress) interface.
7. If the IP output processing queue is empty and the egress I/O controller's transmit (TX) ring has some free space, the dequeuing process transfers the packet directly to the controller's ring. Otherwise, the packet is enqueued in the IP output processing queue.
   a. The dequeuing process may use a number of queuing and scheduling policies (e.g., strict priority, weighted round-robin (WRR), weighted fair queuing (WFQ), etc.) to queue and service packets from IP output processing queue.
8. The egress I/O controller reads the packets in its TX ring and transmits them onto the network.

FIGURE 6.6 Illustration of IP packet forwarding in a centralized software-based routing device.

tasks need to gain CPU control simultaneously, the OS process scheduler gives control to the highest priority process. Note that the scheduler selection algorithm for processes is always optimized to highly favor tasks dealing with packet forwarding.

A very expensive procedure or task in multitasking OSs is performing *context switching* whether this is initiated by a hardware or software interrupt or by process scheduling. When multiple processes share a single CPU running a multitasking OS, the CPU performs a *context switch* (i.e., changes the execution context) by storing the state of a thread or process, so that it can be restored and resume execution at a later point in time. Context switching is a very expensive operation and can have a negative impact on system performance. This is because when control is passed to another process, the CPU has to execute a large number of instructions to save and load all relevant registers and memory maps, update various tables and lists, load the context of the new process, plus other important operations. Furthermore, the memory cache of the CPU may have to be invalidated. Much of the design of OSs particularly in routing devices is to optimize the execution of context switching.

Thus, because of the significant overhead associated with process switching/forwarding, forwarding packets at the interrupt level using a route cache when packets are just received by the router (a forwarding method called *fast switching* in Cisco technology [ZININALEX02]), provides faster forwarding than process switching/forwarding. Although fast switching is software (interrupt level) based, the interrupt handling routines for packet handling and forwarding are implemented in a compact low-level processing language (e.g., assembler) to allow faster packet forwarding and to leave enough time for other CPU tasks like running the routing protocols and processing routing information. Furthermore, by using sophisticated data-processing methods that allow information to be stored and found efficiently (information caching, efficient hash functions, radix trees (or compact prefix trees), etc.), packet forwarding performance is increased.

Process switching/forwarding is essentially platform-independent because the switching/forwarding functions are implemented at the process level in a CPU unlike other architectures (e.g., distributed architectures) in which the functions are mostly platform-dependent. Basically, all router architectures, from centralized to distributed, have a process switching/forwarding component that resides alongside all router control and management functions in a CPU; it is the basic routing component in any routing device. Other than the routing and management protocols, the process switching/forwarding functions sit alongside extensive router troubleshooting and debugging functions. In simple and low-end routing devices, process switching/forwarding is sufficient for implementing routing and packet forwarding.

6.3.1.1.2 *Process Switching/Forwarding Steps*

Figure 6.6 describes in detail the steps involved in process forwarding, however, the main components of the packet processing are highlighted as follows [ZININALEX02]:

- Process forwarding is initiated when the I/O controller of a network interface hardware of a routing device receives a packet and transfers it into the interface's I/O memory.

- The I/O controller buffers the arriving Layer 2 packet and sends an interrupt request (IRQ) to the CPU. The I/O controller interrupts the CPU, alerting it to the reception of a packet in the input (inbound) I/O memory requiring processing.
- The CPU interrupts the process that is currently running in it, executes a context switch, and passes control to the interrupt handler. The interrupt handler proceeds to update its inbound packet counters.
- The interrupt handler takes the Layer 2 packet from the buffer, performs basic checks to verify the packet, and then examines the packet header (e.g., encapsulation type, Network Layer header, etc.) to determine if the packet is an IP packet. If an IP packet, it is placed in the IP input processing queue to be processed by the IP forwarding process.
- The IP forwarding software takes the IP destination address of the packet and performs a lookup in the forwarding table for a matching entry. Upon finding a matching entry, the IP forwarding software retrieves the corresponding next-hop IP address, the outbound interface, and the Layer 2 address associated with the receive interface of the next-hop node. The next-hop node's Layer 2 address can be read from an ARP cache or from the forwarding table if it supports integrating such Layer 2 adjacency information. The forwarding software then updates relevant IP header and Layer 2 frame fields including performing all necessary packet rewrites.
- If the IP forwarding process determines that the IP output processing queue is free and the transmit queue of the I/O controller of the outbound network interface is not full, it enqueues the packet directly on the I/O memory, otherwise it enqueues the packet in the IP output processing queue.
- The I/O controller of the outbound network interface hardware detects the queued packet in the I/O memory, retrieves it, and transmit it out the network interface on its way to the next-hop node. The I/O controller then interrupts the CPU to indicate that the packet has been transmitted. The IP forwarding software then updates the outbound packet counters and frees the I/O memory previously occupied by the transmitted packet.

6.3.1.2   Limitations of the Traditional CPU-Based Forwarding Architectures

The centralized forwarding architectures based on CPUs served the forwarding requirements of earlier networks (which were considered modest at that time). In this architecture, all arriving packets are queued and processed by the CPU which produces poor performance as the traffic arrival rate increases, and as the network routing table size grows. The address lookup times and the associated packet update and rewrite times (and hence the total time to forward a packet) also increase, resulting in poor forwarding rates. Address lookups using longest prefix matching (LPM) is a processing intensive step and can task the CPU as traffic rates and the routing table size increase. Also, recursive routes also demand additional lookups to be performed, further lengthening the overall packet forwarding times.

In the traditional CPU-based forwarding architecture (or process forwarding), all route processing and forwarding functions are implemented in the CPU which has the effect of degrading the performance of the device. All routing protocol packets, management protocol packets, and normal end-user packets are processed by the CPU. Processing IP packets with IP header options further tasks the CPU. Even value-added services such as IP Security (IPSec), Network Address Translation (NAT), Domain Name Service (DNS), and Dynamic Host Configuration Protocol (DHCP), when needed, have to be handled by the CPU. The actual forwarding performance can be significantly degraded when control and routing policies [AWEYA2BK21V1] (i.e., policy routing, packet filtering, packet marking and tagging, traffic policing, traffic shaping, etc.) are configured on the routing device.

Also, proper transmission and reception of routing protocol packets are essential for network stability and avoidance of routing loops. An overloaded CPU can result in routing protocol packets being dropped leading to problems in the performance of the routing protocols and the network in general.

6.3.1.3 Types of Centralized Forwarding Architectures Using the IP Routing Table

Despite the many limitations of the traditional CPU-based architecture, it provides flexibility for the addition of new features. This purely software architecture makes it easy to quickly introduce new features through simple software upgrades. This section describes some example high-level routing device architectures that use centralized forwarding of packets.

Typically, the centralized architectures discussed here use the IP routing table directly for address lookups and packet forwarding (see Figure 6.7) instead of the more compact, resolved, and optimized IP forwarding table (see "Architectures using Topology-Based Forwarding Tables" section below). As discussed in the "Control Plane" section in Chapter 5, the routing table contains relatively a lot more information than the forwarding table, information that is not directly used or relevant for

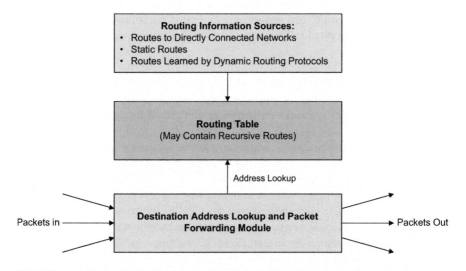

FIGURE 6.7   Centralized address lookups and packet forwarding using a routing table.

actual packet forwarding. The routing table may also contain recursive routes, thereby, causing the router to perform address lookups recursively to find the outbound interface for packets sent on such routes (see the "Recursive Route Lookup in an IP Routing Table" section in Chapter 5).

6.3.1.3.1 Centralized Processor or CPU-Based Architectures

In the centralized processor (CPU)-based architectures (see Figure 6.8), the CPU handles all routing information exchange and forwarding operations. The CPU is responsible for all routing protocol tasks, control and management protocol tasks, device/system management and configuration task, as well as normal packet forwarding. An architecture may use a centralized network processing unit (NPU), also called a network processor, instead of a CPU to improve the packet forwarding performance of the routing device. An NPU is a specialized processor (typically a software programmable device) which has features specifically tailored for networking applications.

6.3.1.3.1.1 Typical Components

This architecture contains a processor which runs the device's operating system (OS), performs packet forwarding functions, and supports various memory types as

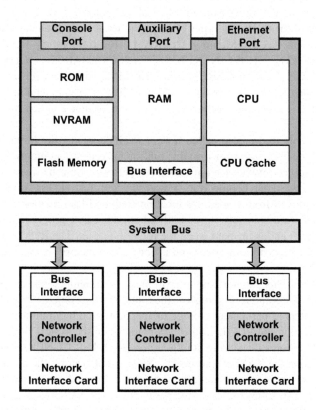

FIGURE 6.8 General hardware components in a router with centralized forwarding.

shown in Figure 6.8. The typical OS in a routing device (e.g., Cisco IOS) is a full-fledged OS with memory management, process scheduling, hardware abstraction, plus many other services related and unrelated to routing and forwarding. The OS supports all the various processes and programs needed for the routing device. It contains protocol-specific code for packet handling.

The OS can be viewed as a combination of processes each performing a specific function or set of functions (control plane and data plane functions). A process is run periodically or is triggered by some event. The OS supports various dynamic routing protocols and mechanisms for installing routes in the routing table, each of which is represented by a software module running in the router. For example, RIP, OSPF, and BGP will each be represented by routing protocol modules that exchange routing information with neighbor routers and install routes in the routing table.

The performance and capabilities of a router's CPU vary and depend on the router platform. The type of centralized processor used depends on the processing requirements of the routing device which also depends on the architecture of the device (number and speeds of the interfaces, switch fabric type (e.g., shared bus, shared memory), etc.). In shared-memory switch fabric architectures, an arriving packet is copied into a memory location that is accessible by both the inbound and outbound network interface processors [AWEYA1BK18]. The typical centralized forwarding architecture supports a shared-bus switch fabric.

In addition to the processor, the typical centralized forwarding architecture employs at least the following types of memory: Read Only Memory (ROM), Non-Volatile Random Access Memory (NVRAM), Flash Memory, Random Access Memory (RAM) [CISCINTRCIOS] [CISC1600ARC] [CISC2500ARC] [CISC2600ARC] [CISC4000ARC] [CISC7200ARC]. To make decisions or to fetch and execute instructions, the router's CPU must have access to data in memory (see also "Memory Components" section in Chapter 3 of Volume 2 of this two-part book):

- **ROM**: This ROM (also called a *Boot ROM*) contains the initial software (called the *bootstrap software*) that runs on the router. It contains the *startup diagnostic code* whose main task is to perform some hardware diagnostics during router bootup (i.e., Power-On-Self-Test (POST)) and to load the router OS from a memory location such as Flash memory. The bootstrap software is usually stored in ROM (e.g., erasable programmable ROM (EPROM)) and is invoked when the router boots up. The ROM is available on a router's processor board and is generally a memory on a chip or multiple chips.
- **NVRAM**: This memory is an extremely fast memory and is persistent across reboots and is used to store the router *startup configuration*. It is used as a writeable permanent storage of the startup configuration. This is the configuration file that the router OS reads when the system boots up. The NVRAM stores the startup configuration (ROUTER CONFIG) and a copy is loaded in the RAM at startup. The NVRAM also stores the *configuration registers* used to specify the router's behavior during the restart or reloading process; it specifies router startup parameters. The NVRAM is also used for permanent storage of hardware revision and identification information, as

well as the MAC addresses of the Ethernet interfaces. The functions of the configuration registers include the following:

o Force the system into the bootstrap software
o Select a boot source and the default boot filename
o Allow the system to recognize break signal from the console
o Set the baud rate of the console terminal
o Control broadcast addresses used by the system
o Load router OS software from a system storage location
o Enable the system to boot from a TFTP (Trivial File Transfer Protocol) server

- **Flash Memory**: An onboard Flash memory is commonly used to store one or more full router OS software images, configuration files, and system information. On some routing platforms, the Flash memory may also store the bootstrap software and the NVRAM data. A removable (portable) Flash memory card may provide a portable storage of the router OS image, configuration files, etc. In Cisco routers, the OS image typically resides in an onboard Flash memory or removable Flash memory card and is loaded into the processor memory (RAM) for execution. On some platforms, the router OS may also run directly from Flash memory to save RAM space. Routers can be designed to either *run from Flash memory* or *run from RAM*. Normally, the Flash memory stores the router OS in compressed form.
- **RAM**: The RAM is usually used for all system operational storage requirements such as router OS system tables and buffers. The RAM (different RAM types available) is used at run time to store the router OS executable code (i.e., running OS image), Layer 2 forwarding tables, Layer 3 routing and forwarding tables, general data structures, system buffers, interface buffers (for packets being received and to be processed, and packets processed and to be forwarded), and the *running configuration* (i.e., the *currently active configuration*). Packet buffers are generally maintained as buffer pools and can be *private* or *public*, *static* or *dynamic*, and buffers are available from the pool in fixed sizes (fixed chunks). Packet buffers are managed by a buffer management process in the OS which is responsible for allocating and reclaiming free buffers. Some routing platforms like the Cisco 4000 and 7200 series routers ([CISC4000ARC] [CISC7200ARC]) have two separate RAM components; one used for the Cisco IOS software code space (i.e., running image), running configuration, routing tables, and data structures, and the other used for the *system buffers*, and *shared interface private buffer pools* and *public buffer pools*. Other platforms like the Cisco 1600, 2500, and 2600 series routers ([CISC1600ARC] [CISC2500ARC] [CISC2600ARC]) have just one large RAM that is partitioned into two parts as shown in Figure 6.9.

The CPU uses various buses for accessing the different components of the router and for transferring instructions and data to or from specified memory addresses. A *CPU bus* may be used for high-speed operations, with direct processor access to RAM,

RAM

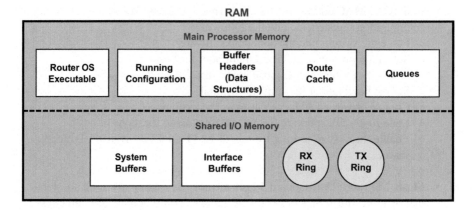

FIGURE 6.9   Memory (RAM) usage.

ROM, NVRAM, and Flash memory. A *system bus* (or *I/O bus*) allows the CPU to individually control other devices such as the network interface cards, and the system management interfaces (i.e., *console, auxiliary*, and *Ethernet* ports as discussed in Chapter 2 of Volume 2). The system management interfaces provide the necessary user interfaces for configuring and managing the device.

The router also supports various *registers*, which are small, fast memory units used for storing special purpose information, such as currently executing instruction, interrupt status, and so on. The location of registers depends on their purpose. For example, the main processor (CPU) contains the instruction register and other control registers. The CPU also contains general purpose registers for integer and floating-point data used in instruction execution. The console interface contains its own status register. Other I/O devices also contain data read/write registers.

Typically, the ROM stores the *startup code* (or *bootstrap program*) that bootstraps the routing device. The bootstrap code initializes the operating system (OS) after power-on or general reset. It loads the OS into the memory (RAM) of the routing device after which the OS will then take care of loading other system software as needed. The startup or bootstrap code checks the system hardware and loads the OS into the RAM. The OS is typically stored in a Flash Memory.

A default behavior may be, the router first tries to boot from the first OS image stored in the onboard Flash memory, if available, and then it tries the removable (external) Flash memory cards. The user may also specify which router OS images or memory locations the system should attempt booting from and the order using appropriate configuration commands (e.g., the **boot system** command in Cisco CLI configuration mode [CISCINTRCIOS]). The system may be configured to attempt booting from an OS image stored in a removable Flash memory in a PCMCIA slot before going to the onboard Flash memory.

When the router OS has been loaded into RAM and is handed control, it copies the system startup configuration which is stored in the NVRAM into a buffer space in the RAM. The OS then passes the configuration to a parser for processing, after which the parser then proceeds to dynamically process corresponding configuration commands. The parser runs at system startup time, when new configuration files are loaded to the router, and when specific CLI commands are added when the router is in the configuration mode (see Chapter 2 of Volume 2). Note that the *startup configuration* is stored in the NVRAM while the *running configuration* is in the RAM.

The network interface cards (generally referred to as line cards in this book) attach to external devices and networks and support the I/O devices such as the network interface controllers (NICs) and the Physical Layer (Layer 1) components that receive and transmit packets. The hardware, and possibly, software components in the line cards provide the low-level network protocol functionalities (Layer 1 and Layer 2 functionalities) that enables the routing device to attach to external devices and networks.

The Layer 1 and Layer 2 components on a line card receive and send packets on an interface using appropriate media-dependent data formatting and transmission techniques. Typically, in this architecture, to ensure data integrity, the NICs check the Frame Check Sequence (FCS) fields for received Layer 2 packets, and calculate the FCS for Layer 2 packets to be transmitted.

The type of communication and the processes involved between the line card I/O controllers and the centralized processor depends on the type of architecture employed in the routing device. In low-end and some mid-end routing devices, the I/O controllers involve the centralized processor by sending an interrupt request as described in Figure 6.6. In high-end routing devices, the line card I/O controllers are able to communicate and pass packets directly over the switch fabric as discussed in the "Distributed Processor or ASIC Based Architectures with Topology Derived Forwarding Tables" section below.

6.3.1.3.1.2   Cisco Router Buffers and Queues

The type of buffers and queues discussed here typically exists in other router types other than Cisco routers [CISCNETS601] [CISCNETS2011] [CISCNETS2111] [CISCNETS2112] [CISCNETS2203]. The *interface buffers* in Cisco routers consist of *interface FIFOs* and *interface receive (RX) and transmit (TX) rings*:

- **Interface FIFO**: This is a very small buffer memory in an interface used to store bits belonging to a packet as they arrive from the physical network medium. These bits are under the control of the interface driver (see discussion below). Typically, these buffers are fixed and not configurable.
- **Interface RX and TX Rings**: These buffers are used by the interface driver on some platforms to receive packets on the inbound (RX) interface and transmit packets on the outbound (TX) interface. Each interface has both RX and *TX rings* which operate in a FIFO manner (FIFO *RX/TX rings*). The *RX ring* is used to store a packet until the interface driver is ready to handle it. The *TX ring* is used to store a packet until the interface can transmit it onto the physical network medium. The interface rings can exist on the network interface card itself, or they can be part of a shared memory on the system (i.e., the *shared I/O memory*). In the latter case, the Cisco IOS software creates the rings on behalf of the network interface controllers and then jointly manages them with the controllers. Typically, these buffers are fixed and are not configurable.

Other than *interface buffers*, the router also supports *internal buffers*. The *internal buffers* consist of *buffer headers*, *shared or main memory system buffers*, and *shared memory interface buffers*:

- **Buffer Headers**: These buffers store *data structures* that contain information about related buffers (e.g., location pointers, buffer size, etc.). The *buffer headers* are mainly used to keep track of buffers and enqueue them for various system processes. The data structures are mostly located in the *main processor memory* (RAM) for all buffers. In some platforms, to speed up processing, the buffer headers or particle headers are stored in *shared I/O memory*.
- **Shared or Main Memory System Buffers**: *System buffers* (located in RAM) are used to store packets that are destined for the processor itself, or packets that are to be handled via process switching in some platforms. The total amount of *system buffers* depends on the available RAM space. The system buffers are configurable and can grow or shrink on demand. These buffers are public and all interfaces can use them. The system buffers in Cisco routers have the following sizes: small (104 bytes), middle (600 bytes), big (1524 bytes), very big (4520 bytes), large (5024 bytes), and huge (18024 bytes).
- **Shared Memory Interface Buffers**: *Interface buffers* are used to store packets that are passed between the interface driver and a forwarding path other than the software process switching path, for example, packets to be forwarded via the route cache (called Cisco fast switching), or via the topology-based FIB (called Cisco Express Forwarding (CEF)). These buffers in Cisco routers are allocated at system startup or after Online Insertion and Removal (OIR). The number of interface buffers depends on the MTU and speeds of the interfaces supported on the router. These buffers are not configurable and are interface-specific.

The processor memory (RAM) is typically logically divided into a *main processor memory* part and a *shared I/O memory* part (see Figure 6.9). The *main processor memory* part holds the router OS executable image, running configuration, *buffer headers* (or data structures), routing tables, and route cache, while the *shared I/O memory* part contains the *system buffers*, *interface buffers*, and possibly, the *RX and TX rings* in some platforms. The *system buffers* of the *shared I/O memory* part is used for temporary storage of packets waiting to be forwarded via process switching, while the *interface buffers* for packets waiting to be forwarded via the route cache. The *shared I/O memory* is shared among all interfaces.

The *shared memory interface buffers* can be further characterized as *particle buffers* or *contiguous buffers*:

- **Particle Buffers**: Incoming packets are segmented into smaller fixed-size data units called *particles* before being stored in the shared I/O memory. The particles are reassembled into their original contiguous packets before being transmitted on the physical network medium. Some routing platforms refer to particles as *cells*. Interface particle buffers are used on some Cisco routing platforms and VIP (Versatile Interface Processor) cards, and are located in the shared I/O memory. These systems allocate packet memory as particles.

- o The size of a particle can be 1024 bytes, 512 bytes, or 128 bytes (128 bytes typically used for multicast/broadcast packets). The particle buffers are not configurable.
- o Similar to the *buffer headers*, particles have associated with them *particle buffer headers* which store information about which particles make up an entire original packet. Particles of an arriving packet are segmented into particle size blocks and stored in free particle buffers.
- o Particle buffer headers can also be cloned (called cloned particle headers). This allows the router to replicate a packet without actually replicating the packet itself for every outbound interface, thereby, significantly improving multicast performance.

- **Contiguous Buffers**: These buffers exist mainly on older Cisco routing platforms. Routers that use contiguous buffers store packets in buffers that are sized with respect to the MTU and speed of the interface media (e.g., 1500-byte MTU for Ethernet). The amount of buffers is based on grouping the requirements of the enabled interfaces on the router.
  - o The use of contiguous buffers can be wasteful in terms of memory usage because arriving Ethernet frames are rarely full size. They are not as efficient as particle buffers when it comes to memory usage, and they are also not efficient during packet replication (for multicast/broadcast purposes).
  - o An arriving packet is treated as a whole unit when contiguous buffers are used. This means when a packet is being replicated, each replica requires a completely new header to be created. In a system using particle buffers, only the particles are replicated and each interface is responsible for reassembling the particles into a copy of the original packet before transmitting it.

Packet buffer pools in Cisco routers (as in other routing platforms) can be created as either *public buffer pools* or *private buffer pools*:

- **Public Buffer Pools**: *Public buffer pools* are created by Cisco IOS software and are available to all interfaces and system processes that have packets to be forwarded. They are used by interfaces that either run out of private buffers or do not support the private buffer function.
- **Private Buffer Pools**: *Private buffer pools* are created during initialization of the Cisco IOS software and are allocated a fixed number of buffers. Each pool is static – new buffers cannot be created on demand for the pool. If a router interface needs a buffer and none is available in the *private buffer pool*, the Cisco IOS software falls back to the *public buffer pool* for the size that matches the interface's MTU. Private buffer pools are created for interfaces to enable them store packets as they arrive from the network medium without relying on the public buffer pools which the rest of the processes and interfaces in the router share. When a packet first arrives on a network interface, it is placed in a buffer in the *RX ring*. The network interface controller then tries to replace this used buffer with a free buffer, either from its *private*

buffer pool, and if this is not possible, from a buffer in the *public buffer pool*. In this case, pulling a buffer from the same sized public buffer pool is called *fallback*. Some routing platforms have interfaces that support *private particle pools*. When such interfaces run short of *private buffers*, they fall back to a buffer a *public particle pool* corresponding to their buffer size.

Queues are used to organize packets stored in memory in the desired order, so that they can be processed relative to other packets. The types of queues used in a system with centralized forwarding can be classified as *system interface queues* and *interface queues*. The *system interface queues* consist of *input hold queues* and *output hold queues*:

- **Input Hold Queues**: These are used to queue packets in the *system buffers* and are to undergo process switching. These queues are located in the *main processor memory* and are configurable on a per interface basis. Note that the *system buffers* are located in the *shared I/O memory part*.
- **Output Hold Queues**: These are used to queue packets in the *system buffers* that have already been handled via process switching and are waiting to be transmitted by the interface driver. These queues are also located in the *main processor memory* and are configurable on a per interface basis.

The *interface queues* consist of *receive queues* and *transmit queues*:

- **Receive Queues**: These queues are used for incoming packets stored in the *interface buffers* and waiting to be forwarded via the route cache (see discussion on route caching and fast switching below).
- **Transmit Queues**: These queues are used for outgoing packets stored in the *interface buffers* and waiting to be transmitted by the outgoing interface driver.

6.3.1.3.1.3   Device Drivers

A device driver provides a software interface to a hardware device in a host system, enabling the host OS and other computer programs running on the system to access the device's hardware functions without having to know precise details of that hardware device. Simply, a device driver allows the host's OS to communicate with the hardware device. Device drivers are hardware-dependent and OS-specific. A device driver provides hardware abstraction as well as serves as a translating interface between the hardware device and the OS and programs that use it.

A router supports various network interfaces through which packets are received and forwarded to their destinations. The router OS has device drivers that support the various interface types. Device drivers provide the following functions:

- Act as the glue between the router OS and the network interfaces.
- Are pieces of the software code that are invoked when arriving packets enter the network interfaces and need to be processed.
- Work with interrupts that are generated when a packet arrives, or when a packet is to be transmitted.

For example, in Cisco routers, device drivers are responsible for the following [CISCNETS2111] [CISCNETS2112]:

- **Inbound packet path**:
  o Manage the reception of a packet from the network interface hardware. Places the packet temporary in a *RX ring* and sends a *receive interrupt* to the CPU to indicate that a packet has been received.
  o Validate the packet (i.e., frame check) as well as identify the contents of the packet. If frame check fails, discard frame and update frame error counters (e.g., runt, giant, CRC).
  o Assign a free *buffer header*. If a free buffer is not available, the incoming packet is dropped and the *ignore counter* is incremented.
  o Move the packet to the *shared I/O memory* (i.e., *interface buffers*).
  o Attempt to forward the packet via the route cache (called *Cisco fast switching*). If this is not possible, prepare the packet for process switching by constructing the packet header.
  o If packet is to undergo process switching, assign a *system buffer* and return the packet buffer (*interface buffer*) to the free list.
  o Enqueue the packet for process switching. If the *input hold queue* is full, drop the packet and increment the *input queue drop counter*.
  o *Receive interrupt* is complete. Return CPU to scheduled tasks.

- **Outbound packet path**:
  o Service the processed packet from the *output hold queue or transmit queue* to the interface *TX ring*.
  o Manage transmission of the packet on the outbound interface hardware.

- **Status/control (housekeeping and statistics collection)**:
  o Return empty buffers to the buffer pool of origin.
  o Collect statistics and perform accounting.
  o Provide statistics and status on request.
  o Change the interface configuration as requested.

6.3.1.3.1.4 Elements of Execution in Some Cisco Routers

Some Cisco routing platforms define the following elements of execution [CISCINTRCIOS] [CISC1600ARC] [CISC2500ARC] [CISC2600ARC] [CISC4000ARC]. Other platforms use different components or mechanisms for booting:

- **ROM Monitor (or ROMMon)**: This is also referred to as *system bootstrap*, *bootstrap code*, or *bootloader*. It contains hardware configuration and system diagnostics code that is run at system startup. The ROMMon performs low-level CPU initialization and POST, and in a system without RxBoot, loads the default Cisco IOS software image into memory. ROMMon does not support the routing code and is only capable of reading the system configuration and accessing the file system. In a system without RxBoot, ROMMon attempts to load the first IOS software image file from Flash memory. The ROMMon is a diagnostic image that is most often used during system recovery procedures (for example, when the user has forgotten the password, or when a wrong/corrupted Cisco IOS software image is loaded). The diagnostic mode provides the user with a limited subset of the Cisco IOS commands. The user can view or modify the configuration register from this mode and can perform a Cisco IOS software upgrade via modem transfer.

- **RxBoot**: This is also referred to as *boot helper image* or *helper Cisco IOS*. This code is a subset of the Cisco IOS software and is used when a valid Cisco IOS image is not present on the router, allowing the router to download a full Cisco IOS image from the network. The boot helper image contains information that allows the system to locate and load a copy of Cisco IOS software according to the settings of the configuration register. The Cisco IOS software image can be located either on an onboard system Flash memory, on a removable Flash memory card, or on a TFTP server in the network. In some platforms, the Cisco IOS software image resides on a removable Flash memory card.

The Boot ROM, typically in an EPROM, is used for permanent storage of the startup diagnostic code (i.e., the ROM Monitor), and the RxBoot.

6.3.1.3.1.5   Cisco Router Startup Process: Boot Sequence

The main job of the CPU (see Figure 6.8) is to load instructions defined in the router OS software from (onboard or removable) Flash memory or from RAM and execute them. The typical startup process (boot sequence) in a Cisco router supporting centralized software-based packet forwarding and using a ROM Monitor and RxBoot can be described as follows [CISCINTRCIOS] [CISC1600ARC] [CISC2500ARC] [CISC2600ARC] [CISC4000ARC]. The bootup sequence when the router first powers up involves the following steps:

Step 1– ROM Monitor:

Upon power up, the ROMMon (which resides in the Boot ROM) takes control of the CPU and performs the following:

- *Configure power-on register settings*: Sets the CPU's control registers and on other devices such as the interface hardware logic for console access (including console settings). Performs configuration register checks.
- *Perform power-on diagnostics*: Performs tests on NVRAM and RAM (by writing and reading various data patterns). This is initial diagnostic tests of memory and other hardware.
- *Initialize the hardware*: Performs initialization of the interrupt vector and other hardware, as well as sizes the various RAM components (memory sizing).
- *Initialize software structures*: Performs initialization of the NVRAM data structures to enable the reading of information about the boot sequence, stack trace, and environment variables (i.e., data structure

initialization). Also, collects information about accessible devices in the initial device table. Performs Flash file system (MONLIB) setup.

Based on the configuration register value in the NVRAM, the router either stays ROMMon, or executes RxBoot from the Boot ROM.

Step 2– RxBoot:

a) *Router has RxBoot* [CISC1600ARC] [CISC2500ARC]:

RxBoot examines the hardware, builds basic data structures, performs interface setup, sets host mode functionality, and performs startup configuration checks. Based on the configuration register value, the router either stays in RxBoot, or executes the Cisco IOS software image file (using default setting or as defined in the startup configuration) from the removable Flash memory card, or from RAM (or from the RAM after the IOS image has been moved there from a TFTP server in the network). The source of the Cisco IOS software image is determined by the configuration register setting. The main Cisco IOS software image then reanalyzes the hardware.

The router configuration file (in NVRAM) may contain the `boot system` commands. For example, the command `boot system flash slot0:router1-sys-122-1a.bin` in the `configuration file forces the RxBoot to search for the file` "router1-sys-122-1a.bin" on the removable Flash memory card called "`slot0:`". The directive provided by boot system command in the router configuration file overrides that in the configuration register. If there is no `boot system` statement, and if the configuration register is set to its default value, then the RxBoot fetches the first IOS image file it finds in its Flash memory. If this fails, RxBoot tries to load an image from Boot ROM.

b) *Router has no RxBoot* [CISC2600ARC] [CISC4000ARC]:

If the router has no separate RxBoot (or *boot helper* image), it needs to have a valid Cisco IOS image in the Flash memory. Next, the ROMMon looks for the Cisco IOS software image in the Flash memory. If the router does not find a valid IOS image in the Flash memory, it will not be able to come up.

Step 3– Cisco IOS:

This stage involves booting the Cisco IOS software image. After the router has successfully located the Cisco IOS software image, it decompresses it and loads it into the RAM. The IOS image then starts to run and performs important functions such as:

- Recognizing and analyzing interfaces and other hardware components.
- Setting up *data structures* such as *Interface Descriptor Blocks* (*IDBs*) in the *main processor memory*. IDBs are used by the Cisco IOS software to describe interfaces and to reference them from a configuration/feature perspective and also from a packet forwarding perspective. IDBs are special control structures that are internal to the Cisco IOS software [CISCIDBLIM12]:

○ An IDB contains information about an interface that includes the IP address, interface state, packet statistics, backplane specific fields, hardware-specific fields (network controller specifics), pointers to external data for extended functionality, and Layer 2 encapsulation [CISCNETS2111]. During the initialization of the router, interfaces and corresponding hardware and software IDBs are created.

○ Two main types of IDBs are supported, *hardware IDBs* and *software IDBs*. A *hardware IDB* represents a physical interface and includes physical ports and channelized interface definitions. A *software IDB* represents a logical sub-interface (e.g., VLAN or Permanent Virtual Circuit (PVC)), or the Layer 2 encapsulation supported (e.g., Ethernet, Point-to-Point Protocol (PPP), High-Level Data Link Control (HDLC), etc.).

○ Each physical router interface consumes a minimum of two IDBs, one hardware IDB for the physical port and one software IDB for the Layer 2 encapsulation. The Cisco IOS software maintains at a minimum, one IDB for each router interface and one IDB for each sub-interface. There is a finite number of IDBs available on a router mainly based on the maximum number of software IDBs that can be supported.

• Allocating *system buffers* and *interface buffers* in the *shared I/O memory*.
• Reading the startup configuration from NVRAM to RAM (i.e., installing the running configuration) and configuring the system.

Note that the RxBoot also performs these functions but does not reanalyze the hardware unless the full Cisco IOS software is executed.

6.3.1.3.1.6   General Layer 3 Packet Processing in a Router with Centralized Forwarding

In the typical CPU-based architecture, when the I/O controller in a network interface receives a Layer 2 packet (Ethernet frame) from the network medium, it places its temporary in an *RX ring* buffer. This frees the interface, allowing it to receive the next arriving Layer 2 packet. Before placing the packet in the *RX ring*, the I/O controller first reads the Layer 2 destination address to check if the packet is destined for the local device plus other basic Layer 2 packet sanity checks.

The I/O controller then sends an *interrupt request* (IRQ) to the CPU (*receive interrupt*) to indicate that a packet has been received. The *interrupt handling routine* (which forms part or the router OS) assumes responsibility of transferring the Layer 2 packet out of the I/O controller's RX ring buffer, performing basic Layer 2 packet verification (plus other packet sanity checks), examining the EtherType field in the frame to determine the upper layer or Network Layer protocol to receive the frame (0x0800 for IPv4, 0x86DD for IPv6, and 0x0806 for ARP, all in hexadecimal notation), and placing the packet in a software *IP input processing queue* (i.e., *input hold queue*).

This CPU-based architecture, generally, does not necessarily have to maintain on a per-interface basis, FIFO queues for the arriving packets. Instead, all arriving packets on all interfaces can be placed in the IP input processing queue. Also, a limit can be set on the number of packets each interface can be stored in the IP input processing queue at any given time, by using a counter that is incremented each time an interface places a packet in the queue. This counter is decremented each time the router completes the processing of a packet and the packet's buffer has been transferred to an interface's output queue or assigned to another process for local delivery.

When the number of packets from a particular interface exceeds a maximum configured limit, the router will start dropping all excess packets, usually via, an active queue management (AQM) mechanism [RFC2309]. Note that in the centralized architecture, all tasks including route processing and packet forwarding are performed by the single CPU, and as a result, can lead to processor overloads when too many packets arrive at the router and the CPU does not have enough processing resources to process them.

Details of the memory and buffer management mechanisms depend on the type of architecture used by the routing device. In some architectures with a single CPU (e.g., Cisco 2500 series routers), all packets are stored in SRAM and are accessible by the I/O controllers and the CPU. In architectures with intelligent line card forwarding processors and a CPU (e.g., the route switch processor (RSP) in the Cisco 7500 series routers), when a packet needs to be processed by the RSP, it is moved from the packet memory in the line card into the main memory of the RSP.

For each packet in the IP input processing queue, the *IP forwarding process* performs the destination address lookup in the routing or forwarding table to determine the outgoing interface and next-hop IP address. The IP forwarding process also determines the Layer 2 (MAC) address of the next-hop to be used for Layer 2 packet rewrites. The IP forwarding process then updates the relevant IP header fields (IP TTL and Checksum), and performs Layer 2 rewrites (source MAC address, destination MAC address, and Ethernet FCS). If the IP output processing queue (i.e., *output hold queue*) is empty and the *TX ring* of the I/O controller has some free space, the processed packet is passed directly to the *TX ring*, otherwise, it is queued in the *IP output processing queue*.

Some router architectures do not discard the arriving Layer 2 frame in which an IP is encapsulated. Instead, the Layer 2 frame information is maintained along with the IP packet because the IP forwarding process needs the Layer 2 information at some steps of the packet forwarding process [ZININALEX02]. In this case, the IP packet is decapsulated only when it is delivered locally.

Other than the traditional FIFO queuing, the IP output processing queue can be organized as a number of priority sub-queues, and advanced scheduling methods such as priority scheduling, weighted round-robin (WRR), and weighted fair queuing (WFQ) used to schedule packets to the *TX ring* of the I/O controller. The particular scheduling method used (*dequeuing process*) is implemented as a specific routine in the router OS which is given control on an interrupt and scheduled by the router OS. This routine examines the output sub-queues according to the scheduling policy configured, and moves packets to the outbound I/O controller's *TX ring* for transmission onto the network.

Other than the CPU or NPU speed, factors such as bus speeds, packet memory speeds, and interface's I/O memory speeds can further limit the performance of a routing device using this architecture. Some improvement in the forwarding rate of the centralized forwarding architecture can be achieved by using a centralized forwarding module based on ASICs as discussed next.

6.3.1.3.1.7 Special Focus: IP Packet Processing in a Cisco Router with Centralized Forwarding (Cisco Process Switching)

Process switching (also called *process forwarding* in this book) is a forwarding method in which the CPU is directly involved in the forwarding decision process for a packet. We assume here that the centralized forwarding process is based on the shared memory architecture. The steps involved in forwarding a packet via Cisco process switching can be described as follows [CISCINTRCIOS] [CISC1600ARC] [CISC2500ARC] [CISC2600ARC] [CISC4000ARC]:

➢ **Receiving a packet**:

1. A packet arrives on the network medium and the network interface controller (precisely, the *interface driver*) detects and copies it into a buffer pointed to by the first free element in the *RX ring*. The interface controller uses the Direct Memory Access (DMA) method to copy packet data into memory.

2. The interface controller changes ownership of the packet buffer (in the receive descriptor) back to the CPU and issues a *receive interrupt* to the CPU. The interface controller does not have to wait for a response from the CPU and continues to receive incoming packets into the *RX ring*.

   Under bursty traffic conditions, the media controller may fill the *RX ring* before the CPU has time to process all the new buffers in the ring, a condition called an *overrun*. When this happens, all incoming packets are dropped until the CPU recovers.

3. The CPU responds to the receive interrupt and attempts to remove the newly filled buffer from the *RX ring* and replenish the ring from the interface's *private buffer pool*. It is important to note that packets are not physically moved within the *shared I/O memory*, instead, only the pointers are changed. The packet is dropped if the interface's *private buffer pool* is full, otherwise, one of the following happens:

   a. The interface's *private buffer pool* has a free buffer available to replenish the *RX ring*: The free buffer is linked to the *RX ring* and the packet now belongs to the interface's *private buffer pool*.

   b. The interface's *private buffer pool* does not have a free buffer available to replenish the *RX ring*: The *RX ring* is replenished by falling back to the *public buffer pool* that matches the interface's MTU. The *fallback counter* is incremented for the *private buffer pool*.

   c. A free buffer is also not available in the *public buffer pool*: The incoming packet is dropped and the *ignore counter* is incremented. In addition, the interface is throttled and all incoming traffic is ignored on the interface for a short period of time.

> **Forwarding the packet**:

After the *RX ring* has been replenished, the CPU begins to forward the packet. We assume Cisco IOS software cannot forward the packet using the route cache (Cisco fast switching) or the optimized topology-based FIB (Cisco Express Forwarding (CEF)).

1. While still in the receive interrupt context, the packet is placed in the *input hold queue* for the *IP input process*, and then the receive interrupt is dismissed. If the input hold queue is full, the packet is dropped and the *input drop counter* is incremented.
2. Note that several other processes run on the CPU. Eventually the packet forwarding process runs, performing a routing or forwarding table lookup to determine the outbound interface, and rewriting IP header and Layer 2 packet fields as needed. Note that the packet still has not been moved from the buffer in which it was originally copied. After the packet has been processed, the Cisco IOS software continues to the packet transmit stage.

> **Transmitting the packet**:

1. After the packet has been forwarded via process switching, it is placed on the *output hold queue* for the input interface. If the *output hold queue* is full, the packet is dropped and the *output drop counter* is incremented.
2. The Cisco IOS software (running in the CPU) attempts to find a free descriptor in the *TX ring* of the output interface. If a free descriptor is available, the Cisco IOS software removes the packet from the *output hold queue* and links the buffer to the *TX ring*. If the *TX ring* is full, the Cisco IOS software leaves the packet in the *output hold queue* until the network interface controller transmits a packet from the *TX ring* and frees a descriptor.
3. The outbound network interface controller polls its *TX ring* periodically for packets that need to be transmitted. As soon as the interface controller detects a packet, it copies the packet onto the network medium and raises a *transmit interrupt* to the CPU.
   a. The interface driver identifies that there is a packet in the *TX ring* waiting to be transmitted, and forwards it onto the physical network medium. The interface driver sends an interrupt to the CPU, requesting that counters be updated and buffers placed back into free pools.
4. The CPU acknowledges the transmit interrupt, unlinks the packet buffer from the *TX ring*, and returns the buffer to the buffer pool from which it originated. The CPU then checks the *output hold queue* for the interface. If there are any packets waiting in the *output hold queue*, the CPU removes the next one from the queue and links it to the *TX ring*. Finally, the CPU dismisses the transmit interrupt.

If an arriving packet is to be forwarded via process switching, the ownership of the buffer passes from the network interface controller to the CPU. If the packet is to be fast switched (via the route cache), ownership is passed either to the *output hold queue* or to the outbound *TX ring*. It is important to note that packets once placed in a buffer are never copied from buffer to buffer; only ownership of the buffer changes using pointers. Once the packet is transmitted, the buffer is returned to its original owner.

6.3.1.3.2   Centralized ASIC-Based Architectures

In the centralized ASIC-based architectures, specialized hardware components are developed and incorporated into the routing device to improve the packet forwarding performance. In these architectures, dedicated centralized forwarding modules mainly implemented in ASIC or using a combination of optimized software and ASIC components, are used to achieve higher forwarding rates compared to the purely centralized processor (CPU)-based architectures. However, it must be noted that bus speeds (in shared-bus architectures), and interface's I/O memory speeds, can still have a significant impact on the overall packet forwarding performance of the routing device. All of these latter factors are needed to achieve higher forwarding rates.

The challenge in designing centralized architectures using CPUs, NPUs, and ASICs is coming out with a design that results in the highest forwarding rates within the limits of the CPU, NPU, or ASIC capabilities, bus speeds, packet memory speeds, interface's I/O memory speeds, and overall system cost (packet forwarding to system cost ratio). The last factor of cost will ultimately determine the final design. As network sizes, interface speeds, and the demand for improved forwarding methods grow, the interest in more improved forwarding architectures also grows.

The discussion up to this point has analyzed the factors that cause performance degradation in the centralized forwarding architecture and the reasons why improved architectures such as distributed forwarding architectures can avoid the shortcomings of the centralized forwarding architecture. In the next two subsections, we discuss a number of architectures that have been developed with the goal of improving the forwarding rates of routing devices.

6.3.2   FORWARDING ARCHITECTURES USING ROUTE CACHES

As networks grew in size and link speeds increased, designers had to rethink the older forwarding architectures to handle faster links and forwarding rates. The concept of route caching was conceived to keep up with increasing data rates. Route caching relies on the temporal and spatial locality exhibited by traffic flows. Figures 6.10 and 6.11 show the typical architectures for route cache maintenance. Both architectures support a route cache update process that is responsible for updating the route cache using the routing information received.

The architecture in Figure 6.11 uses an IP forwarding table (FIB) that has resolved routes and is a well-optimized table created from the IP routing table (which may contain recursive routes as discussed in the "Recursive Route Lookup in an IP Routing Table" section in Chapter 5). The use of a forwarding table results in high forwarding performance because it does not contain recursive routes which can slow down the speed of destination address lookups.

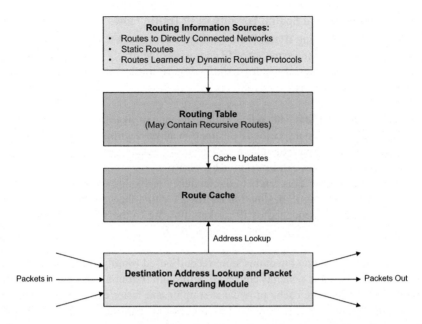

FIGURE 6.10 Address lookups and packet forwarding using route caching: Route cache directly maintained by the routing table.

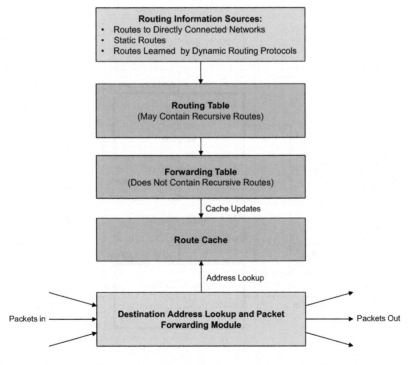

FIGURE 6.11 Address lookups and packet forwarding using route caching: Route cache directly maintained by the forwarding table.

6.3.2.1   Temporal and Spatial Locality of IP Traffic Flows

Locality, when examining IP traffic forwarding, refers to the tendency of a forwarding engine to reference the same set of IP addresses repetitively over a short period of time. The two basic types of locality exhibited by IP traffic are *temporal* and *spatial locality* [HUSSFAULT04]:

- **Temporal Locality**: This refers to the tendency or likelihood of a forwarding engine reusing a specific IP destination address within a relatively short period of time. For example, a sequence of packets carrying the same IP destination address exhibit *temporal locality*.
- **Spatial Locality**: This refers to the tendency or likelihood of a forwarding engine referencing IP destination addresses within the same address range. For example, a sequence of IP packets flowing to the same IP subnet or Virtual LAN (VLAN) exhibits a *spatial locality*.

The route cache-based architectures store recently used routing entries in a fast and convenient lookup table which is consulted before the IP routing or forwarding table (Figure 6.12). The route cache provides a simpler and faster exact matching front-end lookup mechanism that requires less processing than the routing or forwarding tables. If the forwarding process (engine) finds a matching entry during route

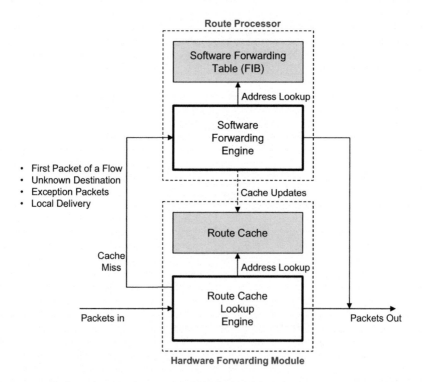

FIGURE 6.12   Packet forwarding using a route cache.

cache lookup, it will forward the packet immediately and not consult the software-based routing or forwarding table. The route cache is populated with information that defines how to forward packets associated with a particular flow. A flow in this case uniquely identifies a stream of packets having the same IP destination address in the network, and each flow entry in the route cache contains sufficient information to forward packets for that flow (Figure 6.13).

The flow entries in the route cache are constructed by routing the first packet in software (via the slower routing or forwarding table lookup process in the route processor), with the relevant values in the forwarded first packet used to create the required information for route cache entry. Subsequent packets associated with the flow are then forwarded using the route cache (usually implemented in hardware using Content Addressable Memory (CAM) devices) based upon the information in the flow entry. For the first packet of a flow to a given IP destination address, the result of the software-based lookup is stored in the route cache as discussed in Chapter 2 and 5. The system then forwards all subsequent packets carrying the same Classless Inter-Domain Routing (CIDR) /32 IP destination address in their IP headers based on the faster route cache lookups.

Forwarding using a route cache is often referred to as "*route once, forward many times*". The route cache entries are created and deleted dynamically as flows start and stop. The routing device may decide to delete certain route cache entries when they are not used for some time (using cache entry aging timers or idle (inactivity) time-outs), or when the route cache memory runs low. The route cache entries for a flow may include information required for QoS and security processing. A route cache entry may contain extra parameters that allow the cache lookup engine to apply packet filtering, mark packets, apply a routing policy, and so on [AWEYA2BK21V1].

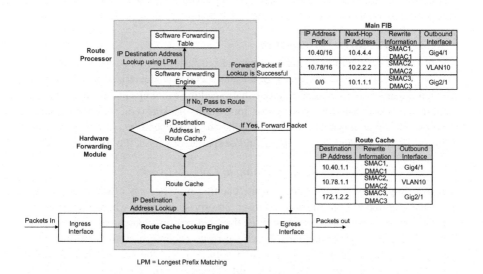

LPM = Longest Prefix Matching

FIGURE 6.13   Illustrating IP packet forwarding using a flow cache.

6.3.2.2   Handling Exception Packets

Exception packets (Figure 6.12) are packets that are received by the routing device and cannot be processed by the normal IP forwarding process and require special attention in a module with more sophisticated processing capabilities such as the route processor. Exception packets include the following:

- Packets too complex for the normal forwarding process such as IP packets containing IP header options, packets that are too large than an interface Maximum Transmission Unit (MTU) and require fragmentation, packets requiring Network Address Translation (NAT), packets requiring encryption, packets requiring tunneling, etc.
- Packets addressed to the router itself, such as routing protocol updates and those belonging to control and management protocols like Secure Shell (SSH), Telnet, and ICMP pings.
- Packets requiring the generation of ICMP error messages, including destination unreachable messages and traceroute responses. The route processor itself generates ICMP error messages when required, and sends them to the appropriate IP source addresses.
- Packets to be forwarded but require additional forwarding information than is available to the forwarding engine, such as packets for which the Layer 2 destination address of the next-hop node for Layer 2 rewrites is unknown. Such packets require the route processor to send ARP request to learn about the Layer 2 address of the next-hope node (see "Switching Within a Subnet" section in Chapter 5 for ARP operations).

The exact nature of exception packets to be handled depends on the routing device architecture, the forwarding engine capabilities, and the structure and contents of the forwarding tables. What may be considered an exception packet in one architecture may not be in another more sophisticated architecture.

6.3.2.3   Route Cache Performance

A route cache is generally a smaller compact but faster memory typically based on a CAM. Generally, the performance of a route cache is characterized by its *hit ratio*, which indicates the percentage of IP destination address lookups in the cache that will yield the correct forwarding information for arriving packets (i.e., successful address lookups). However, it is not hard to see that the hit ratio depends very much on the size of the cache and the degree of temporal and/or spatial locality of the arriving IP traffic.

6.3.2.3.1   Limitations of Route Caching in Core Networks

A forwarding architecture based on route caching may be very efficient at an edge or an aggregation routing device, or in an enterprise network where the IP traffic tends to exhibit more temporal/spatial locality and routing changes occur less often. However, forwarding architectures based on route caching tend to perform poorly when used in the core of service provider networks and the Internet where traffic tends to exhibit less temporal/spatial locality because of the multiplexing of a large number of packets with different destinations and the tendency for more frequent routing changes. When routing changes occur more frequently, the routing device will also have to invalidate corresponding entries in its route more frequently, thereby reducing the cache hit ratio and diminishing the benefits of using a route cache.

Furthermore, route cache entries are (/32) IP destination address based, thereby making route cache-based architectures not scalable when used in the core network routing devices where the number of IP destination addresses can be very large. The memory requirements for the route cache in core networks will have to be very high to avoid memory overflows. Also, increasing the route cache size to allow more entries to be stored can result in higher lookup times (negating the benefits of route caching), and performance degradation.

6.3.2.3.2   Double IP Address Lookups

A reduced or smaller cache hit ratio means the routing device will have to send more arriving traffic to the slower software-based routing or forwarding table for forwarding. The penalty of a lower cache hit ratio and a cache miss is that traffic will have to visit the route cache and then the software-based forwarding table to be forwarded, resulting in extra forwarding delay and lower forwarding rates. With a smaller cache hit ratio, the double lookup makes the forwarding performance of the device poorer than it would have been if only software-based forwarding with no route cache is used.

Furthermore, in addition to having a smaller hit ratio, the system will have to invalidate and replace entries in the route cache with valid ones when routing changes occur, which is an additional penalty paid by the system [HUSSFAULT04]. Given that forwarding via the software-based routing or forwarding table is processor-intensive, a large amount of arriving traffic with a smaller cache hit ratio can easily overwhelm and overload the route processor leading to some service outage.

6.3.2.3.3   BGP-Learned Routes and Recursive Lookups

It is important to note that forwarding via the route cache cannot be used directly with BGP learned routes because BGP specifies only the next-hop addresses of learned routes and not interfaces, thereby requiring the routing device to perform recursive route lookups to determine the outgoing interface for packets traveling on such routes (see Chapter 3 in [AWEYA2BK21V2]). Route caches are meant to provide fast packet forwarding using exact matching of the IP destination addresses of packets. They are not designed or meant to perform recursive lookups (which in itself can slow down packet forwarding rates). This means all recursive lookups have to be done in the route processor via the traditional process forwarding method.

6.3.2.4   Implementing Architectures with Route Caches

The traditional CPU-based architectures (process switching/forwarding) have no mechanisms to recognize that subsequent packets of the same flow can be subjected to the same IP address lookup and forwarding process as the first packet. In these architectures, the first packet and subsequent packets of the same flow are subjected to the same processing intensive CPU forwarding process, even though they are heading to the same destination and have the same forwarding state information. In this section, we examine at a high level, the various architectures that support packet forwarding via route caching.

6.3.2.4.1 Centralized Processor-Based Architectures with Route Caches

In the centralized processor-based architecture using route caches, subsequent packets of each flow are forwarded using route cache entries created by the software forwarding engine (process forwarding) after forwarding the first packets of the flows. The centralized processor-based forwarding using route caches (also called fast cache switching (or forwarding) or fast switching in Cisco terminology) is an enhancement to the traditional process forwarding discussed earlier. The route cache is consulted first each time a packet is received instead of using the processing intensive forwarding feature in the CPU.

In this architecture (Figure 6.14), packet forwarding is performed at the interrupt level using the route cache created by the software-based IP forwarding function in the CPU. Packet forwarding is fast because the route cache data structures and destination address lookup mechanisms are more optimized and also, arriving packets do not have to wait for the IP forwarding process to be scheduled. The route cache is created at the process level when the first packet of a flow is process forwarded; a route cache entry containing the /32 IP destination address as seen in the first packet, the next-hop IP address, the outbound interface, and the Layer 2 parameters of the Layer 2 packet in which the outgoing IP packet is to be encapsulated in (parameters usually maintained in the form of pre-computed Layer 2 packet header and related length).

For example, if an arriving packet with IP destination address 172.16.3.20 matches the network prefix 172.16.3.0/24 which points to the next-hop IP address 10.1.0.2 and outbound interface Gi0/1 in the IP routing or forwarding table, the IP destination

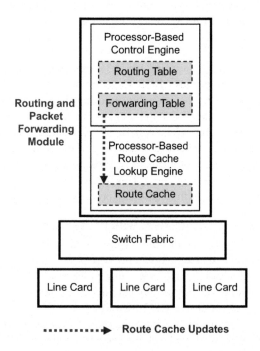

FIGURE 6.14 Centralized processor based architectures with route caches.

address 172.16.3.20, the next-hop IP address 10.1.0.2, and outbound interface Gi0/1 as well as the MAC address rewrite information for the outgoing Layer 2 packet are entered in the route cache. This route cache information is used to match subsequent packets of the same flow without the need to consult the routing or forwarding table. *Note that, in practice, when the route cache contains all Layer 2 rewrite information needed for forwarding a packet, it does not have to contain the next-hop IP address for each entry* (see Figure 6.13).

Using this route cache, all subsequent packets of a flow are forwarded at the interrupt level. Also, using a route cache, the router does not have to perform recursive lookups for subsequent packets of a flow because once the cache entry is created, the outbound interface, the next-hop IP address, and Layer 2 parameters for the next-hop are already known. It is only when a route cache entry does not exist for a given destination address, is the arriving packet queued for process switching/forwarding. The entries of the route cache are organized using special data structures (with efficient data sorting and lookup mechanisms) allowing destination address lookups to be done very fast even though forwarding is done at the interrupt level. In this architecture, special features such as checking ACLs, IP multicast routing, policy routing, and IP broadcast flooding are not implemented at the interrupt level, but rather at the process level.

The steps involved in forwarding a packet in the centralized CPU-based architecture with route caches (see Figure 6.14) can be summarized as follows:

- A packet arrives at a network interface of the routing device and the interface hardware receives and transfers it into the interface's I/O memory (i.e., packet buffer in shared I/O memory). The packet buffer can be pulled from either a public or private buffer pool and is done without interrupting the CPU. The device driver of the interface hardware then interrupts the CPU, indicating to it that a packet is queued in the interface's I/O memory and waiting for processing. The IP forwarding software then proceeds to update its inbound packet counters.
- While still in the receive interrupt context, the IP forwarding software verifies the packet and examines the packet header (e.g., encapsulation type, Network Layer header, etc.) to determine if the packet is an IP packet.
  - If an IP packet, the IP forwarding process consults the route cache to see if there is an entry matching the IP destination address of the packet. If a matching entry is found, the forwarding software retrieves the IP next-hop address, the outbound interface, and the Layer 2 address of the next-hop node's receiving interface from the route cache. The IP forwarding software then updates the relevant IP header fields of the packet, encapsulates the IP packet in a Layer 2 frame, and then performs all relevant Layer 2 frame rewrites.
  - If the IP destination address of the packet is not found in the route cache, the routing device reverts to the traditional process forwarding method as described in the "Packet Forwarding in the Traditional CPU-Based Forwarding Architectures Using the IP Routing Table" section above. After a packet is forwarded via process forwarding, a new entry is

created in the route cache for future use. When process forwarding is invoked upon receipt of a packet, this means this is the first packet of a flow seen by the IP forwarding software. After this, subsequent packets of the same flow will be forwarded via the route cache.

- The packet is sent directly to the outbound interface's I/O controller provided its *TX ring* has some free space, and the outgoing link is not congested. If the *TX ring* is full, the packet is not dropped but placed in the software IP output queue of the network interface to be submitted to the I/O controller when there is space available in its *TX ring*.
- The I/O controller of the outbound network interface detects the queued packet in the I/O memory, retrieves, and transmits it out the network interface on its way to the next-hop node. The network interface then sends a transmit interrupt to the CPU to indicate that the packet has been transmitted. The IP forwarding software then updates the outbound packet counters and frees the I/O memory previously occupied by the transmitted packet.

In the centralized architectures with route caches, particularly, those using shared-bus switch fabrics, the bus speeds, packet memory speeds, and interface's I/O memory speeds can still limit the packet forwarding performance of the routing device.

6.3.2.4.2 Architectures with a Single Dedicated and Centralized Processor or ASIC-Based Route Cache Lookup Engine

These types of centralized processor or ASIC-based architectures with route caches (Figure 6.15) implement the route cache and cache lookup engine using a specialized

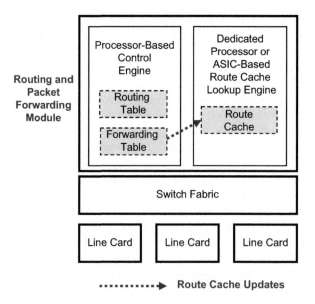

FIGURE 6.15 Architectures with a single dedicated and centralized processor or ASIC-based route cache lookup engine.

processor or hardware components in order to improve packet forwarding performance. In these architectures, a single dedicated centralized processor or ASIC-based forwarding module is used with the goal of achieving higher forwarding rates than the purely centralized processor (CPU)-based architectures with interrupt-level route cache lookup mechanisms (i.e., the Cisco *fast switching* method discussed above in Figure 6.14).

In an ASIC-based implementation, the route cache can be based on a content addressable memory (CAM) architecture allowing faster and exact matching lookups to be performed. The architecture in Figure 6.15 may use a dedicated and specialized processor-based route cache lookup engine (similar to the forwarding method referred to as *autonomous switching* in the Cisco 7000 series routers and *optimum switching* in the Cisco 7500 series [ZININALEX02]) instead of an ASIC-based one.

In this processor-based route cache architecture (e.g., Cisco 7000 series routers), network interface processors or controllers in the line cards communicate over a switch fabric (usually a shared-bus) with a dedicated route cache lookup engine (i.e., the *switch processor* in the Cisco 7000 series router and in other Cisco platforms [ZININALEX02]). The interface controllers pass Layer 2 packets (e.g., Ethernet frames) over the switch fabric to the dedicated route cache lookup engine, while the route cache lookup engine itself communicates with the processor-based *control engine* (also called the *route processor*) over a system bus to obtain routing information in the form of route cache updates.

The route processor runs the router OS which includes the routing protocols, maintains the main routing table, performs all system management and control functions, as well as all necessary process-level forwarding of packets (usually first packets of new flows, and control and management packets). The dedicated route cache lookup engine (i.e., the *switch processor* in the Cisco 7000 series) uses a highly optimized microcode to perform route cache lookups and packet forwarding (a process called *autonomous switching* in Cisco technology). The *switch processor* (or route cache lookup engine) in this architecture is a specialized module dedicated solely to fast-forwarding of packets and does not engage in other router functions (which are left to the route processor).

When the interface controller of a line card receives a packet, it copies it to a shared packet memory (called the *MEMD memory* in the Cisco 7000 and 7500 series routers [ZININALEX02]) on the *switch processor* via the shared-bus switch fabric. The *switch processor* performs a lookup for the IP destination address of the packet. If an entry is found in the cache, the packet is fast-forwarded by the *switch processor* without interrupting the route processor. If no correct entry is found in the cache, the packet is forwarded to the route processor for process-level forwarding using the main routing table information.

After the *switch processor* has determined the next-hop IP address, outbound interface, and associated Layer 2 parameters of the Layer 2 packet in which the outgoing IP packet is to be encapsulated in, the Layer 2 packet is constructed in the shared packet memory. The *switch processor* then signals the corresponding interface controller of the outbound line card to pick up the constructed Layer 2 packet for transmission out of the interface. After forwarding a packet via process switching/forwarding, the route processor always updates the route cache with the corresponding routing information to be used to fast forward other packets going to the same destination.

As with all centralized forwarding architectures especially those based on shared-bus switch fabrics, the bus speeds, and the interface's I/O memory speeds, can still have a significant impact on the overall packet forwarding performance of the routing device. The forwarding rate of the dedicated processor or ASIC-based lookup and forwarding process is still constrained by the shared-bus and the interface's I/O memory speeds. All of these latter factors have to be carefully considered for the routing device to achieve higher forwarding rates. Additional processing is still required in this architecture to handle exception packets which generally cannot be forwarded via the route cache (see "Handling Exception Packets" section above).

6.3.2.4.3 Distributed Processor or ASIC-Based Architectures with Route Caches

In the distributed architectures with route caches, each line has a local route cache and a cache lookup engine as shown in Figure 6.16 to allow packets to be forwarded locally when their IP destination addresses match route cache entries. Packets are forwarded directly to their outbound line cards when route cache entries are matched,

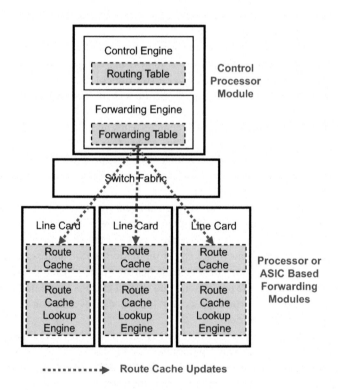

FIGURE 6.16   Distributed processor or ASIC-based architectures with route caches.

but when an entry does not exist for a packet, which happens for the first packet of new flows, the packet is sent to the route processor for the traditional process forwarding. The line cards are designed to have higher port density and modularity with local route cache lookup engines to support faster packet forwarding.

The route cache entries in each line card are created on demand; the first packet of each new flow is always forwarded via process forwarding in the route processor, after which an entry is created in the corresponding line card route cache. In networks with high temporal and spatial locality of traffic, the distributed architectures with route caches can provide very high forwarding rates (see "Temporal and Spatial Locality of IP Traffic Flows" section above).

In the distributed architectures with route caches, exception packets that require additional processing beyond what can be provided by the line card's route cache are forwarded to the route processor (see "Handling Exception Packets" section above). In a distributed processor-based architecture with route caches, high utilization of the line card processor can result in dropped packets and poor forwarding rates. This means a design has to carefully match the processing capability of the line card processor to the aggregate speed of the line card's interfaces to prevent the processor from being overwhelmed during periods of high traffic arrivals.

6.3.2.4.3.1 Example Architecture: Forwarding via the Route Cache in the Cisco 7500 Series Routers with Versatile Interface Processors (VIPs)

This section describes distributed forwarding using route caches in the Cisco 7500 series routers with VIPs [CISC7500ARC] [ZININALEX02]. The Cisco 7500 supports a *route switch processor* (RSP) which performs the functions of the *route processor* and *switch processor* in the Cisco 7000 series as described above. The Cisco 7500 also supports a shared bus called the *CyBus* that connects the line cards to the RSP (Figure 6.17). The VIP is a special line card that was developed to support higher port density and faster local forwarding of packets via route caches. A VIP supports multiple *port adapters* which connect to the attached external network segments. Each VIP has its own CPU, RAM, and packet memory, and runs a specialized version Cisco IOS software image.

The RSP supports both Cisco *process switching* and *fast switching*. When a packet is received by an interface processor in a line card, it is copied into the shared packet memory (*MEMD memory*) on the RSP. The RSP performs destination address lookup either at the interrupt level (using the route cache), or at the process level (using the IP routing table), and then signals the interface processor in the correct outbound line card to pick up the processed packet in the *MEMD* memory.

To start packet forwarding via the route cache in each VIP, the RSP receives first packets of flows (that cannot be forwarded via the route cache) from a VIP, performs IP destination address lookup for those packets, and passes the corresponding routing information to be entered in the route cache of the sending VIP. This is done for each VIP in the system. The RSP also maintains a local copy of the route cache entries sent to the VIPs. The following steps describe the packet forwarding process

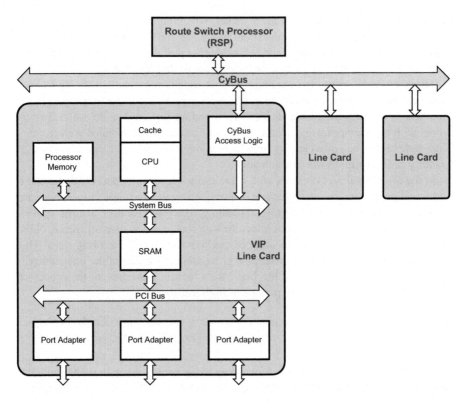

FIGURE 6.17   Architecture of the Cisco 7500 Series routers with versatile interface processors (VIPs).

via the route cache in the Cisco 7500 series with VIPs (Figure 6.17) [CISC7500ARC] [ZININALEX02]:

1. A Layer 2 packet (Ethernet frame) arrives at a port adapter of a VIP and is copied into the shared memory (SRAM) of the VIP.
2. The port adapter notifies the CPU of the VIP about the Layer 2 packet arrival by sending it an interrupt request (IRQ).
3. The VIP's CPU processes the packet and performs an IP destination address lookup in the local copy of the route cache sent by the RSP (i.e., fast switching).
4. If the CPU finds an entry for the packet's IP destination address in the route cache, it rewrites the Layer 2 packet with the appropriate Layer 2 header information. If no entry is found, go to Step 6.
5. If the outbound port adapter is on the same VIP and the transmit (TX) queue of that adapter is not full, the rewritten Layer 2 packet is passed directly to that port adapter. If the outbound port adapter does not reside locally on the VIP or the TX is full, go to Step 6.
6. If forwarding via the local route cache is not possible, the VIP's CPU instructs the CyBus Access Logic to transfer the Layer 2 packet to the MEMD memory of the RSP. From the RSP's perspective, the RSP's MEMD

memory functions as both the TX and RX queues of the interface processors of the port adapters.

7. Upon completing the transfer of the Layer 2 packet into the RSP's MEMD memory, the CyBus Access Logic notifies the VIP's CPU and the CPU in turn notifies the RSP about the presence of the Layer 2 packet.

8. If the RSP determines that the received packet has not already been fast switched by the sending VIP (i.e., a first packet of a new flow), it will perform an IP destination address lookup for the packet to determine the next-hop and outbound port adapter (on the correct outbound VIP), and performs appropriate Layer 2 rewrites. Otherwise, the RSP will skip this step. Note that in a Cisco 7500 configuration with mix VIP and non-VIP line cards, the outbound interface may be simply, a non-VIP line card without local port adapters. In the next steps, we assume a system with all VIP line cards.

   o After the RSP has successfully forwarded a packet that could not be forwarded by a VIP via route cache lookup (i.e., via fast switching), it sends the corresponding routing information to the VIP that sent the packet to be entered its local route cache. Thus, VIP route cache entries are created on demand as first packets of flows are received.

9. The RSP signals the outbound VIP.

10. The CPU of the outbound VIP then instructs the CyBus Access Logic to load the processed Layer 2 packet from the RSP's MEMD memory into the VIP's local SRAM.

11. Upon completing the transfer of the Layer 2 packet into the VIP's SRAM, the CyBus Access Logic notifies the VIP's CPU about the presence of the packet, which then processes the packet.

12. The VIP's CPU then writes a pointer to the Layer 2 packet in the TX queue of the correct outbound port adapter and notifies the port adapter about the presence of the packet in its TX queue.

13. The outbound port adapter retrieves the Layer 2 packet from the TX queue and transmits it onto the external network on its way to the final destination.

When packets are fast switched by the inbound VIP, the RSP does not participate in processing and forwarding packets, but serves only as a CyBus arbiter for inter-process communication. However, in the following cases, the RSP is involved in packet processing and forwarding even if the inbound VIP has a correct entry in its local route cache [CISC7500ARC] [ZININALEX02]:

• If a packet must be forwarded out a local port adapter of the inbound VIP (after fast switching) but the TX queue of that adapter is full, the packet is passed to the RSP to be queued and to share the available bandwidth resources on the outbound interface via one of several configurable queuing policies such as priority queuing, weighted fair queuing (WFQ), and so on.

• If a packet must be forwarded out a port adaptor (after fast switching) but the TX queue of that port adapter in the MEMD memory of its VIP is full, but the queuing strategy on that port adapter is not FIFO but sophisticated like priority queuing, and WFQ, the packet is passed to the RSP.

- If the TX queue of the outbound port adapter is full and the queuing strategy is FIFO, normally the packet would be dropped. However, if the command **transmit-buffers backing-store** is configured on that port adapter, the packet is copied into the DRAM of the RSP and is placed in the software queue of the outbound port adaptor.

Other than offloading packet forwarding tasks from the RSP, VIPs can be configured to provide distributed services such as IP packet fragmentation, IP multicasting, tunneling, access control lists (ACLs) checking, data compression, encryption, and so on [CISC7500ARC] [ZININALEX02].

In a section below, we describe distributed architectures with FIBs (topology-based forwarding tables) in the line cards which avoid the "route once, forward many times" forwarding philosophy and allow routing devices to be highly scalable and resilient.

6.3.2.5 Route Cache Maintenance and Timers

Route cache entries are dynamically created and deleted on demand (i.e., as new packet flows are detected). The following events can lead to the deletion of a route cache entry [ZININALEX02]:

- Some routing changes have occurred and changes were made to the routing table. In this case, all route cache entries affected by the routing changes are invalidated.
- The first packet of a flow has been processed and a new entry needs to be created in the route cache but all memory allocated for the cache has been used up. In this case, the oldest entry in the cache is deleted in favor of the new entry.
- In Cisco routers, 5 percent of the route cache entries are randomly invalidated every minute. This is done so that all route cache entries can be randomly invalidated and refreshed in 20 minutes.

Most route cache-based routing devices including Cisco routers use a number of timers to control invalidation of route cache entries and to prevent cache instability. Cisco routers use the following timers [ZININALEX02]:

- *Minimum Interval*: This specifies the minimum time from the moment a request for a route cache invalidation has been sent and the actual invalidation of the cache. The default setting of this timer is 2 seconds. A cache invalidation request is delayed for at least the *Minimum Interval* when it is first received. Furthermore, all subsequent requests after this are queued and delayed as well.
- *Maximum Interval*: This specifies the maximum time from the moment a request for a route cache invalidation has been sent and the actual invalidation of the cache. The default setting of this timer is 5 seconds.
- *Quiet Interval*: This specifies a quiet time during which if no requests for invalidation have been received, all outstanding requests are processed. The default setting of this timer is 3 seconds.

- ***Threshold***: This specifies the maximum number of invalidation requests that can be ignored (not processed) during the specified quiet time. The default setting is 0.
  - ○ Invalidation requests are executed only if the route cache has been quiet (for the *Quiet Interval*), and the number of requests not greater than *Threshold* is queued during the *Quiet Interval*, or if a request has been waiting for more than *Maximum Interval* to be processed.

A route cache-based architecture may maintain invalidation timers and statistics; invalidation rates may be used to identify excessive invalidation problems in the route cache [ZININALEX02].

6.3.2.6   Exact Matching in IP Route Caches

Each time the first packet of a flow is processed via the route processor (using process switching/forwarding), the routing information corresponding to this packet is entered in the route cache. Each entry of the typical IPv4 route cache contains at a minimum the IP destination address (i.e., the classful IP address as seen in an IP packet), next-hop IP address, and outbound interface/port. The entries in the route cache are destination address-based. They are based on the actual classful IP destination addresses of arriving packets. Because they are fixed and not variable length, address lookups in the route cache are based on exact (destination address) matching similar to lookups in the following devices:

- Ethernet switches using the 48-bit destination MAC address of Ethernet frames
- ARP (Address Resolution Protocol) caches using the CIDR /32 IPv4 destination addresses of packets
- ATM switches using the 8/12-bit VPI (Virtual Path Identifier) and 16-bit VCI (Virtual Channel Identifier) of ATM cells
- MPLS LERs (Label Edge Routers) and LSRs (Label Switch Routers) using the 20-bit MPLS labels of MPLS packets

Unlike the route cache, lookups in the IP routing table (RIB) and forwarding table (FIB) are based on longest prefix matching (LPM).

Compared to LPM, exact matching is easier and efficient in software (e.g., using hashing, binary/multiway search trie/tree, etc.) and hardware (e.g., associative memory (also known as content-addressable memory (CAM)). Figure 6.18(a) illustrates exact matching in a route cache using CAMs, while Figure 6.18(b) illustrates exact matching using hashing.

The size (entries) of a route cache is typically limited and smaller when compared to the size of the routing table and forwarding table. This makes the route cache simple to implement; has a small expected lookup time, and is fast to update. In order to prevent the size of the route cache from becoming unwieldy and to allow it to be refreshed periodically, route caches use a number of timers for aging and invalidating route cache entries as discussed in the "Route Cache Maintenance and Timers" section above.

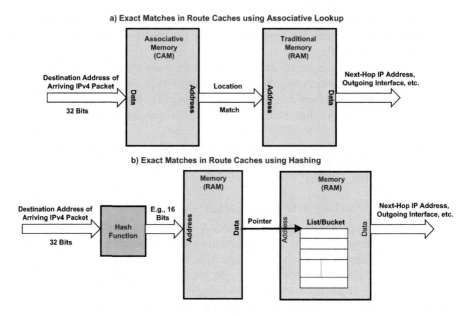

a) Exact Matches in Route Caches using Associative Lookup

b) Exact Matches in Route Caches using Hashing

FIGURE 6.18 Exact matching in IP route caches.

6.3.3 Architectures Using Topology-Based Forwarding Tables

A routing table (also called the Routing Information Base (RIB)) is used to store routing information about directly connected and remote networks. The routing table, which is created by the routing protocols, is not optimized for data plane operations. The entries stored in the routing table consist of information such as the routing protocol that learned a route, metric associated with a route, the administrative distance of a route, and possibly recursive routes, especially those installed by BGP (see "Control Plane" section in Chapter 5).

Although all this information is important to the overall routing process, it is not directly used or relevant to data plane operations. For example, most often the routing table also does not contain the MAC address information for the next hop. This must be determined either via a control plane operation (using ARP), reading the ARP cache or created manually by the network administrator via a router management interface. This means the information in the routing table can be distilled to generate information more relevant for the data plane forwarding operations. This information is stored in the smaller forwarding table (also called the FIB).

Unlike a route cache, the FIB is a topology-derived database that contains the same routes in the RIB. The only difference is that the FIB is a smaller more compact table and contains only information directly relevant for packet forwarding. The RIB contains extra information relevant for mapping out the network topology and discovering destination but not needed for actual packet forwarding.

High-end routers and switch/routers use the optimized lookup table (the IP forwarding table or FIB), which organizes the required routing information for data plane operations in optimized format (e.g., IP destination prefix, next-hop, outbound interface), and may also include a pointer to another optimized adjacency table, which describes the MAC address associated with the various next-hop devices in the network (Figures 6.19 and 6.20). Note that a routing device considers another node to be adjacent if it is directly connected or it can be directly reached over a shared Layer 2 network (e.g., Ethernet network) or a point-to-point Layer 2 network (e.g., Asynchronous Transfer Mode (ATM) network). An adjacent node to a router can be a directly connected host or router, that is, a host or another routing device sharing a common subnet. The optimized lookup tables (FIBs) are created with the goal of achieving high forwarding rates using optimized data structures and specialized lookup algorithms on specialized processors or ASICs engines [AWEYA2001] [AWEYA2000].

These specialized and optimized forwarding architectures, support not only high-performance lookup, but may also possess specialized processor- or hardware-based features that can be used for QoS classification and access security control (using access control lists). These additional features are handled at the same time the normal IP destination address lookups are being performed. ASIC implementations in particular allow these additional features to be turned on without affecting normal packet forwarding performance.

Forwarding using topology-based forwarding tables or FIBs is sometimes referred to as *fast-path forwarding*. Unlike route caches which are not suitable for core routing devices (and for use in core networks), FIBs are designed to handle fast-changing

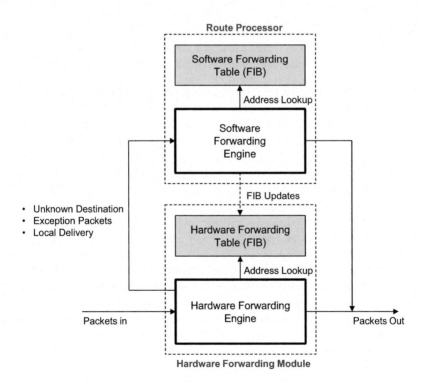

FIGURE 6.19   Packets forwarding using a topology-based forwarding table.

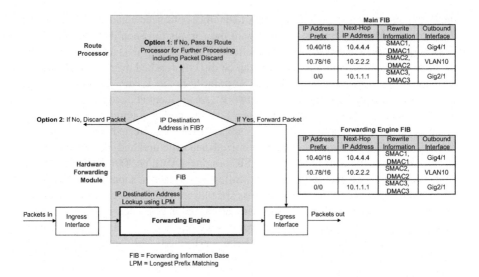

FIGURE 6.20   Illustrating IP packet forwarding using a topology-based forwarding table (FIB).

network conditions (traffic with low temporal and spatial locality), variable length prefixes (Variable-Length Subnet Masks (VLSMs) and Classless Inter-Domain Routing (CIDR)), large routing information (large routes), and supplementary network information used for traffic filtering, marking, tagging, and so on. It should be noted that forwarding via the route cache is a simpler form of fast-path forwarding; forwarding via the FIB and route can are both fast-path forwarding (see Figure 6.5).

6.3.3.1   Implementing Architectures with Topology-Derived Forwarding Tables

Architectures that use network topology-derived forwarding tables (also called FIBs) are flexible and can be implemented in a centralized or distributed environment as discussed in this section. It is important to point out that the architectures described in the "Types of Centralized Forwarding Architectures Using the IP Routing Table" section above can be based on FIBs rather than on the routing table (or RIB) – nothing precludes these architectures from using an FIB.

Unlike route cache-based forwarding architectures, architectures that use topology-derived forwarding provide optimal cache-free forwarding because the FIBs contain the same best routes as in the RIB. The RIB and the FIB are decoupled but they contain the same routing information needed for forwarding packets to their respective destinations. In the topology-based architectures (whether centralized or distributed), packets are forwarded using a pre-computed and highly optimized database, the FIB, which does not contain recursive routes and is derived from the routing table (or RIB).

6.3.3.1.1   The FIB and Resolved Routes

It should be noted that the FIB contains completely resolved routes, that is, recursive routes in the RIB are completely resolved before they are installed in the FIB (see "Resolving Recursive Routes before Installation in the IP Forwarding Table" section in Chapter 5). This allows the forwarding engine to perform address lookups quickly without first resolving recursive routes (i.e., performing recursive lookups). Note that BGP routes specify only the next-hop addresses of the routes and not router interfaces and are therefore installed as recursive routes in the IP routing table (see "BGP-Learned Routes and Recursive Lookups" section above and Chapter 3 in [AWEYA2BK21V2]).

Some FIB entries may be designated as "local delivery" indicating that packets destined to these addresses are to be delivered to the route processor in the routing device for local processing (see "Handling Local Packet Delivery During IP Packet Forwarding" section in Chapter 5). Exception packets, packets with unknown destinations, as well as packets due for local delivery are always sent to the route processor for further processing (Figures 6.19 and 6.20). An FIB entry may also maintain a destination address with multiple next-hop nodes and interfaces which allows the routing device to perform Equal-Cost Multi-Path (ECMP) routing and load balancing.

6.3.3.1.2 Centralized Processor-Based Architectures with Topology-Derived Forwarding Tables

In the centralized architecture with an FIB, the control engine, the forwarding engine, and the FIB all reside on the same processing module (Figure 6.21). This architecture is usually used in low-end routing devices with limited processing in the line cards. Typically, the switch fabric is a bus-based fabric (see Chapter 2 of [AWEYA1BK18]). Some mid-end routers use this architecture but with a shared-memory switch fabric as described in [AWEYA1BK18].

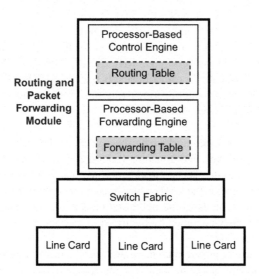

FIGURE 6.21   Centralized processor-based architectures with topology-derived forwarding tables.

The steps involved in forwarding a packet in the centralized processor (CPU)-based architecture with FIBs can be summarized as follows:

- A network interface hardware of a routing device receives a packet and transfers it into the interface's I/O memory (which is a packet buffer in shared I/O memory). The interface hardware then interrupts the CPU, indicating that a packet is queued in the interface's I/O memory and waiting for processing. The FIB forwarding software then proceeds to update its inbound packet counters.
- The FIB forwarding software verifies the packet and examines the packet header (e.g., encapsulation type, Network Layer header, etc.) to determine if the packet is an IP packet.
  - If an IP packet, the FIB forwarding process uses longest prefix matching (LPM) to search the FIB to see if there is an entry matching the IP destination address of the packet. If a matching entry is found, the forwarding software retrieves the next-hop IP address, the outbound interface, and the Layer 2 address of the next-hop node's receiving interface from the FIB. The FIB forwarding software then updates the relevant IP header fields of the packet, encapsulates the IP packet in a Layer 2 frame, and then performs all relevant Layer 2 frame rewrites. The forwarding software then queues the packet on the I/O memory of the outbound network interface.
  - If the IP destination address of the packet is not found in the FIB (unknown destination address), the packet is an exception packet, or the packet is for local delivery, the FIB forwarding software hands the packet to a processing function for further processing (using the traditional process switching/forwarding method as described in the "Packet Forwarding in the Traditional CPU-Based Forwarding Architectures" section above). Optionally, the FIB forwarding software may drop the packet if the IP destination address is unknown (not in the FIB).
  - Examples of packets that are delivered locally in a routing device are routing protocol packets (e.g., RIP, EIGRP, OSPF, IS-IS, BGP), control and management protocol packets (e.g., ICMP, IGMP, SNMP), and packets destined to applications running in the route processor (e.g., SSH, TFTP, FTP, Telnet).

- The outbound network interface hardware detects the queued packet in the I/O memory, retrieves, and transmits it out the network interface on its way to the next-hop node. The network interface hardware then interrupts the CPU to indicate that the packet has been transmitted. The FIB forwarding software then updates the outbound packet counters and frees the I/O memory previously occupied by the transmitted packet.

In the FIB-based forwarding architecture, particularly, those using shared-bus switch fabrics, the bus speeds, packet memory speeds, and interface's I/O memory speeds can still limit the packet forwarding performance of the routing device.

6.3.3.1.3   Centralized ASIC-Based Architectures with Topology-Derived Forwarding Tables

As network sizes and traffic volumes increased in the early days of routing, the short-comings of the centralized CPU-based became more apparent. Designers then started to include centralized ASIC-based forwarding modules to offload the address lookup and packet forwarding tasks from the CPU, and to improve the overall forwarding rate of the routing device (Figure 6.22). The CPU is mainly responsible for all routing protocol tasks, control and management protocol tasks, and device/system management and configuration task (i.e., housekeeping tasks). Once the CPU builds the main FIB, it copies it to the ASIC-based forwarding module and updates it anytime the RIB and main FIB are updated. This architecture is usually used in low- and mid-end routing devices.

In this architecture, packets with IP destination addresses not in the FIB (unknown destination address), exception packets, or packets for local delivery are sent to the route processor for further processing (using the traditional process forwarding method as described in the "Packet Forwarding in the Traditional CPU-Based Forwarding Architectures" section above). Optionally, the ASIC forwarding module may choose to drop the packet if the destination address is unknown (not in the FIB). A designer may decide to incorporate in the ASIC forwarding module other functions such as traffic filtering, packet marking, packet tagging, traffic policing, traffic shaping, and so on.

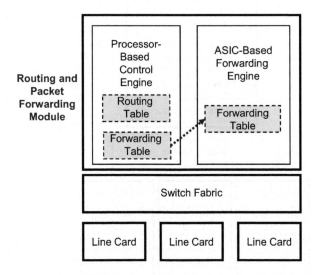

FIGURE 6.22   Centralized ASIC-based architectures with topology-derived forwarding tables.

6.3.3.1.4 Distributed Processor or ASIC-Based Architectures with Topology-Derived Forwarding Tables

As discussed above, the main limitation of the centralized forwarding architectures is that they do not scale well as the link rates, number of line cards, and the aggregate arriving traffic increase. In these architectures, the overall packet forwarding rate is mainly constrained by the speeds of the centralized forwarding engine, the shared bus and various system memory components. Particularly, in centralized forwarding architectures employing shared-bus and shared-memory switch fabrics, other than the bottleneck presented by these switch fabrics, the centralized IP address lookups also present a bottleneck and limits the overall forwarding rate of the device. We also discussed above that, forwarding architectures using route caches do not perform well in core networks due to the reduced temporal and spatial locality of traffic.

For the above reasons, modern high-performance and high-capacity routers use instead distributed forwarding architectures. In these architectures (Figure 6.23), address lookups and related forwarding operations are implemented locally in the line cards in specialized hardware (ASIC) forwarding engines or dedicated processors (sometimes called network processing units (NPUs)). This architecture is usually employed in mid-end and high-end routing devices with enhanced processing in the line cards. Typically, the switch fabric used in high-end routing devices is based on a high-performance single-stage or multi-stage crossbar switch fabric (see Chapter 2 of [AWEYA1BK19]). Figures 6.24 and 6.25 show, respectively, a shared memory-based and crossbar-based architectures with distributed processing in the line cards. Figures 6.26 to 6.29 show different examples of line card designs with local forwarding engines.

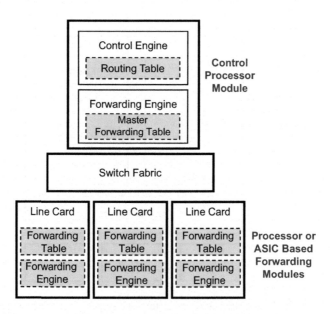

FIGURE 6.23 Distributed processor or ASIC-based architectures with topology-derived forwarding tables.

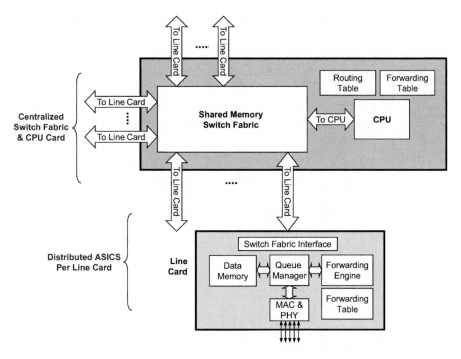

FIGURE 6.24   Shared memory-based architectures with distributed processing.

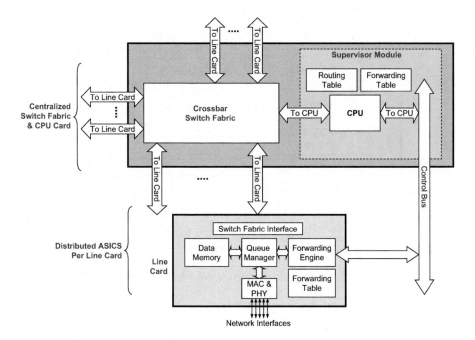

FIGURE 6.25   Crossbar switch-based architecture with distributed processing and a control bus.

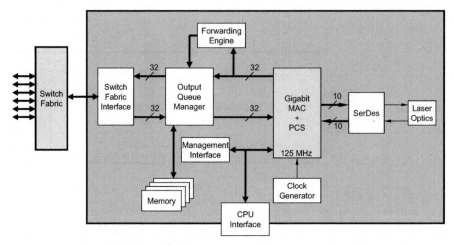

SerDes = Serializer/Deserializer

FIGURE 6.26   Line card design example 1.

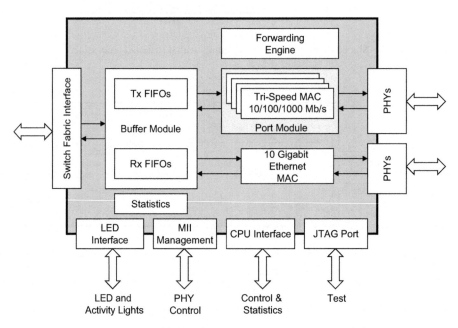

JTAG = Joint Test Access Group

FIGURE 6.27   Line card design example 2.

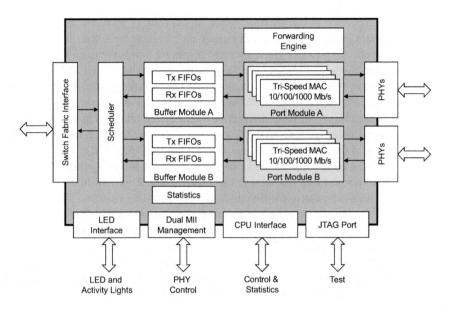

FIGURE 6.28   Line card design example 3.

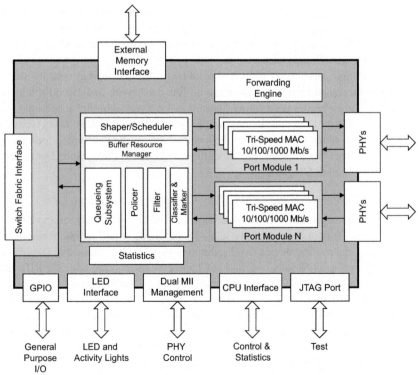

FIGURE 6.29   Line card design example 4.

The route processor is mainly responsible for all routing protocol tasks, control and management protocol tasks, and device/system management and configuration task (i.e., housekeeping tasks). Once the route processor builds the main FIB, it copies it to the line cards, and updates (i.e., synchronizes) all distributed FIBs anytime the RIB and main FIB are updated. Exception packets, packets with IP destination addresses not in a line card's FIB (unknown destination address), or packets destined for the routing device itself (local delivery) are sent by the line card to the route processor for further processing. Optionally, a line card's forwarding engine may choose to drop a packet if the destination address is unknown (not in its FIB).

Routers used in large-scale networks and core networks, including those requiring high port densities and line speeds, use this architecture. Different line cards, with different port densities, can use different capacity forwarding engines. The capacity of a line card's forwarding engine can be tailored to handle just the line speeds and expect traffic load on the line card. A line card may also support other functions for QoS and security processing such as traffic filtering, packet marking, packet tagging, traffic policing, traffic shaping, etc. A line card being somewhat autonomous in the system means a more sophisticated line card design can be used to support in addition more complex functions such as IP packet fragmentation, encryption, data compression, tunneling, multicasting, NAT, and so on.

6.3.3.1.4.1 Device Scalability

Distributed forwarding architectures provide better scalability because each line card is equipped with a forwarding engine that performs address lookups and forwarding operations locally and only on the traffic that arrives to that line card, which is a fraction of the overall arriving traffic.

The distributed forwarding architecture allows the forwarding capacity of the device to be increased simply by adding more line cards with dedicated forwarding engines or adding a higher performing forwarding engine and higher speed links to a line card. The ability to decouple the non-time-critical control plane tasks from the time-critical data plane tasks is one of the key enablers of the distributed forwarding architectures (see "Examining the Benefits of Control Plane and Data Plane Separation" section in Chapter 5).

6.3.3.1.4.2 Decoupling the Forwarding Table from the Routing Table

The ability to decouple the control plane tasks of routing and the data plane tasks of real-time packet forwarding naturally leads to the creation and maintenance of the two databases in a routing device (as discussed in the "Control Plane and Data Plane in the Router or Switch/Router" section in Chapter 5): the routing table (or RIB) and forwarding table (or FIB). This separation allows each database to be optimized and implemented in their respective processing modules. The RIB is constructed and maintained centrally by the route processor (or control engine) and contains directly attached networks, static routes, and routes learned by the dynamic routing protocols.

The FIB is generated from the RIB (a subset of the RIB) and contains the information that is directly relevant for packet forwarding. In the distributed forwarding architecture, the route processor maintains the main or master FIB (generated from the RIB) and distributes copies of the FIB to the line cards for local packet forwarding. Anytime routing changes occur and the RIB is updated, the route processor also updates the distributed FIBs to reflect the RIB updates. An architecture may use an inter-process communication (IPC) mechanism between the route processor and the line cards to ensure that the main FIB and the distributed FIBs in the line cards are always synchronized.

6.3.3.1.4.3   Equal-Cost Multi-Path and Load Balancing

An FIB entry usually points to a single path to a given destination, but in some cases, an entry may point to multiple paths to a destination, allowing ECMP routing to be performed. ECMP also allows a routing device to perform load balancing on the multiple paths to the destination. Load balancing can be performed using either the IP destination address, the IP source and destination addresses, or the five-tuple (IP source address, IP destination address, transport protocol, source transport protocol port number, destination transport protocol port number). Traffic scheduling algorithms that preserve original packet ordering (or sequencing) are generally preferable over per-packet-based scheduling algorithms. RIP and OSPF supports ECMP routing (see [AWEYA2BK21V1] and [AWEYA2BK21V2], respectively), while EIGRP is capable of supporting ECMP routing and Unequal-Cost Multi-Path routing (UCMP) (see [AWEYA2BK21V1]).

6.3.3.1.4.4   Eliminating Route Cache Churn

Unlike a route cache that maintains the routes that have been recently used by the routing device and requires the cache entries to be invalidated when routing changes occur, the entries in an FIB mirror the information in the RIB and are maintained as long as the RIB entries are maintained; FIB entries are invalidated and/or updated only when the RIB is updated. In a dynamic environment, the route cache entries will have to be invalidated and updated more frequently. Also, unlike the route cache, the performance of the FIB does not degrade due to cache churn because it maintains all usable routes learned and stored in the RIB by the system.

6.3.3.1.4.5   Maintaining Layer 2 Adjacency Information

In an integrated implementation, the FIB may contain in addition the Layer 2 addresses of next-hop nodes. The Layer 2 addresses are used for Layer 2 packet rewrites when an encapsulated IP packet is to be transmitted to the next-hop node. Each Layer 2 address is the address of the receiving interface of the next-hop and maps to the IP address of the next-hop's receiving interface. Next-hop Layer 2 addresses (also called Layer 2 adjacencies) are discovered using the ARP or manually configured by the network administrator; manually configured entries are usually static and do not change unless reconfigured or deleted.

Some implementations maintain the Layer 2 adjacencies in a separate database called the Adjacency Table (see "Distributed Forwarding – Express Fast Path Forwarding" section in Chapter 7; see also [AWEYA1BK18] [AWEYA2BK19] [AWEYA2000] [AWEYA2001]). Note that the centralized architecture using an FIB can also maintain the adjacency information in a separate Adjacency Table or integrate this information in the FIB.

6.3.3.1.4.6   Operations in a Generic Distributed Forwarding Architecture

As discussed above, distributed forwarding architectures provide scalable packet forwarding in dynamic network environments. In the distributed forwarding architecture, the process of constructing an RIB, generating the main or master FIB, and distributing the FIB to the line cards (including synchronizing them with the main FIB and RIB) can be summarized as follows:

1. The routing device via the route processor (or control engine) exchanges routing information (via various routing protocols) with neighbor routers to map out the network topology and discover destinations.
2. The route processor uses the routing information learned to build and maintain the RIB. The information in the RIB also includes directly connected networks and static routes.
3. When there exist multiple routes to the same network destination learned by different routing protocols (including static routes), the route processor selects the best route to be installed in the RIB based on the administrative distance assigned to each protocol. The route from the protocol with the smallest administrative distance is preferred and selected for the RIB [AWEYA2BK21V1] [AWEYA2BK21V2].
   a. Note that when multiple routes are learned within any given protocol (e.g., RIP, EIGRP, OSPF, etc.), the best route is based on the route with the lowest routing metric; in this case, all routes have the same administrative distance.
   b. If two or more routes have the same routing metric and the routing device supports ECMP routing, then the route processor can install multiple routes in the RIB (and consequently in the FIB) to enable the device perform ECMP or UCMP routing and load balancing.

4. After the route processor has selected the best routes for the RIB, it also updates the main (master) FIB accordingly to reflect the RIB changes.
5. The route processor then distributes copies of the updated main FIB to the distributed forwarding engines in the line cards.
   a. Given that route processing and route maintenance are centralized and packet forwarding is distributed in the distributed forwarding architecture, the distributed FIBs in the line cards must always be synchronized with the main FIB (and RIB) to allow packets to be forwarded correctly and to provide loop-free packet forwarding in the network.
   b. The contents of the FIBs are always synchronized with those of the main FIB (which is a subset of the RIB). When a line card first starts or is reloaded/restarted, the entire contents of the main FIB are copied to the line card's FIB. Thereafter, incremental changes in the main FIB are copied/synchronized to all of the distributed FIBs. This happens when network routing changes occur which leads to addition/deletion of routes in the RIB, the main FIB, and then the distributed FIBs. Note that a system *restart* on a router is also called a *reload*.

The steps involved in forwarding a packet arriving at a line card can be summarized as follows. We assume that all distributed FIBs in the line cards have been correctly synchronized with the main FIB:

1. A line card receives an IP packet on a network interface and performs Layer 2 and IP level packet verifications to see if the packet should be accepted and processed further (see "Data Plane" section in Chapter 5).
2. The local forwarding engine in the line card extracts the IP destination address from the packet and performs a lookup its local FIB to determine the next-hop node, outbound interface, plus Layer 2 packet rewrite information such as the Layer 2 source and destination addresses. The routing device may have to invoke ARP to determine the Layer 2 destination address of the next-hop node's receiving interface.
3. The forwarding engine updates the IP TTL (Time-To-Live) and IP checksum, performs Layer 2 rewrites, and then forwards the packet to the correct outbound interface. The IP checksum is recomputed because the IP TTL has been updated.
   a. In most architectures, the IP TTL and checksum updates are performed in the inbound interface while the Layer 2 rewrites are performed at the outbound interface.
   b. A Layer 2 rewrite involves setting the Layer 2 source address of the outgoing frame to the Layer 2 address of the transmitting interface, and the Layer 2 destination address of the frame to the Layer 2 address of the receiving interface of the next-hop node.
   c. All Layer 2 rewrites are accompanied by recomputation of the Layer 2 frame checksum. This is because the Layer 2 source and destination addresses have changed; each router hop requires a fresh Layer 2 rewrite and checksum recomputation.

4. The inbound line card then forwards the processed packet directly over the switch fabric to the outbound line card.
5. The outbound line card receives the IP packet and performs the actual encapsulation in the Layer 2 frame, after which the frame is transmitted out the interface on its way to the next-hop node.

If the next-hop node is not the final destination of the packet, it will repeat the above steps until the IP packet reaches its final destination.

6.3.3.1.4.7 Route Processor Redundancy

The control plane which runs in the route processor of a routing device serves as the brain or intelligence of the device. For this reason, in high-performance routing devices, the control plane is typically protected through route processor redundancy, for example, using active and standby processors (see "Node Redundancy and Resiliency" section in Chapter 1 of Volume 2). This is because in a system without route processor redundancy, the single route processor card may fail due to hardware or software faults. However, in a system with active and standby processors, when such failures occur, the system can execute an automatic switchover from the failed active processor to the standby processor.

Given that the control plane has been decoupled from the data (forwarding) plane, a failure of the active processor and the resulting switchover to the standby processor does not disrupt the packet forwarding operations of the data plane (i.e., the forwarding engine(s)). The key requirements for implementing control plane redundancy are the following:

- Decoupling the control plane tasks from the data plane tasks.
- Implementing route processor redundancy. Provide mechanism for detecting active route processor failure and switchover to standby processor
- Providing mechanisms for synchronizing control plane state from the active processor to the standby processor.

Other than providing an effective way for improving the packet forwarding rate and scalability of a routing device, a distributed forwarding architecture also provides an effective way and the architectural framework for implementing control plane redundancy.

6.3.3.1.5 FIB Consistency Checker

As discussed above, distributed forwarding architectures support an RIB that is maintained by a route processor, from which an FIB is derived and copied to the line cards for local packet forwarding. Updating the distributed FIBs may result in inconsistencies in FIB contents because the distribution mechanism for the FIB is mostly asynchronous. Asynchronous database distribution may cause the following types of inconsistencies [CISCFIBCHCK]:

- An FIB on a line card may be missing information such as a particular network prefix.
- An FIB on a line card may contain different information such as different next-hop IP addresses.

Most advanced routing devices such as Cisco routers that support distributed FIBs (Cisco Express Forwarding (CEF)) in line cards use FIB consistency checkers [CISCFIBCHCK] [STRINGNAK07] to address the above database inconsistencies. FIB consistency checkers enables the router to find any FIB inconsistencies, such as a network prefix missing from a line card FIB or route processor RIB.

6.3.3.1.5.1 Passive and Active Consistency Checkers

Cisco routers using CEF-based forwarding support passive and active consistency checkers that run independently to find forwarding inconsistencies [CISCFIBCHCK] [STRINGNAK07]. *Passive consistency checkers* run constantly in the background unless disabled. *Active consistency checkers* are initiated at the console using appropriate commands. When passive consistency checking is used, the following activities occur each minute [STRINGNAK07]:

- Each line card sends one IPC (inter-process communication) message containing FIB consistency checking information by default but the number of messages is configurable.
- The route processor sends one IPC message containing FIB consistency check messages to each line card.
- The route processor compares 1000 network prefixes in the RIB with the local master FIB to ensure that the FIB matches the RIB. This results in 60,000 prefixes per hour.

The number of network prefixes examined in each passive check and the time between passive checks is configurable.

6.3.3.1.5.2 Types of Cisco Consistency Checkers

The following are the consistency checkers that operate on the route processor or the line card [CISCFIBCHCK]:

- **lc-detect consistency checker**: This consistency checker *operates on a line card* and detects missing network prefixes on the line card FIB. The checker determines network prefixes missing from the line card FIB and passes the information to the route processor for confirmation. If network prefixes are missing in the line card FIB, the line card cannot forward packets for those prefixes. If the route processor finds that it has the relevant entry in its RIB for a missing FIB network prefix, an inconsistency is detected, and an error message is displayed. The route processor then sends a signal back to the line card confirming that the network prefix is an inconsistency.
- **scan-lc-rp consistency checker**: This consistency checker *operates on a line card* and searches through the line card FIB for a configurable time period and sends the next *n* number of network prefixes to the route processor. The route processor does an exact lookup in its local master FIB, and if it finds that a prefix is missing, it reports an inconsistency. The route processor then sends a signal back to the line card for confirmation. The time period and the number of network prefixes scanned are configurable.
- **scan-rib-ios consistency checker**: This consistency checker *operates on the route processor* and compares the centralized RIB to the local master FIB and provides the number of entries missing from the FIB.
- **scan-ios-rib consistency checker**: This consistency checker *runs on the route processor* and compares the local master FIB to the centralized RIB and provides the number of entries missing from the RIB.

6.3.3.1.5.3 Consistency Checking Process

The FIB consistency checking process has two phases [STRINGNAK07]:

- Constructing, transmitting, and comparing the FIB
- Handling a detected inconsistency

The consistency checking process can be described as follows:

1. The FIB consistency checker running on the route processor constructs a *consistency check IPC message* by walking through the local master FIB. For each FIB entry, the checker inserts a description of the entry including a checksum into the IPC message.
2. When the checker builds a full IPC message, it transmits it to all line cards.
3. The FIB consistency checker on the line card compares the information received in the consistency check IPC message from the route processor with the same entries in the local FIB, including comparing the checksum computed locally with the checksum in the IPC message computed by the route processor.
4. If the line card finds that any entry does not match, it creates a new IPC message containing the local information about this mismatched entry and transmits this to the route processor.
5. When the route processor receives the IPC message from the line card, it reexamines the local master FIB looking for a mismatch. If the route processor finds that the data received is a mismatch after local FIB search, it builds a new IPC message containing the correct information and transmits it to the line card. If after three consistency checks (this is to allow some time for any pending updates to be completed), the line card and the route processor FIBs continue to be inconsistent, the route processor marks the line card as *inconsistent*.
6. If the route processor marks a line card as inconsistent, the router may run the *auto-repair* routine on the line card FIB, if this is enabled. To run auto-repair, the router waits 10 seconds to allow all current consistency checks to complete.
7. At the end of the 10-second wait, the FIB epoch is incremented. This, in turn, causes the route processor to walk through the local master FIB, generating updates for every entry to each line card. As the route processor generates these updates, it also purges old FIB information. The router uses a *hold-down timer* to prevent multiple auto-repairs from running concurrently.

Cisco CEF uses the following processes to manage the CEF data structures and CEF operation [STRINGNAK07]:

- **FIB Manager**: This process manages the installation of network prefixes in the master FIB, including dynamic allocation of new memory chunks as needed, as well as handling FIB statistics.
- **Adjacency Manager**: This process manages the adjacency tables, including managing interface states, enabling and disabling protocols, and maintenance of a per-interface tree.
- **Update Manager**: This process keeps track of which entries in the FIBs need to be updated. The update process is responsible for passing updates from line cards to the route processor, which allows the line cards to regulate the rate at which FIB information is transferred.

6.4   ROUTING BETWEEN VLANs

A virtual LAN (VLAN) facilitates the configuration of end-systems and Layer 2 (Ethernet) switches according to logical rather than physical network topologies. Using VLANs, a network administrator can combine any collection of LAN segments within an internetwork into an autonomous user group, which appears as a single LAN broadcast domain. VLANs logically segment the network into different broadcast domains so that packets are switched only between users and ports within the same VLAN.

Modern enterprise network design methodologies recognize the business benefits of separating functional groups within the network using multiple VLANs. Separating functional groups surely has design benefits such as increasing network performance and efficiency, and also allowing security access controls to be applied between each functional group (VLAN). Typically, a VLAN corresponds to a particular IP subnet, although that does not necessarily have to be the case.

6.4.1   WHAT IS A VLAN?

A network manager could assign VLANs on a per-port, protocol, IP subnet, or IEEE 802.1Q tag basis. Possible configurations could include:

- **Port-based VLANs** to allows the network manager to group "specific port traffic" into different broadcast domains, eliminating broadcast storms and by maintaining distinct Spanning Tree domains. Port-based VLANs indicate that all frames that originate from a port belong to the same VLAN.
- **Protocol-based VLANs** to enable the network manager to easily and transparently group similar protocols into defined VLANs.
- **IP subnet-based VLANs** to allow the network manager to map each IP subnet into a distinct VLAN. In this case, the assignment of an IP address to a network interface (e.g., via the Dynamic Host Configuration Protocol (DHCP)) automatically places that interface in a defined VLAN.
- **IEEE 802.1Q standard VLAN tagging** to enable the creation of VLANs that cross switch boundaries. IEEE 802.1Q specifies how Ethernet frames can be explicitly tagged to define the VLANs to which they belong.

VLANs can also be used to provide simple and effective isolation of traffic for numerous individual customers or peer groups, as well as to support the delivery of multiple services beyond connectivity. Since each VLAN represents a unique broadcast domain and can be configured to be non-routable (i.e., a single CIDR /32 IP address is not assigned to an entire VLAN), a high degree traffic isolation is achieved. Tagged VLANs play an important role in building Layer 2 switched networks that use configurations of redundant switches to achieve high availability.

The main benefits of VLANs (particularly, using IEEE 802.1Q-based VLANs) are summarized as follows:

- **Broadcast containment**: A VLAN prevents the propagation of broadcast and multicast traffic beyond its boundaries; such traffic is confined to the VLAN itself and cannot spread to other VLANs and networks.

- **Easy administration of logical groups**: A VLAN allows end-systems that are not even physically located on the same LAN segment to communicate as if they are on the same LAN segment.
- **Improved security**: The use of VLANs creates firewalls between VLANs and allows communication between VLANs only through a router (which serves as a point of enhanced security implementation).
- **Managing VLAN members**: VLANs simplify VLAN member adds, moves, and changes.

6.4.2 IEEE 801.1Q

IEEE 802.1Q standard defines procedures for supporting VLANs on an Ethernet network [IEEE802.1Q05]. The standard defines a VLAN tagging system for Ethernet frames and the accompanying procedures to be used by bridges and switches in handling tagged frames. The standard also defines a mechanism for implementing QoS prioritization scheme commonly known as IEEE 802.1p.

In addition, IEEE 802.1Q defines the Generic Attribute Registration Protocol (GARP), now replaced by the Multiple Registration Protocol (MRP). MRP (added as the amendment IEEE 802.1ak to the IEEE 802.1Q standard) is a generic registration framework. MRP like GARP, allows bridges, switches, or other similar devices to be able to register and de-register attribute values, such as VLAN identifiers and multicast group membership across a LAN.

IEEE 802.1Q specifies a tag that is placed at a defined spot in an Ethernet MAC frame (Figures 6.30, 6.31, and 6.32). The 4-byte tag field is placed between the source MAC address and the Type/length fields of the original Ethernet frame. The IEEE 802.1Q tag consists of the following parts:

- **Tag Protocol Identifier (TPID)**: The TPID is a 2-byte field that is set to a value of 0x8100 in order to identify the frame as an IEEE 802.1Q-tagged frame. As illustrated in Figures 6.30, 6.31, and 6.32, the TPID field is located at the same position as the EtherType/Length field in untagged Ethernet frames, which allows tagged frames to be distinguished from untagged frames.
- **Tag Control Information (TCI)**: The TCI field consists of a Priority Code Point (PCP) (also called User Priority), Drop Eligible Indicator (DEI) (previously called Canonical Format Indicator (CFI)), and VLAN Identifier (VLAN ID) as shown in Figure 6.31.
  - *Priority Code Point (PCP)*: The PCP (or User Priority) is a 3-bit field which specifies the IEEE 802.1p priority class of service (i.e., indicates the frame's priority level). These values are used by network devices to prioritize different classes of traffic (e.g., voice, video, data, etc.).
  - *Drop Eligible Indicator (DEI)*: The DEI is a 1-bit field which may be used separately or with the PCP to indicate a frame's eligible to be dropped by a network device in the presence of congestion.
  - *VLAN Identifier (ID)*: The VLAN ID is a 12-bit field specifying the VLAN to which the Ethernet frame belongs.

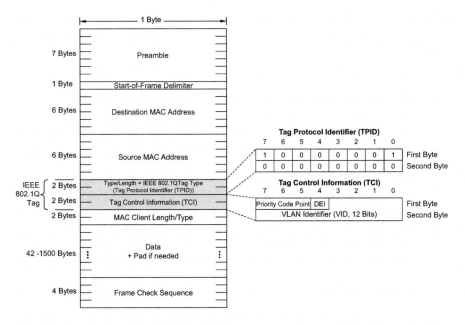

FIGURE 6.30 VLAN tagged Ethernet frame format.

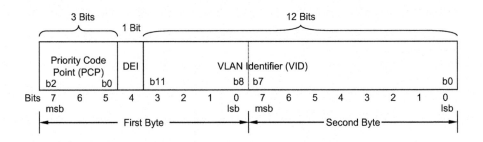

DEI = Discard Eligibility Indicating

FIGURE 6.31 VLAN tag TCI (TAG Control Information) field format.

6 Bytes	6 Bytes	2 Bytes	2 Bytes	2 Bytes	42-1500 Bytes	4 Bytes
Destination MAC Address	Source MAC Address	TPID =0x8100	TCI	Type/ Length	Data	FCS

FIGURE 6.32 VLAN protocol identifier (VPID) on Ethernet.

In the 12-bit VLAN ID field (Figures 6.30 and 6.31), the hexadecimal values of 0x000 and 0xFFF are reserved. All other values may be used as VLAN IDs, allowing up to 4,094 VLANs.

- **The null VID (0x000):** The reserved value 0x000 indicates that the Ethernet frame does not belong to any VLAN. It indicates that the tag header contains only priority information; no VLAN identifier is present in the frame. This VID value is not to be configured as a Port VID or a member of a VID set, or configured in any filtering database entry, or used in any management operation.
- **The default VID (x001):** On bridges, VLAN 1 (the default VLAN ID) is often reserved for a management VLAN but this is vendor-specific. The default VID value is used for classifying frames on the ingress port of a bridge. The Port VID value of a port can be changed by management.
- **Reserved for implementation use (0xFFF).** This VID value is reserved and is not to be configured as a Port VID or member of a VID set, or transmitted in a tag header. When used, this VID value is used to indicate a wildcard match for the VID in management operations or filtering database entries.

As illustrated in Figure 6.33, IEEE 802.1Q adds a 4-byte field between the source MAC address and the EtherType/Length fields of the original Ethernet frame. This leaves the minimum Ethernet frame size unchanged at 64 bytes but extends the maximum frame size from 1,518 bytes to 1,522 bytes.

IEEE 802.1Q tagged frames have a minimum payload of 42 bytes, while untagged frames have a minimum payload of 46 bytes. Two bytes of the 4-byte IEEE 802.1Q

Untagged Ethernet Frame

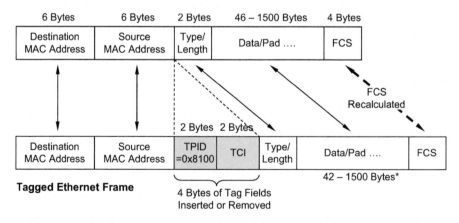

TPID = Tag Protocol Identifier
Note:
*Data Field Reduced by 4 Bytes

FIGURE 6.33 Mapping between untagged and tagged Ethernet frame formats.

field are used for the TPID, while the other two bytes are used for the TCI. The TCI field is further divided into PCP, DEI, and VID. Inserting the IEEE 802.1Q tag into an Ethernet frame changes its contents, thus, requiring recalculation of the 4-byte FCS field in the Ethernet frame trailer.

The Maximum Transmit Unit (MTU) is the size (in bytes) of the largest protocol data unit that a protocol layer can pass onto another entity. Standard Ethernet frame (MTU) is 1500 bytes (Figure 6.33). This does not include the Ethernet header and trailer fields (which take up 18 bytes), meaning the total Ethernet frame size is actually 1518 bytes. Thus, the MTU size refers only to the Ethernet payload. The Ethernet frame size refers to the whole Ethernet frame, including the header and the trailer.

A "baby giant" frame refers to an Ethernet frame with size up to 1600 bytes, and a "jumbo" frame refers to an Ethernet frame with size up to 9216 bytes [CISCBABY]. Changing the maximum Ethernet frame size from 1518 bytes to 1522 bytes to accommodate the four-byte VLAN tag, may be problematic to some network devices (that do not understand IEEE 802.1Q tagged frames). Some network devices that do not support the larger frame size (i.e., tagged frames) will process the tagged frame successfully but may report them as "baby giant" anomalies [CISCBABY].

A network can be constructed to have segments that are VLAN-aware (i.e., IEEE 802.1Q conformant) where frames include VLAN tags, and segments that are VLAN-unaware (i.e., only IEEE 802.1D conformant) where frames do not contain VLAN tags. When a frame enters the VLAN-aware segment of the network, a tag is added (see Figure 6.33) to represent the VLAN membership of the frame. Each frame must be distinguishable as belonging to exactly one VLAN.

The VLAN ID field in the IEEE 802.1Q tag is 12 bits long, meaning up to 4,096 VLANs can be supported. While this number is adequate for most smaller networks, there are many networking scenarios where double-tagging (IEEE 802.1ad, also known as provider bridging, Stacked VLANs, or simply QinQ or Q-in-Q) needs to be supported. Double-tagging can be useful for large networks and Internet service providers, allowing them to support a larger number of VLANs, in addition to other important benefits. A double-tagged frame has a theoretical limitation of 4096×4096=16777216.

6.4.3   INTER-VLAN ROUTING

Inter-VLAN routing is required when hosts in different VLANs want to communicate with each other. From an IP perspective, each VLAN behaves like an IP subnet. For an IP subnet to communicate with remote IP subnets, IP routing is required similar to inter-VLAN routing. Supporting inter-VLAN routing provides several benefits to the network operator, some of which include the following:

- Inter-VLAN routing reduces broadcast domains, since broadcast traffic cannot cross routing devices, thereby, increasing network performance and efficiency.
- Inter-VLAN routing allows the network operator to implement access control and security between VLANs.

- Inter-VLAN routing increases network manageability by creating smaller "troubleshooting domains", where the effect of a faulty network interface card (NIC) sending uncontrolled traffic is isolated to its specific VLAN and does not propagate to other VLANs and the entire network.
- Inter-VLAN routing allows multilayer network topologies to be built that are much more scalable and can support more efficient Layer 3 routing and redundant paths in the network. This is more efficient than flat Layer 2 topologies that rely on only spanning trees.

The role of VLANs and benefits of inter-VLAN routing of course, mean that inter-VLAN routing should not degrade network performance, as users expect high performance from the network. Performance is a major consideration for inter-VLAN routing within a network. Hardware-based Layer 3 forwarding within networks is one effective method to overcome the performance limitations of software-based Layer 3 forwarding.

6.4.4 IMPLEMENTING INTER-VLAN ROUTING

VLANs divide a Layer 2 broadcast domain into multiple separate broadcast domains. Each VLAN becomes its own individual broadcast domain (logically similar to an IP subnet). Only interfaces on the Layer 2 switches belonging to the same VLAN can communicate without an intervening device – a router. Interfaces assigned to separate VLANs require a router to communicate with each other. Routing between VLANs can be accomplished in one of several ways as described below.

6.4.4.1 Using an External Router

Inter-VLAN routing can be accomplished using an external router that has a separate physical interface for each VLAN. The external router can be attached to multiple Layer 2 switches to provide inter-VLAN connectivity as depicted in Figures 6.34 and 6.35. This option is the least scalable solution for inter-VLAN routing and is not very practical in a network environment with a large number of VLANs.

6.4.4.2 Using a One-Armed Router

Another way for two different VLANs to communicate is via an external router that has a single link to a Layer 2 switch over which all inter-VLAN traffic can be routed. Figure 6.36 shows a simple configuration for inter-VLAN routing where a router is connected to one of the switch ports. This router is sometimes referred to as a "*one-armed router*" since it receives and forwards traffic on the same port. A one-armed router has a single interface which it uses to route traffic between multiple VLANs (providing inter-VLAN routing). Inter-VLAN Traffic goes into the router and out of it through the same port. This router literally "hangs off" one of the ports of the Layer 2 switch. This method is also known as a "*router-on-a-stick*", "*one-arm bandit routing*", or "*lollipop routing*".

The one-armed router must support either IEEE 802.1Q or Cisco ISL (Inter-Switch Link) trunking encapsulations, and the switch port must be configured as a trunk. *Trunking* is a way to carry traffic from several VLANs over a point-to-point

FIGURE 6.34   Using an external router that has an interface to each VLAN.

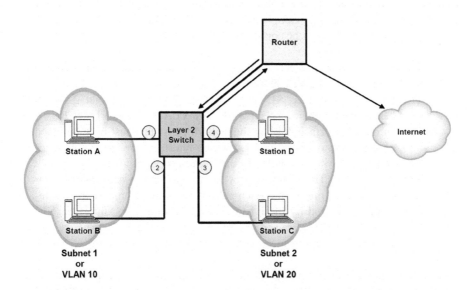

FIGURE 6.35   Layer 2 switch with external router for inter-VLAN traffic and connecting to the Internet.

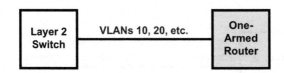

FIGURE 6.36   Using a one-armed router (Router-on-a-stick).

link between the two devices. Two ways in which Ethernet trunking can be implemented are IEEE 802.1Q and ISL (a Cisco proprietary protocol).

Trunking (through IEEE 802.1Q and ISL) allows multiple VLANs to operate independently across a single link between two switches or between a switch and a router. Trunking allows multiple VLANs to operate simultaneously on a single link (a trunk link). A port on a switch normally belongs to only one VLAN, where any traffic received or sent on this port is assumed to belong to the configured VLAN. A trunk port, on the other hand, is a port that can be configured to send and receive traffic for many VLANs. It accomplishes this by tagging VLAN information to each frame. Also, trunking must be active on both sides of the link connecting the two devices; the other side must be expecting frames that include VLAN information for proper communication to occur. The single interface on the one-armed router is configured as a trunk link and is connected to a trunk port on the Layer 2 switch.

The one-armed routing functionality can be provided in an external routing device or in some cases within the same Layer 2 switch (i.e., implemented internally within a switch/router as described below) to avoid using an external router and to free up another switch ports. When the router acts as a one-armed router and is connected to a Layer 2 switch, the same port (linking the switch) may support multiple VLANs.

In the one-armed routing example in Figure 6.37, to enable inter-VLAN communication, three elements must be configured [BALCH2009]:

- The interface "Interface_ST" on the Layer 2 switch must be configured as a trunk port to carry the VLAN traffic to and from the one-armed router.
- The interfaces "Interface_S1" and "Interface_S3" on the Layer 2 switch must be assigned to their respective VLANs, VLAN 10, and VLAN 20.
- The interface "Interface_R1" on the one-armed router must be subdivided into *sub-interfaces* one for each VLAN. Sub-interfaces are virtual interfaces that are created when a single physical interface is divided into multiple logical interfaces. The sub-interfaces use the parent physical interface for sending and receiving traffic. Each sub-interface must support the frame-tagging protocol used by the Layer 2 switch's trunk port "Interface_ST" (IEEE 802.1Q or ISL). Frames sent to and from the Layer 2 switch across the trunk port have to be tagged with the VLAN ID the frame belongs to.

Host devices in each VLAN will point to their respective sub-interfaces on the one-armed router. The IP address on each sub-interface represents the default IP gateway for each VLAN. In reality, the host devices have no knowledge of the sub-interfaces or even that there is a one-armed router present, and as a result must each be

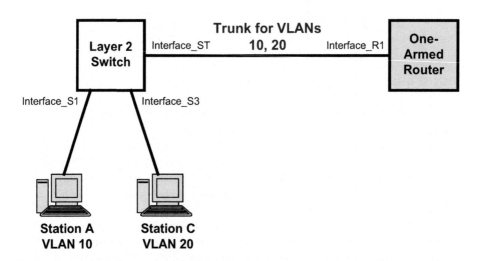

FIGURE 6.37   One-armed router (Router-on-a-stick).

configured with a default gateway corresponding to the IP address of a sub-interface on the one-armed router. For example, Station A's default gateway could be sub-interface 172.16.1.1 and Station C's sub-interface could be 172.16.2.2 (with all these sub-interfaces on the one-armed router). These functions allow the one-armed router to perform inter-VLAN communication on behalf of the Layer 2 switch.

The one-armed routing approach is appealing because of its simplicity. The number of VLANs that can be supported is not limited by the number of ports available as in the case when a traditional router is used. The one-armed router also removes the need for multiple wiring from the router to the switches. However, there are inherent limitations to this approach [BALCH2009]:

- The trunk carrying VLAN traffic between the Layer 2 switch and the one-armed router (Figures 6.36 and 6.37) may have insufficient bandwidth for each VLAN as all routed traffic will need to share the same router interface.
- Depending on the design and capacity of the one-armed router, there may be an increased load on it route processor and forwarding engine, to support the IEEE 802.1Q or ISL encapsulation taking place in it. As traffic from all VLANs must travel through the trunk link, it may become a source of congestion; inter-VLAN traffic travel over the trunk twice.
- If the one-armed router fails, there is no backup path for inter-VLAN traffic. The router may become the bottleneck in the network. The use of Link Aggregation (IEEE 802.3ad), where multiple links are aggregated to create the trunk, can mitigate the single link trunk bottleneck; the single interface is combined with other interfaces.

A more efficient alternative for inter-VLAN routing is to use a switch/router (i.e., a combined Layer 2/Layer 3 switch).

6.4.4.3 Using a Switch/Router

One approach that has become widely popular for inter-VLAN routing while at the same ensuring that the performance of the network is not degraded is to use switch/routers (integrated Layer 2/3 switches). We have seen in previous chapters that switch/routers are essentially Layer 2 switches with a Layer 3 routing function that is designed to specifically route traffic between IP subnets or VLANs in a network when routing is required. Switch/routers provide a number of benefits for inter-VLAN routing over traditional standalone routers as discussed below [MENJUS2003]:

- **Higher Port Density**: As discussed in the previous chapters, switch/routers are typically designed to have enhanced features beyond the traditional software-based routers and, hence, have higher port densities. By implementing routing and forwarding engines in software, traditional software-based routers typically have much lower port densities.
- **Network Deployment Flexibility**: Given that switch/routers allow the integration of Layer 2 switching plus Layer 3 routing and forwarding functions in a single platform, a switch/router can be configured to operate as a normal Layer 2 switch when routing is not required, or as a Layer 3 routing

and forwarding device when routing is required. It also supports these functions when connected, for example, to multiple VLANs where the Layer 2 switching functions handle intra-VLAN communication, while the Layer 3 routing and forwarding functions handle inter-VLAN communications.

- **Performance versus Cost**: Switch/routers offer a much more cost-effective approach for delivering high-speed inter-VLAN routing than conventional routers because the normally separate Layer 2 and Layer 3 functions are integrated into one device. High-performance aggregation and core routers are typically more expensive than switch/routers. Switch/routers are relatively less expensive as they are targeted specifically for inter-VLAN routing, where Ethernet access technologies are dominant and used in high densities. This situation creates a more attractive environment for switch vendors to develop high-performance switch/routers, as vendors can develop specialized hardware chips (ASICs) that route traffic between Ethernet networks, without having to be constrained by the complexities of also supporting WAN technologies such as ATM, SONET/SDH, Packet over SONET (POS), etc.

The following port types can be configured on switch/routers [BALCH2009]:

- **Switch Ports**: These are Layer 2 ports on the switch/router on which the MAC addresses of user devices are learned. A switch port can either be an access port attached to an end-user device, or a trunk port from another Layer 2 or 3 device.
- **Layer 3 Ports or Routed Ports**: These are routing ports on switch/routers handling Layer 3 traffic. The Layer 3 port is assigned an IP address when configured. A Layer 3 port behaves like a physical router interface on the traditional router.
- **Switched Virtual Interfaces (SVI) or VLAN Interfaces**: An SVI is a virtual routed interface that connects the Layer 2 forwarding (or bridging) function on the switch/router (i.e., a Layer 2/3 forwarding device) to the Layer 3 routing function on the same device. An IP address is assigned to the SVI connecting the switch/router connects to the corresponding VLAN. The attached VLAN itself is treated as an interface on the switch/router. An SVI is a logical routed interface and is referenced by the VLAN number it serves. Each SVI on the switch/router is assigned an IP address and allows an entire VLAN to be connected. SVIs have become the most common method for configuring inter-VLAN routing on switch/routers. An SVI only becomes alive or comes online only when the VLAN is created and at least one port is active in the VLAN.

In addition to Layer 2 switching, a switch/router supports both a Layer 3 forwarding engine and a routing engine (i.e., the route processor). In the route cache forwarding method, an IP packet must first be routed (i.e., forwarded via the routing engine), allowing the forwarding engine to cache the IP packets IP destination address and outgoing port. After this cache entry is created, subsequent IP packets of the cached flow are forwarded (via the forwarding engine) and not routed, thereby reducing latency. This concept is often referred to as *"route once, forward many times"*.

A more advanced packet forwarding method as described above is using a forwarding engine combined with a network topology-based optimized lookup table (i.e., an FIB). This method addresses some of the disadvantages of the route cache-based method because it does not cache routes, thus, there is no danger of having stale routes in the cache if the routing topology changes. The topology-based method contains the (Layer 3) routing engine (which builds the routing tables from which a forwarding information base (FIB) is constructed) and the Layer 3 forwarding engine (that forwards packets based on the FIB).

The routing engine builds the routing table dynamically via routing protocols (such RIP, EIGRP, OSPF) and manually when the network manager enters static routes. The routing table is then reorganized into a more efficient table, the FIB. The most relevant information in the routing table that is useful for actual packet forwarding is distilled into the FIB. The Layer 3 forwarding engine then utilizes the FIB for packet forwarding.

Additionally, the forwarding engine may maintain a separate adjacency table which contains the MAC addresses of the next-hop nodes. The adjacency table information may be integrated in the FIB. Entries in the adjacency table are made as new neighboring routers or end systems are discovered using the ARP. This process of discovering the neighbors using ARP or other means like "snooping" (see discussion below) is sometimes referred to as "gleaning" the next-hop MAC address. When fully constructed, the FIB contains the following information needed for forwarding packets:

- The destination IP network prefix (and mask)
- The next-hop MAC address
- The interface or port on the router or switch/router that leads to the next-hop device

The Layer 3 function on the switch/router can configure to support an SVI for each VLAN – each SVI is assigned an IP address. Each SVI IP address will serve as the default gateway for end users on the corresponding VLAN. By assigning an IP address to each SVI, the SVI will be added to the routing table maintained by the switch/router as directly connected routes, thereby allowing routing of packets.

The switch/router can be used to provide inter-VLAN routing as shown in Figures 6.38 and 6.39. The switch/router supports Layer 2 switching and so it forwards traffic between Stations A and B at Layer 2 since these stations are in the same subnet (VLAN). However, communication between Stations A and C (which are in different VLANs) has to be facilitated by the switch/router which forwards the traffic at Layer 3 (routing).

Station A sends an IP packet addressed to the MAC address of the switch/router, but with an IP destination address equal to Station C's IP address. The switch/router rewrites the MAC header of Station A's frame with the MAC address of switch port 3 (Figure 6.39) and forwards the frame to Station C after performing the IP forwarding table lookup, decrementing the TTL, recalculating the IP checksum and inserting Station C's MAC address in the outgoing frame's destination MAC address field.

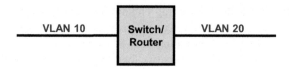

FIGURE 6.38   Using a switch/router.

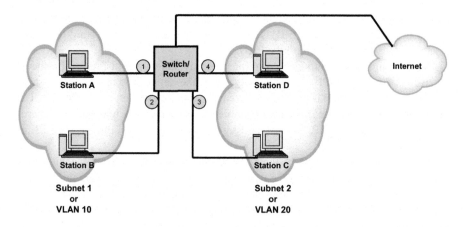

FIGURE 6.39   Switch/router connected directly to the Internet.

For intra-VLAN forwarding, the switch/router has the capability of simply employing the MAC address learning process used in transparent bridges (switches) to determine on which ports the stations (MAC addresses) are located. However, for inter-VLAN
forwarding (routing) between directly attached VLANs as shown in Figure 6.39, a
number of methods are available for the switch/router to determine the ports on which
the IP addresses and MAC addresses of stations (involved in the inter-VLAN communication) are located. Note that when the switch/router performs learning at Layer 2, it
only knows Station C's MAC address. Other methods exist, but the following are some
methods for learning Station C's address on directly attached networks:

- **Address learning using ARP (Address Resolution Protocol)**: A device
  uses ARP to find the MAC address of another device from its known IP
  address. ARP also maintains a cache (table) in which MAC addresses are
  mapped to IP addresses. The switch/router can perform an ARP operation
  on all the IP Subnet 2 ports for Station C's MAC address and to determine
  Station C's IP-to-MAC mapping and the port on which Station C lies.
- **Address learning through IP header snooping**: Here the switch/router uses
  IP "*snooping*" techniques to learn the MAC/IP address relationships of end
  stations when the switch/router communicates in a network with routers and
  end-stations that exist in the network. This process involves snooping into the
  IP header of the MAC frames on a port and determining the IP source address
  of a source on that port from the IP source address field of the received packet.

Using this method, the switch/router can determine Station C's IP-to-MAC address mapping by snooping into the IP header upon receiving any MAC frame from Station C. Note that when the switch/router performs normal Layer 2 switching, it learns the MAC addresses on the ports. IP header snooping allows it to learn, in addition, IP-to-MAC address mapping. A network device operating in this mode is sometimes referred to as a *"Layer 3 Learning Bridge"*.

Although for Layer 3 forwarding, the switch/router can be configured with the ports (MAC and IP addresses) corresponding to each of the subnets (VLANs), this option is very laborious to carry out.

Configuration of inter-VLAN routing on a switch/router can be performed as follows:

- First, create the required VLANs
- Then routing must be (globally) enabled on the switch/router
- Next, each VLAN SVI (or VLAN interface) is assigned an IP address

These IP addresses (i.e., the SVIs) will serve as the default gateways for the clients on each VLAN. By adding an IP address to an SVI, those networks will be added to the routing table as directly connected routes, allowing routing to occur.

REVIEW QUESTIONS

1. When a host A in one IP subnet communicates with another host B in another IP subnet and sends IP packets encapsulated in Ethernet frames through the default gateway router, what MAC address is written in the destination MAC address field of the frames sent by the host A?
2. What is the difference between slow-path forwarding (also called process switching) and fast-path forwarding in a routing device?
3. What are the main limitations of the traditional CPU-based forwarding architectures?
4. What type of information is normally stored in the ROM (e.g., EPROM), NVRAM, Flash Memory, and RAM of the typical centralized forwarding architecture?
5. Explain the purpose of the interface FIFOs and interface receive (RX) and transmit (TX) ring buffers in a router.
6. Explain the difference between particle buffers and contiguous buffers.
7. Explain the difference between public buffer pools and private buffer pools.
8. Explain the main difference between temporal and spatial locality of IP traffic flows.
9. Explain briefly what an exception packet is. Which component in a router is responsible for handling exception packets? Give at least three examples of packets that are considered exception packets.
10. Explain why route cache-based forwarding architectures are not very effective or suitable for routing in core networks.
11. What are the main parameters that make up a route cache entry?

12. What is double IP address lookup in a route cache based forwarding architecture?
13. Explain why exact matching lookups are used in route caches and not longest prefix matching (LPM) lookups. Apart from route caches, what other devices use exact matching lookups?
14. Explain why lookups in a route cache do not involve recursive route lookups.
15. Why does an FIB not contain recursive routes?
16. Why does an FIB still require LPM lookups?
17. Explain why distributed forwarding architectures using FIBs are generally more efficient that route cache based architectures.
18. What are the conditions that can lead to the invalidation and deletion of a route cache entry?
19. What is the purpose of an FIB Consistency Checker in a routing device?
20. Explain briefly the purpose of the Priority Code Point (PCP), Drop Eligible Indicator (DEI), and VLAN Identifier (VLAN ID) fields in a tagged Ethernet frame.
21. What is IEEE 802.1Q trunking?
22. What is a one-armed router? What are its advantages and limitations?
23. What is a sub-interface?
24. What is a switched virtual interface (SVI), also called a VLAN interface?

REFERENCES

[AWEYA1BK18]. James Aweya, *Switch/Router Architectures: Shared-Bus and Shared-Memory Based Systems*, Wiley-IEEE Press, ISBN 9781119486152, 2018.

[AWEYA2BK19]. James Aweya, *Switch/Router Architectures: Systems with Crossbar Switch Fabrics*, CRC Press, Taylor & Francis Group, ISBN 9780367407858, 2019.

[AWEYA2BK21V1]. James Aweya, *IP Routing Protocols: Fundamentals and Distance Vector Routing Protocols*, CRC Press, Taylor & Francis Group, ISBN 9780367710415, 2021.

[AWEYA2BK21V2]. James Aweya, IP Routing Protocols: *Link-State and Path-Vector Routing Protocols*, CRC Press, Taylor & Francis Group, ISBN 9780367710361, 2021.

[AWEYA2000]. J. Aweya, "On the Design of IP Routers. Part 1: Router Architectures," *Journal of Systems Architecture (Elsevier Science)*, Vol. 46, April 2000, pp. 483–511.

[AWEYA2001]. J. Aweya, "IP Router Architectures: An Overview," *International Journal of Communication Systems (John Wiley & Sons, Ltd.)*, Vol. 14, Issue 5, June 2001, pp. 447–475.

[BALCH2009]. Aaron Balchunas, "Multilayer Switching", *Switching Guide*, Router Alley, 2014.

[CISC1600ARC]. Cisco Systems, "Cisco 1600 Series Router Architecture", *Document ID: 5406*, October 10, 2002.

[CISC2500ARC]. Cisco Systems, "Cisco 2500 Series Router Architecture", *Document ID: 5750*, September 5, 2002.

[CISC2600ARC]. Cisco Systems, "Cisco 2600 Series Router Architecture", *Document ID: 23852*, October 11, 2002.

[CISC4000ARC]. Cisco Systems, "Cisco 4000 Series Router Architecture", *Document ID: 12758*, November 6, 2002.

[CISC7200ARC]. Cisco Systems, "Cisco 7200 Series Router Architecture", *Document ID: 5910*, October 11, 2002.

[CISC7500ARC]. Cisco Systems, *Inside Cisco IOS Software Architecture*, Chapter 6: "Cisco 7500 Routers", Cisco Press, July 2000.

[CISCBABY]. Cisco Systems, "Understanding Baby Giant/Jumbo Frames Support on Catalyst 4000/4500 with Supervisor III/IV", *Document ID: 29805*, March 24, 2005.

[CISCFIBCHCK]. Cisco Systems, *IP Switching Cisco Express Forwarding Configuration Guide*, Chapter "Configuring CEF Consistency Checkers", January 20, 2018.

[CISCIDBLIM12]. Cisco Systems, "Maximum Number of Interfaces and Subinterfaces for Cisco IOS Routers: IDB Limits", *Document ID:15096*, May 24, 2012.

[CISCINTRCIOS]. Cisco Systems, *Internetworking Technologies Handbook*, 4th Edition, Chapter "Introduction to Cisco IOS Software", October 31, 2003.

[CISCNETS601]. Cisco Systems, "Cisco Router Architecture", Cisco Networkers 1998, *Session 601*.

[CISCNETS2011]. Cisco Systems, "Catalyst Switch Architecture and Operation", Cisco Networks 2003, *Session RST-2011*.

[CISCNETS2111]. Cisco Systems, "IOS Router Operation and Architecture", (Part 1), Cisco Networks 2003, Session RST-2111.

[CISCNETS2112]. Cisco Systems, "IOS Router Operation and Architecture", (Part 2), Cisco Networks 2003, Session RST-2112.

[CISCNETS2203]. Cisco Systems, "Router Switching Performance Characteristics", Cisco Networks 2000, Session 2203.

[HUSSFAULT04]. I. Hussain, *Fault-Tolerant IP and MPLS Networks*, Cisco Press, 2004.

[IEEE802.1Q05]. IEEE Standard for Local and Metropolitan Area Networks: Virtual Bridged Local Area Networks, IEEE Std 802.1Q-2005, May 2006.

[IEEE802.3]. IEEE 802.3-2018 - IEEE Standard for Ethernet.

[MENJUS2003]. Justin Menga, "Layer 3 Switching", *CCNP Practical Studies: Switching (CCNP Self-Study)*, Cisco Press, Nov. 26, 2003.

[RFC2309]. B. Braden, D. Clark, J. Crowcroft, B. Davie, D. Estrin, S. Floyd, V. Jacobson, G. Minshall, C. Partridge, L. Peterson, K. K. Ramakrishnan, S. Shenker, J. Wroclawski, L. Zhang, "Recommendation on Queue Management and Congestion Avoidance in the Internet," *IETF RFC 2309*, April 1998.

[STRINGNAK07]. N. Stringfield, R. White, and S. McKee, *Cisco Express Forwarding, Understanding and Troubleshooting CEF in Cisco Routers and Switches*, Cisco Press, 2007.

[ZININALEX02]. Alex Zinin, *Cisco IP Routing: Packet Forwarding and Intra-Domain Routing Protocols*, Addison-Wesley, 2002.

7 Review of Multilayer Switching Methods
Switch/Router Internals

7.1 INTRODUCTION

As discussed in previous chapters, LAN switches implement Layer 2 functionality, while routers implement Layer 3 functionality. However, some LAN switches used in the access layer of enterprise and service provider networks implement, additionally, Layer 3 functionality. The resulting platform in this case then takes the form of a switch/router, where the functions of a LAN switch and a router are merged. A switch/router (also referred to as a multilayer switch) performs three major functions: Layer 2 and Layer 3 packet forwarding, route processing, and advanced network service processing (Quality of service (QoS), security, multicasting, tunneling, multiprotocol label switching (MPLS), etc.).

Route processing is typically performed in software on the route processor (also called the routing engine). Layer 2 forwarding which involves relatively simple operations is typically performed in hardware. Compared to Layer 2 forwarding, Layer 3 forwarding in routers and switch/routers involves more intensive data plane operations than is required in Layer 2 forwarding. As discussed in Chapter 6, Layer 3 forwarding can be performed in software, where an operating system application is responsible for the data plane operations, or in hardware, where a hardware chip (or ASIC) designed specifically for Layer 3 forwarding is responsible for the data plane operations.

Performing Layer 3 forwarding in software provides more design flexibility because code can be written that uses a common processor to perform the specific data plane operations in addition to the advanced network services required. Performing Layer 3 forwarding in hardware increases packet forwarding performance because all operations are performed by a function-specific chip, leaving the route processor free to perform other duties. However, using hardware Layer 3 forwarding provide less design flexibility and is more expensive because a hardware chip has to be designed to handle the various forwarding features desired. In some cases, separate chips must be designed for each major feature, for example, for separate ASIC for Layer 2 Ethernet forwarding, IP forwarding, MPLS forwarding, etc. This chapter reviews some of the specific methods used for packet forwarding in switch/routers including their internal mechanisms.

7.2 MULTILAYER SWITCHING METHODS

There are numerous approaches to multilayer switching and all strive to meet the same objectives, but each approach differs in its execution. A number of these technologies are proprietary in nature. Some of the widely accepted approaches used nowadays are the following [CISCCAT6500] [CISCIOSR12.2] [CISCKENCL99] [FOUNLAN00]:

- **Front-end processor approach**: This is an older (legacy) approach (now disused) that was used in the early days of IP routing when routers were mainly software-based. This architecture was used to boost the performance of Layer 2/Layer 3 forwarding in campus and enterprise networks. In this architecture, a multilayer switch (switch/router) sits in front of conventional software-based router, off-loading and forwarding IP traffic before it reaches the router. The switch/router acts as a "virtual router", that is, a front-end processor that off-loads traffic sent to existing routers, in some cases up to 80% offload possible. Using mainly standard protocols, the front-end processor is used to accelerate IP subnet-to-subnet traffic without relying too much on the router. In regards to the network topology, the router is not even aware of the presence of the switch/router, and other network devices think they are communicating directly with the router. The switch is conceptually like a front-end processor of the router, handling simpler communications (like IP subnet-to-subnet forwarding) while the more expensive software-based router sitting behind it resolves larger matters (like IP subnet to Internet forwarding). The front-end switch/router is transparent to other network devices – they are only aware of the router it is directly attached to. One method the switch/router uses to learn IP forwarding addresses on the fly is by monitoring broadcast traffic and ICMP Route Discovery Protocol messages [RFC1256].
- **Conventional standalone switch/router approach**: This architecture aims at high-performance Layer 2/Layer 3 forwarding for data centers and server farms, large campus networks, enterprise networks, service provider networks, as well as wide area network backbones. In this architecture, a switch/router performs standard Layer 3 IP routing and Layer 2 switching concurrently – a true switch and router in a single box. For example, the switch/router may support Gigabit Ethernet and multi-gigabit Ethernet interfaces carrying traffic in the network access, aggregation or backbone. Other routers in the network see the switch/router as just another router, Layer 2 switches see it as another Layer 2 switch, while host sees it as either a Layer 2 switch or router (depending on the destination of the traffic). A typical switch/router supports various Layer 3 IP protocols like RIP, EIGRP, OSPF, IGMP, VRRP, etc. The device may support 100 Mb/s Ethernet, Gigabit Ethernet (GE), 10 GE, in some cases, multi-gigabit Ethernet (e.g., 25 GE, 40 GE, 100 GE) interfaces with advanced features such as traffic prioritization, QoS, and access security control (using access control lists (ACLs)).

Some of the multilayer switching approaches are described in detail in the following sections. The data plane implementation is a key factor in determining the packet forwarding speed of the overall switch/router. A high-performance data plane (e.g., using specialized ASICs) allows the switch/router to support full line-rate forwarding and low, predictable latency for both Layer 2 or Layer 3 traffic, irrespective of the advanced Layer 2/Layer 3 services that are configured.

7.3 UNDERSTANDING THE FRONT-END PROCESSOR APPROACH WITH FLOW-BASED MULTILAYER SWITCHING

We describe in this section, an example of the front-end processor approach for multilayer switching which is now disused and has only historic significance now [CISCCAT6500] [CISCIOSR12.2] [CISCKENCL99] [MENJUS2003]. The architecture uses flow-based switching technology to route inter-VLAN traffic. In reality, this flow-based technology is based on route caching as discussed in Chapters 2, 5, and 6. This architecture represents the first hardware-based Layer 3 forwarding mechanism supported on Cisco Catalyst switches such as the Catalyst 5000/5500 [CISCKENCL99] and Catalyst 6000/6500 [CISCCAT6000].

7.3.1 BASIC ARCHITECTURE

The front-end processor approach consists of the following two main components (Figure 7.1):

- **Multilayer Switching–Route Processor (MLS-RP)**: The MLS-RP implements the control plane of the multilayer switch (switch/router), i.e., the route processing functions based on routing protocols such as RIP and OSPF. The MLS-RP constructs and maintains the routing table and is responsible for updating the routing table as network topology changes occur. The MLS-RP constitutes the routing engine of the switch/router.
- **Multilayer Switching–Switching Engine (MLS-SE)**: The MLS-SE implements the data plane of the switch/router, i.e., the packet forwarding functions. When IP packets are received and require routing, the MLS-SE is responsible for determining the next hop and egress interface information for the packet, and then rewriting the correct information in the outgoing

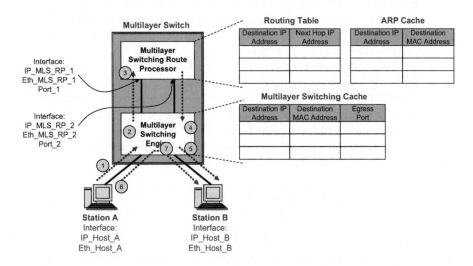

FIGURE 7.1 Flow-based multilayer switching: Components on the same platform.

Ethernet frame carrying the IP packet as required, and forwarding the frame to the correct egress interface to the next hop. The MLS-SE represents the forwarding engine in the switch/router.

The MLS-RP maintains the routing table and is also responsible for communicating the control plane information to the MLS-SE to be used for packet forwarding. The MLS-RP and MLS-SE communicate via the MLS Protocol (MLSP), a Cisco proprietary protocol that uses multicast Ethernet frames for the communication [CISCCAT6500] [CISCIOSR12.2] [CISCKENCL99]. Events such as routing topology changes and access control list (ACL) configuration changes are communicated to the MLS-SE by the MLS-RP via MLSP messages.

Figure 7.1 shows the high-level architecture of the multilayer switching and how packets are forwarded using flows. In this figure, the MLS-RP and MLS-SE are implemented on the same platform. A flow in this case is defined as a stream of packets having the same forwarding characteristics and heading to the same network destination. In the flow-based switching architecture, flows represent the entries (forwarding information) in the multilayer switching cache (or route cache) located on the MLS-SE (Figure 7.1). A flow may be defined based upon any of the following information:

- **Destination IP address (destination-only mask)**: This represents a unique (single) destination IP address. All packets sent to this unique destination IP address are considered as belonging to a single flow. For example, all packets that are sent from any source IP address to a destination IP address of 192.168.4.2 represent a unique flow.
- **Source and Destination IP address (source-destination only mask)**: This represents a unique combination of source and destination IP addresses. All packets sent between a specific source and a destination IP address are considered as belonging to a single flow. Several flows might exist for the same destination IP address but each flow is differentiated by its source IP address. For example, all packets sent from source IP address 192.168.2.1 to destination IP address 192.168.4.2 represent a single flow, with the return packets representing another flow.
- **Full flow (full 5-tuple mask)**: This represents a unique combination of source and destination IP addresses, as well as source and destination Layer 4 (i.e., TCP or UDP) ports. A full flow represents all packets associated with a specific source IP address, destination IP address, source TCP/UDP port, and destination TCP/UDP port. For example, a VoIP connection between 192.168.1.1 and 192.168.2.1 would constitute a separate flow from an HTTP connection between the same two hosts.

The MLS-SE constructs the required frame rewrite information for Layer 3 forwarding for each new flow by allowing the MLS-RP to perform the normal routing/forwarding table lookup (to determine the next-hop IP address and egress port) for the first packet of that flow. As discussed in Chapter 2, the frame rewrite information consists of the source and destination MAC addresses which are rewrite information required for forwarding the frame to the next-hop node. The MLS-SE learns (via the MLSP) the required rewrite information for the source and destination MAC addresses of a framed IP packet after its next-hop has been determined by the MLS-RP; the appropriate rewrite information is then stored in the multilayer switching cache maintained by the MLS-SE (Figure 7.1).

The front-end processor approach is designed to support a distributed architecture, meaning the various components do not need to be located on the same physical device as is the case in Figure 7.2. Figure 7.2 shows the topology where the two components are not implemented on the same device. In Figure 7.2, the multilayer switching device (the front-end processor or virtual router) provides the high-speed Layer 3 forwarding required for inter-VLAN communication. The multilayer switch (switch/router) here acts as the MLS-SE (see Figure 7.1), which is responsible for the data plane operations required for Layer 3 forwarding.

The router in Figure 7.2 provides routing and control plane operations, that is, initially routing the first packet of each flow sent through the multilayer switch, allowing the multilayer switch to learn the required MAC address rewrite information for Layer 3 forwarding of subsequent packets of the same flow. The router acts as the MLS-RP (Figure 7.2), which is responsible for making routing decisions for the first packet of a new flow. The router has a physical Ethernet interface that connects as an IEEE 802.1Q trunk to the multilayer switch. Two virtual interfaces are required on the trunk to provide inter-VLAN routing between VLANs.

The Cisco Catalyst 6000/6500 Supervisor 1A with Policy Feature Card (PFC) can act only as an MLS-SE when used in conjunction with the Catalyst 6000/6500 Multilayer Switch Feature Card (MSFC) [CISCCAT6000] [CISCKENCL99]. In this configuration, the MLS-SE (the PFC) and MLS-RP (the MSFC) do not communicate over IP, instead they communicate via an internal bus. However, the MSFC can also

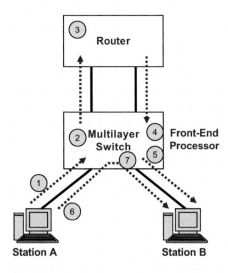

FIGURE 7.2   Flow-based multilayer switching: Components on separate platforms.

function as an MLS-RP when used with other MLS-SEs such as the Catalyst 5000 with NetFlow Feature Card (NFFC) [CISCKENCL99].

Newer generations of Cisco Catalyst Layer 3 switches as well as current high-performance switch/routers are all based on distributed forwarding architectures using distributed forwarding information bases (FIBs) (we use FIB and forwarding table interchangeably throughout the book). Distributed forwarding using FIBs offers significant improvements over route cache-based forwarding (also referred to as flow-cache based forwarding), the most notable being that the first packet in a flow does not need to be forwarded via process switching by the control plane routing component, as is the case with the multilayer switching method discussed here.

With distributed forwarding, all packets (including the first) belonging to a flow are Layer 3 forwarded using optimized lookup algorithms and the FIBs. This is an important feature in environments where many new flows are being established continuously (e.g., an Internet service provider environment). This is because large amounts of new flows in the route cache-based forwarding approach can potentially reduce forwarding performance due to the continuous cache updates required to accommodate the new flows.

Distributed forwarding (discussed later in this chapter) allows the appropriate information required for the data plane operations of the Layer 3 forwarding process (e.g., MAC address rewrites on an Ethernet network and determining the egress port through which a routed frame should be sent) to be stored in a compact data structure optimized for fast lookups. The route cache-based architecture uses a flow-based caching mechanism, where packets must first be forwarded via process switching by the MLS-RP to generate flow entries in the cache.

Environments where thousands of new flows are created per second can cause the MLS-RP to become a bottleneck. Distributed forwarding using FIBs was developed to eliminate the performance penalty associated with forwarding the first packet via process switching by the control plane. This new architecture allows the forwarding table used by the Layer 3 forwarding engine to contain all the necessary Layer 3 forwarding information before any packets associated with a flow are received.

7.3.2 FLOW-BASED MULTILAYER SWITCHING PACKET PROCESSING STEPS

This section describes the steps involved in forwarding packets in a multilayer switching system based on the front-end processor approach and flow-based switching. The processing steps described below involve a host Station A sending packets to another host Station B on a different IP subnet through a default gateway which is the MLS-RP (see Figures 7.1 and 7.2).

Step 1 – Station A sends an IP packet addressed to Station B on a different subnet
- The IP packet is addressed to Station B which is located on a different IP subnet from Station A. So, Station A addresses the Ethernet frame (encapsulating the IP packet) to be sent to its configured default gateway (MLS-RP) for routing. The frame has a destination MAC address *Eth_MLS_RP_1* which belongs to Port_1 of the MLS-RP which is the interface facing or leading to Station A.

Step 2   – MLS-SE marks the first framed IP packet as a candidate frame for Layer 3 forwarding

- The MLS-SE receives the Ethernet frame sent by Station A and checks the destination MAC address of this first frame. Given that the destination MAC address of the frame is the MAC address of the MLS-RP (*Eth_MLS_RP_1*), the MLS-SE understands that this frame contains an IP packet that requires routing and immediately marks the frame as a *candidate frame* for Layer 3 forwarding. It should be noted that any frame requiring Layer 3 forwarding always carries the destination MAC address of a routing device.
- The MLS-SE examines the destination IP address in the encapsulated IP packet and performs a lookup in the multilayer switching cache for a flow entry that is associated with this packet. The MLS-SE discovers that there is no entry for this packet because this is the first packet sent by Station A to Station B, so the packet is sent to the MLS-RP for routing.
- The MLS-SE writes an *incomplete* (or *partial*) flow entry in the multilayer switching cache, which includes only partial information (e.g., source and destination IP addresses) that identifies the particular flow this packet belongs to at this stage of the forwarding process.

Step 3   – The MLS-RP performs normal IP routing table lookup on the first framed IP packet

- The MLS-RP receives the framed IP packet, extracts the destination IP address, and performs a lookup in its local IP forwarding table. The result of the table lookup shows that the destination (Station B) is locally attached (i.e., directly attached to one of the switch's interfaces).
- The MLS-RP then checks its local ARP cache to determine the MAC address of Station B (*Eth_Host_B*). If the ARP cache does not contain an entry for Station B, the MLS-RP sends an ARP request for the MAC address associated with Station B (*Eth_Host_B*). After obtaining Station B's MAC address, the MLS-RP generates a new Ethernet frame to transport the IP packet to its intended destination. This new frame is sent back to the MLS-SE for forwarding to the destination.

Step 4   – The MLS-SE writes the destination MAC address of the routed IP packet into the incomplete flow entry in the multilayer switching cache

- The MLS-SE receives this first framed IP packet (of the flow) from the MLS-RP after it has completed the necessary IP routing and forwarding table lookup process. The MLS-SE then writes the destination MAC address of this framed IP packet into the *incomplete* flow entry in the multilayer switching cache that was initially created in Step 2.
- The MLS-SE also examines its local Layer 2 forwarding (bridge) table to determine the egress port on the MLS-SE associated with the destination MAC address (*Eth_Host_B*) and registers this information into the flow entry as well.

- The flow entry information in the multilayer switching cache is now ready for use in future rewriting of MAC address information and forwarding of packets sent to Station B without having to forward the packets to the MLS-RP. The flow entry now has complete information needed for forwarding subsequent packets of the flow without having to involve the MLS-RP.

Step 5 – ML-SE forwards Ethernet frame to Station B
- After rewriting the necessary source and destination MAC address information in the frame, the MLS-SE forwards the frame out the appropriate egress port to Station B

Step 6 – Station A sends another IP packet to Station B
- From the destination MAC address in the frame (*Eth_MLS_RP_1*), the MLS-SE sees that the frame is destined to the MLS-RP and is therefore a candidate for Layer 3 forwarding. The MLS-SE examines the destination IP address (*IP_Host_B*) in the encapsulated IP packet and performs a lookup in the multilayer switching cache for an entry that matches the flow entry completed in Step 4.
- The MLS-SE finds a matching flow entry and rewrites the source and destination MAC addresses of the frame. The MLS-SE on its own performs other necessary Layer 3 forwarding operations such as decrementing the IP TTL and recomputing the IP header checksum, rewriting the source and destination MAC addresses of the outgoing frame, and recomputing the Ethernet frame checksum.

Step 7 – ML-SE forwards subsequent Ethernet frames belonging to the created flow to Station B
- Finally, the MLS-SE forwards the rewritten frame out the correct egress port as in Step 5 to Station B. The MLS-SE Layer 3 forwards all subsequent IP packets from Station A to Station B as described in Step 6 and Step 7 as long as the flow entry created in Step 4 is valid or has not been aged out.

As illustrated in Figure 7.3, the first packet in a flow from a host is routed through the MLS-RP, generating the relevant forwarding information which then allows the MLS-SE to maintain the appropriate information for Layer 3 forwarding of subsequent packets to the destination host.

The MLS-RP and MLS-SE also communicate regularly (using the MLS Protocol (MLSP)). With this, if the MLS-SE detects that an MLS-RP has stopped functioning, the MLS-SE can flush the appropriate flow entries in the multilayer switching cache. This feature is particularly important in a multilayer switching environment with redundant route processors where two or more MLS-RPs provide inter-VLAN routing because it ensures that the redundant MLS-RP can be used if the primary MLS-RP fails. The redundant MLS-RP works with the MLS-SE to continue routing Layer 3 packets.

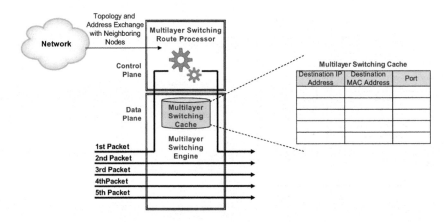

FIGURE 7.3   Basic flow-based switching architecture.

7.3.3   MULTILAYER SWITCHING USING A ROUTE CACHE AND ACCESS CONTROL LISTS

In the multilayer switching architecture discussed above, the multilayer switching cache stores the flow-based information required for packet rewrites performed during the Layer 3 forwarding process. The flow configured in a multilayer switching system depends on the requirements of the network. For example, if the network manager requires only simple routing where the routing decisions need to be made based only on the destination IP address of each packet, flows can be set up based on destination IP address only.

However, if there is the requirement to use access control lists (ACLs) on a routed interface through which a packet would normally travel, then the multilayer switching system can use source and destination IP address or full flow information depending on the granularity of the ACLs [MENJUS2003]. Figure 7.4 shows an example topology of a multilayer switching system with an extended ACL configured on an interface on the MLS-RP.

The processing steps that occur in the multilayer switching system with an extended ACL in Figure 7.4 are described below [CISCCAT6500] [CISCIOSR12.2]. Here, we assume again that Station A sends packets to another Station B on a different IP subnet through the default gateway which is the MLS-RP. An inbound ACL is implemented on the ingress port of the MLS-RP to control traffic.

Step 1   – Station A attempts to establish an HTTP connection to Station B on a different subnet
- To establish the HTTP connection, Station A sends an IP packet with a source IP address *IP_Host_A*, destination IP address *IP_Host_B*, protocol number 6 (TCP), destination TCP port 80, and a random source TCP port 1111 to Station B. Because Station B is not on the same IP subnet as Station A, the frame containing the IP packet is addressed to the configured default gateway MLS-RP (which has MAC address *Eth_MLS_RP_1* as illustrated in Figure 7.4).

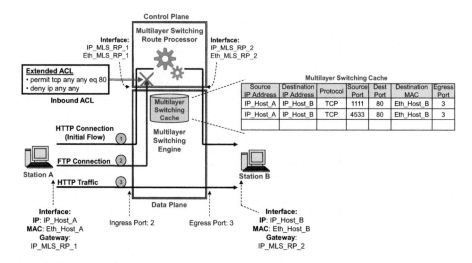

FIGURE 7.4 Flow-based switching with Access Control List (ACL).

- The MLS-SE receives the framed IP packet from Station A and because no flow entry exists in the multilayer switching cache that matches the received packet, the MLS-SE marks it as a candidate packet requiring Layer 3 forwarding, and forwards it to the MLS-RP.
- The MLS-RP receives the packet and inspects its header parameters against the inbound ACL it maintains and permits the packet because the packet is a TCP packet with a destination port of 80. The MLS-RP performs an IP forwarding table lookup to determine the packet's next-hop and sends the packet back to the MLS-SE.
- The MLS-SE receives the routed IP packet and writes a complete flow entry in the multilayer switching cache.

Step 2 – Station A next attempts to establish an FTP connection to Station B

- Station A next sends a packet carrying an FTP message to Station B. The multilayer switching cache on the MLS-SE now contains a full flow mask (for the HTTP flow already processed), but the FTP packet sent does not match the flow entry created in Step 1 and hence is forwarded to the MLS-RP.
- The MLS-RP examines the packet header contents against the inbound ACL and drops the packet because it is not an HTTP packet. We assume the ACL is configured to allow only HTTP traffic.
- This results in the MLS-SE not seeing a returned routed FTP packet from the MLS-RP that includes the required information to complete the entry for the flow. This means the MLS-SE cannot complete the *incomplete* (or *partial*) flow entry created when the FTP packet was first received by the MLS-SE.

- Any subsequent FTP connection request packets or any other non-HTTP traffic are always forwarded to the MLS-RP which then drops the packets because no complete flow entries were ever created on the multilayer switching cache by the MLS-SE.

Step 3 – All traffic associated with the HTTP connection established in Step 1 is Layer 3 forwarded by the MLS-SE
- All packets associated with the HTTP connection established in Step 1 are forwarded directly by the MLS-SE because a complete flow entry exists in the multilayer switching cache.
- However, if a new HTTP connection is established between Station A and Station B, a new flow entry must be created by the MLS-SE as explained in Step 1, because the source TCP port is different for each connection.

The configuration in Figure 7.4 uses full flow entries because the MLS-RP must be able to permit traffic based on specific combinations of source and destination IP addresses, and source and destination TCP/UDP ports as defined in the ACL it maintains. If a destination IP address only or destination-source IP address flow was configured on the MLS-RP, the MLS-SE would not differentiate HTTP packets from FTP packets and FTP packets would be incorrectly Layer 3 forwarded (or permitted) to Station B.

7.3.4 MULTILAYER SWITCHING CACHE TIMERS

The multilayer switching cache in the MLS-SE is a finite resource that can only maintain a finite number of flow entries. If the cache becomes fully utilized, the MLS-SE will be unable to register new entries for new flows. Which means any new flows that require routing cannot be Layer 3 forwarded directly by the MLS-SE. Instead, they have to be routed normally via the MLS-RP which acts as a one-armed router (or router-on-stick) as discussed in Chapter 6. This is similar to the router-on-stick topology discussed in Chapter 6 where the MLS-SE represents the Layer 2 switch and the MLS-RP represents the one-armed router [CISCCAT6500] [CISCIOSR12.2] [MENJUS2003].

For example, the multilayer switching cache described in references [CISCCAT6500] [CISCIOSR12.2] and [MENJUS2003] has a capacity of 128 K (i.e., 128 × 1024) entries; however, when the number of entries exceeds a threshold of 32 K, the MLS-SE may have to send some flows to the MLS-RP for forwarding. To prevent the multilayer switching cache from exceeding 32 K entries, the MLS-SE operates two timers. These timers are both used to age out idle flow entries in the cache after a configurable period of time has elapsed:

- **Multilayer switching fast-aging timer**: First, each flow is configured with a *packet-send threshold* (a configurable parameter) which specifies the number of packets a flow can send in a given time period. This timer is used to age out flows that have not sent packets exceeding the configured packet threshold within a configured *fast-aging time period*. The network manager can configure the fast-aging timer as 32, 64, 96, or 128 seconds, and configure a packet threshold of 0, 1, 3, 7, 15, 31, or 63 packets. For example, if the fast-aging timer is set to 32 seconds and a packet threshold to 15, any flow that does not send more than 15 packets within 32 seconds is aged out [CISCCAT6500] [CISCIOSR12.2] [MENJUS2003].

- **Multilayer switching aging timer**: This timer (which we refer to in this chapter as the *flow aging timer*) is used to age out idle flows, that is, flows that have not been active or have not sent a single packet during a specified *aging timer interval*. If one or more packets are sent in a flow, the aging timer is reset. The multilayer switching aging timer has a default setting of 256 seconds and the aging time can be configured in 8-second increments between 8 and 2032 seconds [CISCCAT6500] [CISCIOSR12.2] [MENJUS2003].

7.3.5   MLS-RP to MLS-SE Communications

As stated above, the MLS-RP and MLS-SE communicate using the Cisco proprietary protocol, MLS Protocol (MLSP) [CISCCAT6500] [CISCIOSR12.2]. This multilayer switching architecture uses the MLSP as an internal communication mechanism between the MLS-RP and MLS-SE to ensure that the Layer 3 forwarding process on the MLS-SE has the correct information to perform the forwarding operations. The information used by the MLS-SE has to be refreshed periodically. The following two major events can cause the multilayer switching cache to invalid its entries:

- **Routing topology changes**: If a routing topology change occurs, the flow information in the multilayer switching cache may need to be updated to reflect the changes. In the event of a routing topology change, the MLS-SE flushes the multilayer switching cache to ensure that packets are not Layer 3 forwarded in error.
- **ACL configuration**: The multilayer switching cache also needs to be updated in the case where an ACL is applied to an interface (or modified) on the MLS-RP. In the event an ACL is applied to an interface on the MLS-RP, the MLS-SE also adjusts the flow mask to that specified by the MLS-RP to ensure that traffic cannot bypass the ACL. For example, the flow mask would be changed when an extended ACL is applied to a previously unfiltered interface. As illustrated in Figure 7.4, a full flow mask is required when extended ACLs are used, so the MLS-RP sends an MLSP message to the MLS-SE instructing it to flush its multilayer switching cache and update the flow mask. The MLS-RP communicates the required flow mask to the MLS-SE, which ensures that anytime an ACL is applied, the appropriate flow mask is immediately used on the MLS-SE ensuring packets are not incorrectly permitted if they have been denied by the ACL.

The MLS-RP and MLS-SE use MLSP messages (sent in multicast Ethernet frames) to communicate with each other. The MLS-RP sends MLSP messages to the MLS-SE if any of the above events occur. This is to indicate to the MLS-SE that it should flush the multilayer switching cache it maintains and possibly modify the flow mask used. The MLSP messages are also used to verify that components in the multilayer switch are still alive and functioning, which is done via the exchange of hello packets [CISCCAT6500] [CISCIOSR12.2]. By default, these messages are configured to be sent every 15 seconds.

In a configuration where multiple MLS-RPs are installed for redundancy purposes, the MLS-SE must be able to differentiate between each MLS-RP. A less effective approach is to do the differentiation based on the MAC address of each MLS-RP. Note that each interface on an MLS-RP is assigned a unique MAC address. However, let us take the case where there are thousands of flow entries in the multilayer switching cache. If each of the entries associated with an MLS-RP that has just stopped functioning needs to be flushed, it would be a lengthy process searching through the multilayer switching cache based on a 48-bit MAC address value. So, an approach that facilitates faster cache purges is preferable. To achieve this, each MLS-RP is assigned an 8-bit XTAG value which serves as an identifier and index for each MLS-RP [CISCIOSR12.2]. The shorter XTAG allows the MLS-SE to differentiate between the flow entries associated with each MLS-RP in the multilayer switching cache, facilitating faster cache purges when needed.

7.4 IP MULTICAST MULTILAYER SWITCHING

This section discusses the forwarding of IP multicast packets in multilayer switches that are directly connected to VLANs or IP subnets. We focus on the scenario where IP multicast multilayer switches forward IP multicast packet flows between VLANs/IP subnets, thereby off-loading the more processor-intensive, multicast packet routing on network routers [CISCCAT6500] [CISCIOSR12.2]. Similar to the unicast forwarding method discussed above, the multicast multilayer switch acts as a front-end processor to a traditional multicast router, offloading as much as possible, traffic heading to the router. This multicast forwarding architecture was used when routers were mainly based on software forwarding.

The packet forwarding function in this architecture is moved into the connected multilayer switch whenever a path exists between a multicast source and members of a multicast group. Packets that do not have a path to reach their destinations are still forwarded in software by (other) routers. The discussion starts with a recap of the main elements of IP unicast forwarding in Cisco multilayer switches [CISCCAT6500] [CISCIOSR12.2] [CISCKENCL99].

The unicast and multicast multilayer switching methods discussed up to this section are no more in use due to advances in switch/router architectures as discussed in the previous chapters and in the "The FIB, Fast-Path Forwarding, and Distributed Forwarding Architectures" section below. We include these legacy architectures in this book only to highlight how switch/routers (or multilayer switches) have evolved over the years.

7.4.1 Unicast Multilayer Switching in Cisco Catalyst 5000, 6000, and 6500 Series Switches

IP provides a connectionless service and forwards each IP packet independently of other packets. However, the actual traffic in a LAN segment, IP subnet, or a network in general, consists of many end-to-end conversations, or flows, carrying data between end-users or applications. As defined earlier, a full flow (or simply, a flow) is a unidirectional sequence of packets sent from a particular source to a specific destination that share the same Network Layer protocol (IP) and Transport Layer information (TCP or UDP port).

Packets in a communication from a client to a server and from the server to the client constitute separate flows. Flows can also be defined based on only IP addresses where only a destination IP address or both source and destination IP addresses can be used to define a flow. When only the destination IP address is used to identify a flow, IP traffic from multiple users or applications to a particular destination constitutes a single flow.

The main elements considered here for multilayer switching (MLS) (already described above) are the MLS-SE, MLS-RP, and MLSP, which is the protocol running between the MLS-SE and MLS-RP to enable multilayer switching. The MLS-SE maintains the multilayer switching cache for the Layer 3-forwarded flows. The cache also includes entries for traffic statistics (for traffic flows) that are updated at the same time packets are forwarded. After an entry in the multilayer switching cache is created for a flow, packets identified as belonging to that flow can be Layer 3-forwarded based on the cached information. The multilayer switching cache maintains flow information for all active flows; entries for inactive flows are aged out periodically. When an entry for a flow ages out, the flow statistics for that flow can be exported to a network management system called a flow collector application.

The forwarding of IP unicast packets in the Cisco multilayer switches is summarized by the following steps [CISCCAT6500] [CISCIOSR12.2] [CISCKENCL99]:

Step 1: Initialization of the multilayer switching process
- Using the MLSP, the MLS-RP multicasts its MAC address and VLAN information to all MLS-SEs on the chassis (i.e., the MLS-RP's MAC addresses used on different VLANs and the MLS-RP's routing and ACL changes).
- When the MLS-SE receives an MLSP *hello* message indicating initialization of the MLS process, the MLS-SE programs itself with the MLS-RP MAC address and its associated VLAN identifier in addition to any relevant routing and ACL information required.

Step 2: Station A transmits first packet to MLS-RP intended for Station B on another network
- Let us assume Station A and Station B are located on different VLANs and Station A initiates a data transfer to Station B. When Station A sends the first packet to the MLS-RP, the MLS-SE recognizes and flags this packet as a *candidate packet* for Layer 3 forwarding because

the MLS-SE has already been programmed with the MLS-RP's destination MAC address and VLAN through the MLSP.

- The MLS-SE extracts the Layer 3 flow information from the packet
(such as the destination IP address, source IP address, and Transport
Layer protocol port numbers), and forwards the first packet to the
MLS-RP. The MLS-SE creates a *partial* or *incomplete* entry for this
Layer 3 flow in the multilayer switching cache.

- The MLS-RP receives the (candidate) packet, performs a lookup in
its local forwarding table to determine how to forward the packet, and
applies services such as ACLs and class of service (CoS) policy to the
packet. The MLS-RP rewrites the MAC header of the frame carrying
the packet by rewriting the destination MAC address of the frame to
be the MAC address of the receiving interface of Station B, and the
source MAC address to be its own MAC address.

**Step 3: The MLS-RP forwards the processed candidate packet back to the
MLS-SE**

- When the MLS-SE receives the packet, it recognizes the source MAC
address as that of the MLS-RP, and that the packet's flow information
matches the flow for which it recently set up a candidate entry.

- The MLS-SE flags this MLS-RP processed packet as an *enabler
packet* and completes the partial flow entry (established by the candidate packet) in the multilayer switching cache.

**Step 4: MLS-SE forwards all subsequent packets in the flow using the entry
in the multilayer switching cache**

- After the flow entry has been completed in Step 3, all IP packets
belonging to the same flow from Station A to Station B are Layer 3
forwarded directly by the MLS-SE, bypassing the MLS-RP.

- After the routed path between Station A and Station B has been established by the MLS-RP, all subsequent packets from Station A have
their headers appropriately rewritten by the MLS-SE before they are
forwarded to Station B. The rewritten information includes Layer 3
header updates (decrementing the IP TTL and recomputing the IP
header checksum), new source and destination MAC addresses for the
outgoing frame, and recomputing the Ethernet frame checksum.

The above multilayer switching processing is unidirectional which means for Station
B to communicate with Station A, another Layer 3 (routed) path needs to be created from Station B to Station A. The multilayer switching process allows the multilayer switch to enforce ACLs on every packet of the flow while forwarding packets.
Additionally, the multilayer switching process allows route topology changes and the
addition of ACLs to be reflected in the multilayer switching cache and the forwarding
process [CISCIOSR12.2].

Let us consider the case where an ACL has been configured on the MLS-RP to
deny communication from Station A to Station B. Let us assume that Station A

initiates communication with Station B by sending the first packet to the MLS-RP. The MLS-RP receives this packet and checks its ACL to see if this packet is permitted to proceed to Station. If the ACL is configured to deny this packet, it will be discarded. Because the first packet does not return from the MLS-RP to the MLS-SE, a multilayer switching cache entry is not created by the MLS-SE, thereby, preventing other similar packets from Station A from getting to Station B; no flow entry is created in the cache.

Let us consider another case where ACLs are configured on the MLS-RP while flows are already being Layer 3 forwarded within the MLS-SE. The MLS-SE (through the MLSP) learns about the introduction of the ACLs and immediately enforces the ACLs for all affected flows by purging them [CISCIOSR12.2]. Similarly, when the MLS-RP detects a routing topology change, the appropriate multilayer switching cache entries are deleted in the MLS-SE.

7.4.2 Understanding IP Multicast Multilayer Switching

Unlike an IP unicast flow, an IP multicast flow is a unidirectional sequence of packets sent from a multicast source to members of a multicast group (identified by a multicast address carried in the destination IP address field of the multicast packets). Multicast flows are based on the IP address of the multicast source device and the address of the IP multicast group (i.e., the multicast group address).

7.4.2.1 IP Multicast Multilayer Switching Network Topology

The IP multicast multilayer switching process described here (which is based on Cisco multicast multilayer switching (MLS) technology) requires a specific network topology for the multicast forwarding process to work correctly [CISCCAT6500] [CISCIOSR12.2]. In this topology, the multicast traffic from the source is received on a Layer 2 (LAN) switch, traverses a trunk link to a router, and returns to the switch over the same trunk link for the switch to forward the traffic to the destination multicast group members.

The basic forwarding architecture consists of a Layer 2 switch connected to an internal or external router through an IEEE 802.1Q or Cisco ISL (Inter-Switch Link) trunk link. ISL is Cisco's VLAN encapsulation protocol and is supported only on some Cisco platforms. It is offered as an alternative to the IEEE 802.1Q standard, the widely used VLAN tagging protocol.

Figure 7.5 shows the basic configuration when IP multicast MLS is not deployed. Figure 7.6 shows the configuration when IP multicast MLS is deployed (assuming a complete multicast flow has been established in the multicast multilayer switch). The architecture (Figures 7.5 and 7.6) consists of a Layer 2 switch, a directly connected external router, and multiple VLANs (or IP subnets).

The network in Figure 7.5 does not have the IP multicast multilayer switching feature enabled. In Figure 7.5, the multicast traffic flow is represented by the dash arrows from the router to each multicast group in each VLAN. In this configuration, the router (which is based on older software-based router architecture technology) must replicate the multicast data packets from the source to the multiple VLANs. If there are N VLANs, the router has to replicate N streams of multicast traffic.

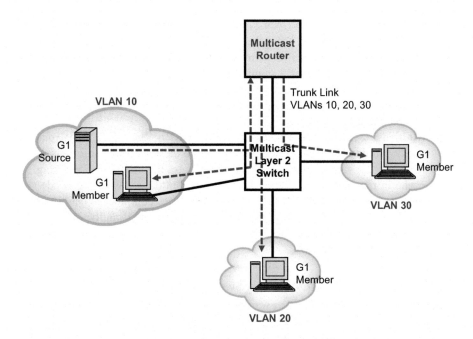

FIGURE 7.5 IP multicast multilayer switching not deployed.

With this setup, the router can be easily overloaded or overwhelmed with forwarding and replicating multicast traffic if the multicast input rate increases or the number of outgoing interfaces carrying multicast traffic increases. The forwarding architecture in Figure 7.6 prevents this potential problem by modifying the Layer 2 switch hardware and having it forward the multicast data traffic directly to the VLANs having multicast group members. In this multicast MLS method, multicast control packets will still have to be sent between the router and Layer 2 switch.

7.4.2.2 Cisco IP Multicast MLS Components

The IP multicast MLS network topology (in Figure 7.6) has two main components [CISCIOSR12.2]:

- **Multicast MLS-Switching Engine (MMLS-SE)**: When adequately programmed with the correct forwarding instructions, the MMLS-SE is able to forward directly both Layer 2 and Layer 3 traffic. The MMLS-SE is represented by the Layer 2 switch in Figure 7.6. To forward multicast traffic, the MMLS-SE employs a *Layer 2 multicast forwarding table* that it maintains to determine which switch ports Layer 2 multicast traffic should be forwarded on. The entries in the Layer 2 multicast forwarding table are generated by enabling CGMP (Cisco Group Management Protocol), IGMP (Internet Group Management Protocol) snooping, or GMRP (GARP Multicast Registration Protocol) on the switch. The Layer 2 multicast forwarding table entries map the destination multicast MAC addresses to outgoing switch ports for a given VLAN. Note that the Layer 2 multicast forwarding table (which is different from the *Layer 3 multicast multilayer switching cache*) is used in conjunction with CGMP (or IGMP) snooping and GMRP, that is, when these Layer 2 multicast related protocols are enabled. Note that ISL and CGMP are Cisco proprietary protocols that were used in Cisco platforms before IEEE 802.1Q and IGMP, respectively, became the widely accepted industry standards.

- **Multicast MLS-Route Processor (MMLS-RP):** The MMLS-RP runs the IP multicast routing protocols to generate the forwarding information needed to forward multicast packets, and updates the multilayer switching cache in the MMLS-SE. When IP multicast multilayer switching is adequately initialized in the MMLS-SE, the MMLS-RP continues to handle all non-IP-multicast traffic while transferring the tasks of IP multicast traffic forwarding to the MMLS-SE. The MMLS-RP is represented by the multicast router in Figure 7.6.

7.4.2.3 Layer 3 Multicast Multilayer Switching Cache

Other than the Layer 2 multicast forwarding table, the MMLS-SE maintains another forwarding table which is the *Layer 3 multicast multilayer switching cache* used to identify individual IP multicast flows. The entries in the multilayer switching cache

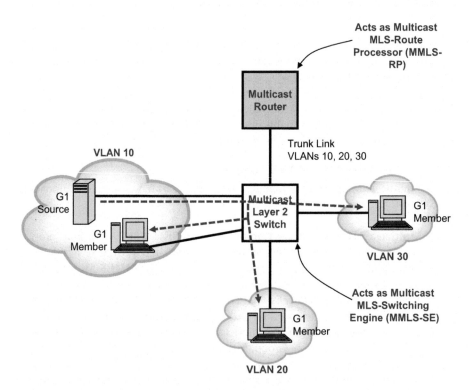

FIGURE 7.6   IP multicast multilayer switching deployed.

are of a different form than those in the Layer 2 multicast forwarding table. Each entry in the multilayer switching cache is structured in the following form: {source IP address, multicast group IP address, source VLAN ID}. This translates into a single flow mask that is designated as *source destination vlan* [CISCIOSR12.2].

As discussed above, the maximum size of the multilayer switching cache (in the Catalyst 6500) is 128K and is shared by all multilayer switching processes on the switch such as the IP unicast MLS [CISCIOSR12.2]. However, when the number of cache entries exceeds 32K, there is a high chance that a flow will not be forwarded by the MMLS-SE but instead passed to the router (MMLS-RP) for forwarding.

The MMLS-SE populates the multilayer switching cache using the forwarding information passed on from the MMLS-RP via the MLSP. The MMLS-RP communicates with other routers participating in IP multicast routing to learn about multicast traffic flows and trees in order to generate the routing information needed for its multicast routing table. The multicast routing table is then distilled to create the forwarding information for the multilayer switching cache in the MMLS-SE.

The router (MMLS-RP) and Layer 2 switch (MMLS_SE) in Figure 7.6 exchange information using the *multicast MLSP*. Whenever the MMLS-RP receives traffic for a new multicast flow, it updates its multicast routing table and forwards the new flow information to the MMLS-SE using the MLSP. Furthermore, when an entry in the multicast routing table is aged out, the MMLS-RP deletes that entry and forwards the updated information to the MMLS-SE.

The MMLS-SE maintains in its multilayer switching cache only information that apply to active multilayer switched flows. After the entries in the multilayer switching cache are created, multicast packets identified as belonging to an existing flow can be Layer 3 forwarded by the MMLS-SE based on the cache entry for that flow. To forward multicast packets out the correct interfaces, the MMLS-SE maintains for each cache entry, a list of outgoing interfaces for the destination IP multicast group. The MMLS-SE uses this list (of interfaces-to-destination IP multicast group address mapping) to determine which switch port and VLANs a given multicast flow should be replicated on.

As discussed above, the IP multicast multilayer switching process supports a single flow mask, designated as *source destination vlan*. With this, the MMLS-SE maintains one multicast multilayer switching cache entry for each {source IP, destination group IP, source VLAN}. This multicast *source destination vlan* flow mask is different from the IP unicast multilayer switching *source destination ip* flow mask (which has elements {source IP address, destination IP address}). This is because, for IP multicast multilayer switching, the source VLAN is included as part of the entry [CISCIOSR12.2]. The source VLAN serves as the multicast Reverse Path Forwarding (RPF) interface for the multicast flow (see multicast RPF discussion in Chapter 5).

7.4.2.4 Layer 3-Switched Multicast Packet Rewrite

After performing the required lookup operations in the multilayer switching cache, the MMLS-SE performs a frame rewrite based on information learned from the MMLS-RP and stored in the multicast multilayer switching cache. Then the multicast packet is Layer 3 forwarded out the switch port to a destination multicast group.

Let us consider an example where Server A transmits a multicast packet addressed to IP multicast group G1. We assume members of multicast group G1 are on VLANs other than the source (Server A's) VLAN. As required by IP forwarding, the MMLS-SE must perform a rewrite on the frame encapsulating the IP multicast packet when it replicates the frame to the destination VLANs. The multicast packet the MMLS-SE receives is pre-formatted in a manner similar to the example shown in Figure 7.7.

The MMLS-SE rewrites the frame as follows:

- Decrement the IP header Time-to-Live (TTL) value by one and recompute the IP header checksum
- Replace the source MAC address in the Layer 2 frame header (which is the MAC address of Server A) with the MAC address of the MMLS-RP (this MAC address is stored in the multicast multilayer switching cache entry for the flow).
- Recompute the Layer 2 (Ethernet) frame checksum

The outcome of the frame rewrite process is an IP multicast packet that appears to have originated from the router (MMLS-RP). The MMLS-SE then replicates the rewritten frame (carrying the multicast packet) onto switch ports that lead to the appropriate destination VLANs, where it is forwarded to members of IP multicast group G1. The frame format after the MMLS-SE performs the necessary rewrite is shown in Figure 7.8.

Note that when sending an IP unicast packet in an Ethernet frame to a router, the destination MAC address is set to the MAC address of the router. However, when sending an IP multicast packet in an Ethernet frame (regardless of the destination

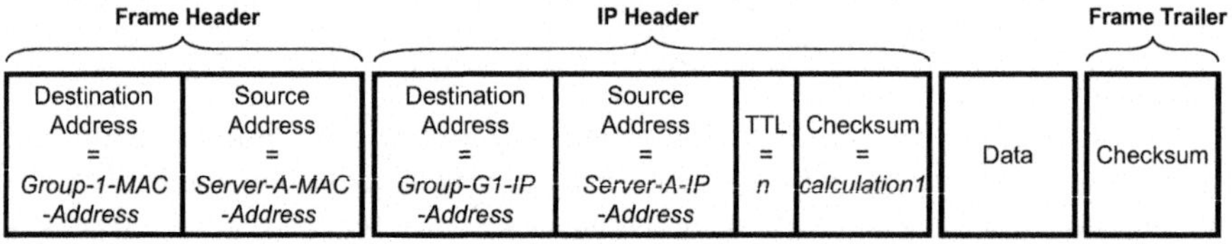

Frame Header		IP Header					Frame Trailer
Destination Address = Group-1-MAC -Address	Source Address = Server-A-MAC -Address	Destination Address = Group-G1-IP -Address	Source Address = Server-A-IP -Address	TTL = n	Checksum = calculation1	Data	Checksum

FIGURE 7.7 Packet format when the multicast multilayer switch receives a multicast packet (only relevant fields shown).

Frame Header		IP Header					Frame Trailer
Destination Address = Group-1-MAC -Address	Source Address = Multicast-Router -MAC-Address	Destination Address = Group-G1-IP -Address	Source Address = Server-A-IP -Address	TTL = n -1	Checksum = calculation2	Data	Checksum

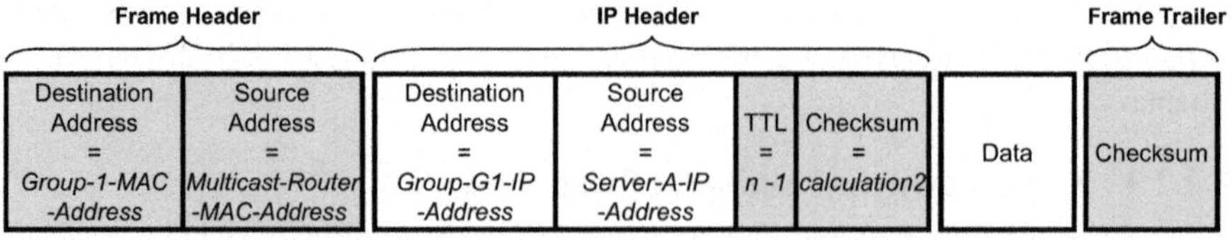

FIGURE 7.8 Packet format after the multicast multilayer switch packet rewrite (only relevant fields shown).

type, whether a router interface or another host on the same IP subnet or VLAN), the destination address is always set to the MAC address obtained after converting the IP multicast group address to a corresponding Ethernet multicast MAC address. Reference [CISCWILLB03] describes a mechanism for mapping an IP multicast group address to an Ethernet multicast MAC address. *For multicast packets, the destination IP address and destination MAC address are both always multicast addresses.*

7.5 THE FIB, FAST-PATH FORWARDING, AND DISTRIBUTED FORWARDING ARCHITECTURES

Using a standalone multilayer switch (or switch/router) in place of a system with a front-end processor as discussed above, provides efficient unicast and multicast packet forwarding. In particular, as discussed briefly above (and in earlier chapters), distributed forwarding using network topology information on line cards is a technology that provides Layer 3 forwarding based on a topology map of the entire network rather than on cached routes. The topology map structured in the form of an FIB is distributed to each line card on the chassis, allowing the line card to make autonomous packet forwarding decisions without involving the centralized CPU serving as the route processor (or routing engine) of the system (Figure 7.9).

Distributed forwarding optimizes network performance and scalability of networks with large and dynamic traffic patterns, such as the Internet, and networks with many short-lived Web-based applications, or interactive sessions. This technology, when used within the Internet and in large core networks, provides scalable packet forwarding as discussed in Chapter 6.

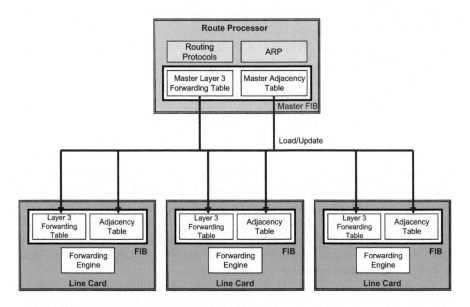

FIGURE 7.9   Distributed forwarding architecture.

7.5.1 BENEFITS OF DISTRIBUTED FORWARDING

One of the key benefits of distributed forwarding in the switch/router or router is rapid routing convergence. Since the FIB is distributed to all line cards, whenever a route disappears or is added, the main FIB in the route processor is updated with that information and the same routing information is sent to the line card FIBs. The line cards receive the new network topology information (from the routing protocol running in the system) very quickly and reconverge around a failed link. In order for all this to happen in larger systems, the control plane and data plane have to have a scalable performance to ensure that the FIBs provided to the line cards accurately reflect the current topology of the network. Fast reconvergence is difficult to achieve in a system based on route cache forwarding. Using a distributed forwarding architecture with FIBs ensures that the line card forwarding capacity is not degraded due to stale routing information in a route cache.

Distributed forwarding architectures offer highly sustainable level of forwarding consistency and stability in large, dynamic networks. This is because the distributed FIBs contain the same routes as the route processor's FIB, therefore eliminating the potential for cache misses that occur in route caching schemes. In such demand caching schemes, if a route is not found in the route cache, the system has to rely on the route processor's routing table to determine the next-hop IP address and outbound interface, and then a route cache entry is added for that destination. Because the route cache information is derived from the routing table, routing changes cause existing cache entries to be invalidated and then reestablished to reflect any network topology changes.

In networking environments that frequently experience significant routing activity (such as the Internet and enterprise and service provider network backbones), the route cache-based system can cause traffic to be forwarded via the routing table maintained by the route processor (*Process Switching* in Cisco terminology) as opposed to via the route cache (called *Fast Switching*). During major network convergence or flux, network performance would thus be suboptimal. Distributed forwarding (using optimized lookup algorithms and the FIBs) obviates the Process Switching/Fast Switching scenario because the FIB is topology-driven rather than traffic-driven; distributed forwarding performance is largely independent of and unaffected by network size or dynamics.

7.5.2 DISTRIBUTED FORWARDING ARCHITECTURE

As discussed in Chapters 2 and 5, the functions involved in routing IP traffic in routing devices can be split into two distinct planes of operation – the control plane and the data plane. The control plane deals with determining the topology of a network to obtain information about where each destination IP address in the network is. Each destination IP address in the routing device is associated with an outbound interface and a next-hop IP address, which represents the next best router closest to the local node that leads to the destination.

The data plane deals with the physical operations that are required to actually forward a packet to the next-hop; this refers to the operation of determining the next-hop IP address and outbound (egress) interface, decrementing the IP header TTL, recomputing the IP header checksum, rewriting the appropriate Layer 2 addressing

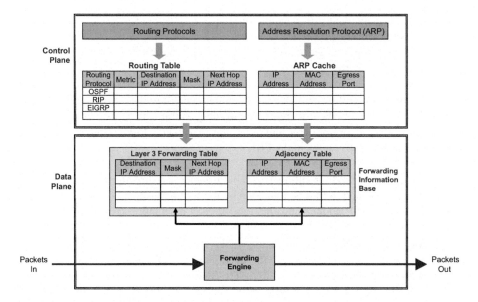

FIGURE 7.10 Components of the distributed forwarding architecture.

information on the outgoing packet, recalculating the Layer 2 packets checksum, and then forwarding the packet out the appropriate egress interface to the next-hop.

Most high-performance switch/routers and routers employ a distributed architecture in which the control path and data path are decoupled and operate relatively independently (Figure 7.10). The control path software, which includes the routing protocols, runs on the route processor, while most of the data packets are forwarded directly by the line cards over the switching fabric. Each line card includes a forwarding engine that handles all packet forwarding.

The main tasks of the control plane in the distributed forwarding architectures include:

- Running the routing, control, and system management protocols. Extracting routing and control information from arriving routing updates, updating the routing and forwarding tables, and then conveying the routing information to the line cards.
- Handling special packets sent from the line cards to the route processor:
  o ARP request and replies
  o SNMP messages
  o ICMP messages
  o IP packets requiring a response from a route processor (TTL has expired, MTU is exceeded, fragmentation is needed, etc.)
  o IP broadcasts that will be relayed as unicast (e.g., DHCP requests)
  o Routing protocol updates
  o Packets requiring special processing like encryption
  o Packets requiring Network Address Translation (NAT) processing

- Collecting the data path information, such as traffic statistics, from the line cards to the route processor
- Managing the internal housekeeping tasks and system environment monitoring and control

Distributing routing information from the route processor to the individual line cards results in high-speed FIBs lookups and forwarding as illustrated in Figure 7.11.

As illustrated in Figures 7.9 and 7.10, the distributed forwarding architecture maintains two data structures in the FIB:

- **Layer 3 Forwarding Table**: This table is generated directly from the routing table (which in turn is populated by the routing protocols) and contains the next-hop IP address information for each destination (IP route) in the network.
- **Adjacency Table**: The adjacency table specifies the MAC address and egress interface associated with each next-hop IP address. The MAC address information is obtained via the ARP process or manual configuration (by the network administrator). The next-hop MAC address column represents the destination MAC address of the next hop router, which is the address used to rewrite the destination MAC address in the outgoing Layer 2 packet.

In Figures 7.9 and 7.10, the routing table and ARP cache are both control plane entities, meaning both are generated and maintained by the route processor (i.e., control or routing engine). From these tables, the Layer 3 forwarding table and adjacency tables are created, which are essentially data structures optimized for fast lookup by the data plane processor (i.e., the forwarding engine). In Figures 7.9 and 7.10, the Layer 3 forwarding engine uses the Layer 3 forwarding table and adjacency table to determine the next-hop device's IP address and MAC address

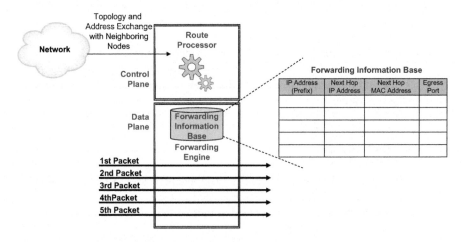

FIGURE 7.11 Distributed forwarding operation.

for a packet. This provides the required information for MAC address rewrites in departing Layer 2 packets.

The Layer 3 forwarding table is extracted from the routing table maintained by the control plane. The main elements of this table (the destination IP address prefixes, next-hop IP addresses, and outbound interfaces) are extracted from the routing table (the remaining information in the routing table is not directly useful for the Layer 3 forwarding process). The adjacency table holds the MAC address details that are used to rewrite the destination MAC address of each outgoing packet. This table is maintained using the ARP cache (which in turn is populated via the control plane using ARP requests) or via manual configuration.

The MAC address of the switch/router's transmitting interface(s) itself must also be known to the Layer 3 forwarding engine, which is used to rewrite the source MAC address of the outgoing Layer 2 packet. The source MAC address is always the same and does not need to be included in the Layer 3 forwarding table and adjacency table. It should be noted that all of the information contained in the Layer 3 forwarding table and adjacency table is the same information contained in the routing table and the ARP table (cache). The Layer 3 forwarding table and adjacency table exist purely for organizing the relevant information required for Layer 3 forwarding into a structure that is optimized for fast lookups by the data plane.

Although there are many different implementations of distributed forwarding, they all share one common feature; they all implement multiple Layer 3 forwarding engines so that simultaneous Layer 3 forwarding operations can occur in parallel, thereby, boosting overall system performance (Figures 7.9 and 7.10). The preferred implementation uses multiple FIBs distributed across multiple line cards installed on the system chassis. Each line card in this architecture has its own dedicated specialized processor- or hardware-based Layer 3 forwarding engine and FIB, allowing multiple Layer 3 data plane operations to be performed simultaneously on a single chassis. The main route processor of the switch/router is responsible for generating a central master forwarding table and adjacency table, and distributing these tables to each line card supporting distributed forwarding.

7.5.3 DISTRIBUTED FORWARDING WITH INTEGRATED HARDWARE PROCESSING

In order to achieve high scalability together with carrier-grade resiliency, switch/routers use various forms of fully distributed architectures with all packet forwarding decisions made by distributed forwarding engines, and all control functions handled by the switch/router's centralized route processor CPU(s). In some high-end switch/routers, the control plane (just like the data plane processing) is also distributed across multiple processors (i.e., route processor units) on the line cards. Distribution of control plane processes greatly increases the processing capacity of the control plane, providing high scalability route processing power.

In some designs, a CPU is dedicated to each of three functional areas [FOR10FTOS08]: Layer 3 processes, Layer 2 processes, and management/control processes. Each of these processes is handled by a dedicated processor. The modularity of these individual control plane processes allows them to be readily partitioned among the three route processors. Resiliency is also significantly improved because failure of a process on a CPU will not affect the other processes on the remaining two CPUs (except processes that require current information from the failing process).

The switch/router in [FOR10FTOS08] has three processors on each Route Processor Module (RPM) and one processor on every line card. The line card CPUs perform local control functions such as network traffic sampling, aggregation, and reporting using sFlow [RFC3176]. The modularity of the operating system in [FOR10FTOS08] allows for future system redesign and configuration where more centralized control plane processes can be designed and distributed across the line card CPUs. The traditional switch/router design has only a single CPU that performs all control plane and management functions.

Switch/router and router vendors can achieve better price/performance by using hardware integration and advanced silicon (ASICs, field-programmable gate array (FPGA), etc.) to enhance normal routing and packet forwarding functions. In the past, traditional software routers use expensive and relatively slow processors to perform Layer 3 functions. In recent years, vendors have cast these IP routing forwarding functions (particularly, the data plane functions) into ASIC, essentially creating a "router-on-a-chip". With this approach, a separate centralized processor (the route processor) still handles the routing, control, and system management protocols. New multi-gigabit switch fabrics also play a key role in speeding traffic in and out of the Layer 3 forwarding process, replacing the bus-based and shared-memory backplanes of traditional routers.

Typically, high-end, high-performance switch/router architectures completely separate the data plane from the control plane. In many designs, the data plane, uses ASICs exclusively to implement all packet processing functions; all such functions are in hardware. In some designs, ternary content addressable memories (TCAMs) on the line cards perform packet classification while other ASICs handle buffering and traffic management functions including other functions such as traffic policing, traffic shaping, queue scheduling, ACLs, statistical sampling of traffic using sFlow [RFC3176], and congestion control. This approach is adopted to ensure that even with the complete range of QoS, and traffic management and monitoring services enabled in a system, full line-rate packet forwarding performance is still maintained.

With the right integrated hardware processing, the control processor (route processor) is removed from the normal Layer 3 forwarding path. Through tight integration of hardware and software resiliency features, the switch/router could be designed to deliver high performance and high availability. Traditional routers send each data packet to a single processor for next-hop address lookup and packet modification, such as IP header TTL and checksum) update. In these older router architectures, such functions are controlled by software and must be invoked one packet at a time before the processor sends data to the outbound port for delivery to the next-hop.

Some switch/routers implement cache-based techniques to try to shortcut this by not handling every subsequent packet in a data stream as described above. An integrated hardware processing approach using FIBs instead maintains the standard packet-by-packet handling of data, but trades the router's slow processor and software for an ASIC that can handle Layer 3 forwarding functions, packet modifications and updates on the fly. With this, packet forwarding is done more quickly as data is received from the network.

IP address lookup is another major challenge that has always demanded attention in switch/router and router design. IP address lookup is the single most time-consuming activity that switch/routers and routers must perform – they have to comb through very lengthy forwarding tables to correctly map destination IP addresses to next-hops one packet at a time. To address this problem, some vendors design custom-built ASICs to streamline the address lookup task. These ASICs use hardware-based address lookup logics (sometimes with integrated Layer 2/Layer 3 address resolution capabilities) rather than software-based lookup algorithms that cannot sustain wire speed performance.

7.5.4   Benefits of Multilayer Switching with Integrated Hardware Processing

Other providing dramatic performance improvements, an integrated hardware approach achieves equally significant cost savings. Designers favoring ASIC-based architectures argue that, with integrated hardware-based packet processing and forwarding on all interfaces, full line-rate performance can be maintained even as the switch architecture evolves to require additional protocol formats and encapsulations (e.g., tunneling). They argue that, by replacing the multiple processors of a traditional router (with its accompanying software development and integration cost) with ASIC-based forwarding engines, the hardware-based processing architectures cost less to build. Overall systems costs are further reduced by integrating both Layer 2 and Layer 3 functionality in one ASIC, which in turn eases network management and overall administrative complexity.

To conclude, the rationale behind using multilayer switches (or switch/routers) is that, hardware integration of Layer 2 and Layer 3 IP processing functions on a single platform further helps users simplify their network infrastructure. For example, by replacing one or more switches and a traditional router with a single multi-gigabit switch/router, multiple interfaces can be configured as separate Layer 2 switching domains (VLANs) and still allow inter-VLAN communication and communication to the wider Internet to take place. This configuration significantly increases performance – routing or Layer 2 forwarding is performed only when required – and provides new levels of flexibility through the more efficient use of router interfaces and network addresses.

REVIEW QUESTIONS

1. What are the functions of the Multilayer Switching–Route Processor (MLS-RP) in the front-end processor approach for packet forwarding?
2. What are the functions of the Multilayer Switching–Switching Engine (MLS-SE) in the front-end processor approach for packet forwarding?
3. What is the purpose of the MLS Protocol (MLSP)?
4. What constitutes a full flow in multilayer switching?
5. What is a candidate frame (packet) in multilayer switching?
6. What is a enable frame (packet) in multilayer switching?
7. What is an incomplete (or partial) flow entry in the multilayer switching cache?

8. Why are the entries of the multilayer switching cache periodically aged and deleted?
9. What is the purpose of the fast-aging timer in multilayer switching?
10. What is the purpose of the flow aging timer in multilayer switching?
11. What major events can cause the multilayer switching cache to invalid its entries?
12. What is the difference between a unicast flow and a multicast flow in multi-layer switching?
13. What is the purpose of the Layer 2 multicast forwarding table in multilayer switching? How are the entries of this table generated?
14. What is the purpose of the Layer 3 multicast multilayer switching cache in multilayer switching? How are the entries of this cache generated?
15. In multilayer switching, what goes into the source and destination MAC address fields of a received frame?
16. In multilayer switching, what goes into the source and destination MAC address fields of an outgoing frame?
17. What is the difference between the Layer 3 forwarding table and the adjacency table? How are the entries of these tables generated?
18. In the distributed forwarding architecture, name four examples of special packets that are handled solely by the route processor (i.e., the routing or control engine).

REFERENCES

[CISCCAT6000]. Cisco Systems, Catalyst 6000 and 6500 Series Architecture, *White Paper*, 2001.
[CISCCAT6500]. Cisco Systems, Cisco Catalyst 6500 Architecture, *White Paper*, 2007.
[CISCIOSR12.2]. Cisco Systems, "Multilayer Switching Overview," *Part 4 in Cisco IOS Switching Services Configuration Guide*, Release 12.2.
[CISCKENCL99]. Kennedy Clark, *Cisco LAN Switching*, CCIE Professional Development Series, Cisco Press, 1999.
[CISCWILLB03]. Beau Williamson, *Developing IP Multicast Networks*, Volume 1, Cisco Press, 2003.
[FOR10FTOS08]. Force10 Networks, FTOS: A Modular Switch/Router OS Optimized for Resiliency & Scalability, *White Paper*, 2008.
[FOUNLAN00]. Foundry Networks, The Convergence of Layer 2 and Layer 3 in Today's LAN, *White Paper*, 2000.
[MENJUS2003]. Justin Menga, "Layer 3 Switching," *CCNP Practical Studies: Switching (CCNP Self-Study)*, Cisco Press, November 26, 2003.
[RFC1256]. S. Deering, Editor, "ICMP Router Discovery Messages", *IETF RFC 1256*, September 1991.
[RFC3176]. InMon Corporation's sFlow, "A Method for Monitoring Traffic in Switched and Routed Networks", *IETF RFC 3176*, September 2001.

8 Quality of Service in Switch/Routers

8.1 INTRODUCTION

This chapter explains the basic Quality of Service (QoS) control mechanisms available in a typical switch/router. Switches, switch/routers, routers, and networks, in general, are increasingly being called upon to transport unprecedented volumes of increasingly diverse traffic. Real-time application traffic (e.g., multimedia conferencing, broadband streaming video, online computer gaming, Virtual Reality (VR)/Augmented Reality (AR), factory automation, online electronic transactions, network storage, and cluster/Grid interconnect) is highly sensitive to latency and jitter (more appropriately called latency variation or packet delay variation (PDV)).

Enterprise data applications, such as Enterprise Resource Planning (ERP) and Customer Relationship Management (CRM), generate traffic that often requires high priority transfer to protect them from packet loss. Other applications, such as data backups and network stored video, generate traffic that is not very sensitive to latency or packet loss, but are bandwidth intensive.

To adequately address growing business demands, enterprises and service providers continually upgrade the bandwidth of their networks to accommodate the growing volumes of aggregated traffic. In doing so, they also need to address the ability of the network to support a diverse number of applications with different service requirements, many requiring predictable and/or guaranteed levels of service.

To simply QoS control, applications with similar or close service requirements are typically grouped into service classes. Within each class, the level of service needs is generally defined in terms of bandwidth, delay (latency), packet level "jitter" (more appropriately called latency variation or packet delay variation (PDV)), packet loss, and service availability. Thus, in order to deliver predictable service levels and meet required Service Level Agreements (SLAs), switch/routers normally employ well-defined set of QoS mechanisms.

To meet existing and emerging network and end-user service requirements, switch/routers support advanced Layer 2 features, Layer 3 routing capabilities, and QoS control mechanisms that include traffic prioritization and rate-limiting. The extensive feature set supports network end-user requirements ranging from basic connectivity to high-definition broadband streaming audio/video applications among others. The reality of today's networks is that different applications expect to be given the QoS appropriate to their differing needs and requirements.

DOI: 10.1201/9781003311249-8

8.2 NETWORK REQUIREMENTS FOR REAL-TIME TRAFFIC TRANSPORT

Enterprises and service providers that are moving their networks to handle real-time traffic like broadband streaming video also have to adapt their networks to handle a high mixture of latency-insensitive traffic (like TCP traffic) and latency-sensitive real-time traffic. Deployments of real-time applications generally involve implementing some changes and improvements in the existing network infrastructures that have been in many cases optimized for traditional data applications.

In general, the most significant aspects of the changes in the network include using non-blocking and high forwarding architectures, as well as providing adequate network bandwidth, end-to-end QoS, and telecommunication (telecom) grade network availability.

8.2.1 ADEQUATE BANDWIDTH

The network must have enough bandwidth provisioned in it to ensure that existing and anticipated real-time traffic meets their QoS requirements while not totally locking out non-real-time traffic. As discussed in this chapter and Chapter 9, various QoS control and traffic engineering mechanisms are available that allow the network engineer to allocate bandwidth to the different traffic flows; both real-time and non-real-time traffic. The network engineer normally ensures that real-time traffic is allocated a fair and manageable percentage of the overall link bandwidth. Typical, the network uses traffic prioritization and rate-limiting mechanisms to limit real-time traffic to no more than a specified percentage of the capacity of a link.

Most often, the bandwidth over the access layer of the network is the biggest concern and not as much as in the present-day network backbone or core. This is because as the price of Ethernet bandwidth continues to fall, desktop connections have transitioned from 100 Mb/s Ethernet to Gigabit Ethernet, while aggregation and core links of the network have transitioned from Gigabit Ethernet links to 10, 25, 40, 100, and even higher Gigabit Ethernet links. It is observed that with continuing improvements in Ethernet technology and cost, as well as the use of optical fiber and wavelength-division multiplexing (WDM) technology, there is generally ample bandwidth available in the backbone to support an increasingly rich array of end-user applications. In many cases, providing adequate bandwidth to the high volume of bandwidth hungry applications at the network access is of greater concern.

8.2.2 NETWORK ELEMENTS WITH NON-BLOCKING AND IMPROVED SWITCHING AND FORWARDING PERFORMANCE

The network elements (e.g., switches, switch/routers, routers, multiplexers) must be capable of providing adequate bandwidth for the application and network control traffic that pass through them. To do this, the switching and routing elements of the network as well as other network access elements such as firewalls and gateways must have the internal capacity to forward traffic at full line rates in spite of variations in traffic patterns. Real-time sessions should not be interrupted by a shortage of forwarding resources within the network infrastructure. The internal packet forwarding engines, switch fabric, and network interfaces must be provisioned with enough bandwidth to allow end-user applications derive the maximum benefit from the network.

The switch and router architectures used in today's networks must be capable of delivering consistent latency and PDV with minimum packet loss under all traffic conditions. Basically, a non-blocking architecture is one that has the forwarding capacity to support concurrently traffic from all ports when the ports are operating at full port bandwidth capacity. Non-blocking should hold for the expected minimum packet sizes. A truly non-blocking architecture does not impede packet forwarding in any way, enabling data to flow at wire-speed through the device. Non-blocking architectures provide the most consistent and predictable performance possible, irrespective of traffic patterns and packet sizes. Although many platforms claim to offer non-blocking switch backplanes, they do not offer a true non-blocking path through the device from port to port. Often, individual line cards or some other elements like the switch fabric will have some form of blocking. Occasionally, the line cards do not support enough processing resources and may not be capable of forwarding packets at wire speed. Architectures that use centralized forwarding engines may not have enough resources to handle traffic from all line cards at high traffic load conditions.

Furthermore, to provide predictable performance, line-rate packet forwarding performance should still be non-blocking and not diminish even when all additional Layer 2 and Layer 3 services are enabled, including the full range of traffic management/control and QoS services supported by the device. For example, running large Access Control Lists (ACLs) for Layer 2 or Layer 3 traffic should not degrade the packet forwarding rates, which is a requirement for real-time traffic (latency, PDV, and packet loss should not get worse).

8.2.3   END-TO-END QOS

While provisioning ample bandwidth in the network reduces the probability of congestion, this may not eliminate it entirely because of the bursty nature of traffic generated by both individual and aggregated data applications. Congestion in networks poses the most serious threat to real-time applications. When congestion occurs, the queuing delays in network device buffers can easily exceed the latency limits required by applications. Buffer overflows during periods of network congestion are also the leading cause of packet loss. By configuring appropriate end-to-end QoS mechanisms in the network, it is possible to control queuing delay and congestion-related packet loss.

The most important aspect of end-to-end QoS is to ensure that all the network elements on the path from source to destination provide a consistent level of service to end-user applications. Application traffic entering the network is appropriately classified and prioritized, allowing the network nodes to treat traffic according to their QoS requirements; real-time application traffic can be classified and allocated appropriate network resources. Real-time traffic is given relatively higher priority than non-real-time traffic which typically comes from TCP applications. In many cases, real-time traffic is placed in traffic classes that map to strict priority queues on network interfaces. Strict priority ensures that the streaming voice and video traffic, for example, will be forwarded before any other traffic that is waiting in a network queue.

8.2.4   HIGH TELECOM GRADE NETWORK AVAILABILITY

Users of the legacy voice-centric telecom networks are quite familiar with their very high ("five-9" or 0.99999) availability where a user picks up the phone and expects a dial tone to be immediately there. The current trend in today's telecom networks which are based on packet technologies (e.g., Ethernet, IP, MPLS) is to design them such that they support five-9 availability to enable real-time communications to be handled as seen in the legacy telecom networks. Networks with availability greater than 99.99% are necessary in order to meet user expectations that have been set by the legacy telephony system consisting of enterprise TDM PBXs (private branch exchanges) and the PSTN (public switched telephone network).

8.2.5   REMARK ON NETWORK QoS FOR REAL-TIME APPLICATIONS

From the above discussion, we see that deploying today's real-time applications requires fairly significant changes within the network infrastructure, plus the added challenge of managing end-to-end service levels for the higher number of disparate classes of traffic. Real-time application traffic requires a switched packet network infrastructure that features non-blocking nodes, network topologies with effective QoS mechanisms to ensure service quality, and high degree of reliability/resiliency to ensure service availability.

8.3   ELEMENTS OF QoS IN SWITCH/ROUTERS

QoS in a packet network is not a specific algorithm or technique, but rather the application of a number of distinct, well-defined traffic control operations, on a network-wide basis to achieve a desired end-user service result. QoS comprises several technologies which help control bandwidth, network delay, delay variation, and packet loss in a network, and are mostly triggered when the network becomes overloaded or congested. Specifically, QoS is a distinct set of operations, which occur, in series, and may consist of traffic classifying, queuing, packet forwarding, metering, shaping, and re-tagging so that a desired set of behaviors can occur on a network-wide basis.

As discussed in this chapter, QoS in network devices is based on multiple mechanisms and involves several operational steps. The implementation of QoS in contemporary Ethernet switches involves distinct stages. Switches that support QoS support the following stages described in no specific order:

- **Traffic classification from packet header information**: This stage includes explicitly identifying an arriving packet using, for example, using IEEE 802.1p priority tags, or IP packet header Differentiated Services Code Point (DSCP) field tagging information. This stage may also involve implicit classification by looking into the packet content at Layer 2 (Ethernet frame header), Layer 3 (IP header), Layer 4 (UDP/TCP port numbers), or higher protocol layer content. *Traffic classification* is sometimes referred to as *traffic coloring*.

- **Assigning the classified traffic to one of multiple queues to service the traffic classes**: This stage ensures that each classified packet is assigned to the proper queue based on its priority. The network device may offer N queues per interface, allowing N levels of traffic prioritization. In simple designs, N typically ranges from 4 to 8 queues, while in some high-end architectures N may be up to 128 queues or even more.
- **Use of appropriate queue management algorithms that allocate forwarding bandwidth among classes**: This stage includes the use of mechanisms such as traffic policing, traffic shaping, and specific traffic scheduling algorithms such as Strict Priority, Weighted Round Robin (WRR), Deficit Round Robin (DRR), Weighted Fair Queuing (WFQ), and so on.
- **Adequate internal switching capacity**: This is to ensure that the network node itself including the forwarding engines and switch fabric does not present a bottleneck to traffic flows.
- **Traffic management**: Traffic management may be invoked to select which traffic will be subjected to rate limiting and/or packet loss using algorithms such as traffic shaping or smoothing, Random Early Detection (RED), and Weighted Random Early Detection (WRED).
- **Packet remarking (rewriting)**: Packets may be remarked or reclassified with new priorities at a network node due to prevailing node/network conditions and sent to downstream nodes. The process of rewriting is the ability of the network node to modify the Class of Service (CoS) field in the Ethernet frame header or the DSCP field in the IPv4 header (see discussion below). This is particularly important when forwarding untagged traffic to a VLAN support VLAN tagging and IEEE 802.1p, or translating IEEE 802.1p priority marking to IP header DSCP field priority tagging in a network node.

Modern day switch/routers support extensive QoS and traffic management capabilities including standard-based ones such as the IEEE 802.1p and IP DSCP specifications. With service-aware QoS capabilities in the network nodes, enterprise and service providers are in a better position to honor end-user traffic.

8.4   TRAFFIC CLASSES

It is easier and more practical to classify IP-based traffic into distinct traffic classes and manage these classes rather than classify each individual application or user in the network. This concept of using traffic classes makes the problem of QoS much more manageable. Both the IEEE and IETF have defined mechanisms for supporting the differentiation of traffic based on Layer 2, Layer 3, and higher protocol layer data into eight distinct classes within a network domain.

Some of the important classification mechanisms are described below. However, for any QoS services to be applied to the traffic in today's networks that are mainly based on Ethernet and IP, there must be a way to tag or prioritize an IP packet or an Ethernet frame. The CoS fields discussed below are used to achieve this.

8.4.1 IEEE 802.1p/Q

IEEE 802.1p specifies a CoS mechanism for prioritizing VLAN tagged Ethernet frames. This is done using a 3-bit field called the Priority Code Point (PCP) (or user priority) within VLAN tagged Ethernet frames as defined by IEEE 802.1Q (see Figure 8.1). IEEE 802.1p (which is only a technique and not a standard or amendment published by the IEEE) specifies an Ethernet frame priority value from 0 to 7 is used to differentiate traffic.

The 3-bit PCP field in the IEEE 802.1Q header added to tagged Ethernet frames provides eight different classes of service. IEEE 802.1p was incorporated into the IEEE 802.1Q standard which specifies the VLAN tag may be inserted into Ethernet frames. As shown in Figure 8.1, the IEEE 802.1Q tag is placed between the 6-byte source MAC address field and the 2-byte Type/Length field.

The IEEE 802.1Q tag is recognizable by VLAN-tag aware Ethernet switches and does not require the switch to parse any field beyond the Ethernet frame header. Basically, as stated above, IEEE 802.1p defines 8 priority levels (0–7) and is only a mechanism for tagging packets with a priority value at Layer 2 and does not define how tagged packets should be treated in a network. The way traffic should be treated when assigned any particular IEEE 802.1p priority value is undefined and left to user/vendor implementation. However, some broad recommendations have been made by the IEEE on how users/vendors can implement these traffic classes as shown in Table 8.1.

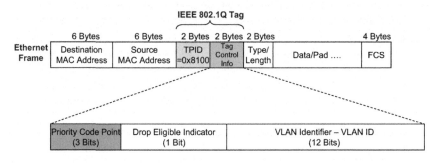

FIGURE 8.1   IEEE 802.1p: LAN Layer 2 QoS/CoS protocol for traffic prioritization.

TABLE 8.1
IEEE 802.1p Priority Values and Associated Traffic Types

Priority Code Point (PCP)	Priority	Binary	Traffic Type
1	0 (lowest priority)	001	Background
0	1 (default priority)	000	Best Effort
2	2	010	Excellent Effort
3	3	011	Critical Applications
4	4	100	Video
5	5	101	Voice
6	6	110	Internetwork Control
7	7 (highest priority)	111	Network Control

Ethernet switches may support the creation of VLANs based on port groupings on a single switch or based on IEEE 802.1Q tags for VLANs that may extend across multiple switches. With IEEE 802.1Q, a tag header is added to the Ethernet frame immediately after the destination and source MAC address fields.

Since the VLAN ID field in the IEEE 802.1Q tag is 12 bits long, up to 4,096 VLANs can be created as discussed in Chapter 6. While this number is likely to be adequate for most smaller networks, some high-end switches also support VLAN stacking (or double tagging defined in IEEE 802.1ad), where a second VLAN tag is added to the frame, expanding the number of possible VLANs to over 16 million.

8.4.2 IETF TYPE-OF-SERVICE

The 8-bit Type-of-Service (ToS) field was originally defined as part of the IP packet header [RFC791]. Figure 8.2 shows the location of the ToS field in the IP header. Similar to the IEEE 802.1Q tag in Ethernet frames, the IP header was defined to contain a field that specifies a priority value for an IP packet. The ToS is now an obsoleted IP header mechanism for providing packet prioritization and is replaced with a 6-bit DSCP field [RFC2474] and a 2-bit Explicit Congestion Notification (ECN) field [RFC3168]; the original 8-bit ToS field has been replaced by these two fields. ECN provides an optional end-to-end mechanism for signaling network congestion to network devices without dropping packets. The optional ECN feature (the two least significant bits) may be used between two ECN-enabled nodes when the underlying network infrastructure supports this feature.

Prior to being deprecated, the ToS field was defined to have a 3-bit Precedence subfield, 3-bit ToS subfield, and 2 bits unused. In practice, only the upper 3-bit IP Precedence subfield was ever used. Similar to IEEE 802.1p, the higher the IP Precedence field value, the higher the priority of the IP packet. The three most significant bits of the ToS field (i.e., the IP Precedence subfield) yield eight priority values (Figure 8.2). This provides for eight levels of IP Precedence or priority levels. The newly defined DSCP field provides a method for assigning priority values to IP packets. The 6-bit DSCP field (which coincides with the 6 most significant bits of the

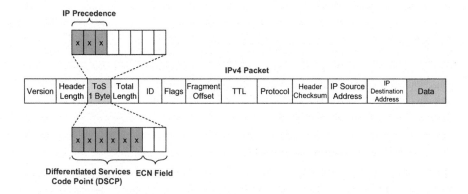

FIGURE 8.2   Reading IP Precedence and DSCP from the ToS byte.

TABLE 8.2
IP Precedence Values and Associated Traffic Types

Precedence Value	Binary	Precedence Name	Recommended Use
0 (lowest priority)	000	Routine or Best Effort	Default marking value
1	001	Priority	Data applications
2	010	Immediate	
3	011	Flash	Call signaling
4	100	Flash Override	Video conferencing and streaming video
5	101	Critic	Voice
6	110	Internetwork Control	Network control traffic (such
7 (highest priority)	111	Network Control	as routing, which is typically precedence 6)

ToS field) yield 64 different priority values (Figure 8.2). It should be noted that the upper 3 bits of DSCP field were defined to give values that maintain compatibility with 3-bit IP Precedence values.

IP Precedence was the only industry-accepted packet prioritization mechanism for traffic prioritization in IP routers. Its prioritization features are somewhat similar to IEEE 802.1p, but done at Layer 3 by routers and Layer 3 switches and generally not by Layer 2 switches. IP Precedence uses priority values 0 to 7, much like IEEE 802.1p, but within the IP packet header (as opposed to at IEEE 802.1Q tag which is implemented at Layer 2). In fact, IP Precedence is the only part of ToS that was ever really implemented in some routing platforms, particularly, Cisco routers. Table 8.2 lists the eight different IP Precedence values defined in [RFC791].

In the networking industry, the ToS is sometimes mistakenly referred to as the IP Precedence. In addition to the 3 bits used for the IP Precedence, there were additional definitions to the other bits in the ToS field. One of the other bits in the ToS byte, when set, would indicate that the packet should get low delay, high throughput, and others. These old semantics of ToS, other than the IP Precedence, are mostly ignored and were never deployed.

Also, the actual deployment of IP Precedence in the Internet and private networks was not done as was originally intended in the IETF standards but frequently left to the needs of the specific network service provider or router vendor. Generally, a Precedence value of 0 means a packet should receive only best-effort service. The other values, however, meant different things to different providers or vendors; use mostly depended on the service provider or router vendor and what methods were used for QoS control in a network.

8.4.3   IETF DIFFERENTIATED SERVICES

Differentiated services (DiffServ) is a more recent QoS architecture defined by the IETF [RFC2474] [RFC2475] [RFC3260]. DiffServ specifies a simple and scalable coarse-grained, class-based mechanism for traffic forwarding in IP networks.

It defines a mechanism for classifying and marking packets as belonging to a specific CoS. It includes the concept of marking IP packets with priority values that allow routers to classify the marked packets into traffic classes, each associated with certain forwarding behaviors in the network. Using DiffServ-based classification, each packet is placed into a limited number of traffic classes (or queues), compared to the fine-grained, flow-based architecture (Integrated Services (IntServ) architecture [RFC1633]) where network traffic is differentiated based on the requirements of an individual flow. DiffServ allows each router on the network to differentiate traffic based on the limited range of packet markings and traffic classes. Each traffic class can be managed differently, allowing routers to provide preferential treatment for higher-priority traffic on the network.

A *DiffServ domain* consists of a group of routers that implement a common, administratively defined set of policies based on DiffServ traffic markings. Traffic entering a DiffServ domain is subjected to a well-defined set of classification and conditioning policies. Traffic may be classified based on different parameters, such as source IP address, destination IP address, Transport Layer protocol type (TCP or UDP), source port number, and destination port number. A traffic classifier in a domain may honor the DiffServ markings in arriving packets or may elect to override or ignore those markings. Additionally, traffic in each class may be conditioned by subjecting it to traffic policing and shaping. Typically, all traffic classification and policing as well as other traffic conditioning are done at the boundaries between DiffServ domains.

As discussed above, packet marking is done using bits in the IP header field formally designated the ToS field, and now redefined as the DSCP field (Figure 8.2). The term DSCP is commonly used to describe the value with which an IP packet is marked. The DSCP contains 6 bits and is written into the left-most (most significant) 6 bits of the previous ToS byte of the IP header (see Figure 8.2) [RFC2474] [RFC2475]. The ECN occupies the least-significant 2 bits.

The 3 most significant bits of the DSCP field have been defined to give values that are backward compatible with the old IP Precedence, while the remaining bits are not backward compatible with the semantics of the other bits in the old IP header ToS byte. For backward compatibility, the 8 values obtained from the upper 3 bits of the DSCP field correspond to the 8 old IP Precedence values; they were defined to mean the same as the old IP Precedence values. Since the IP Precedence is now represented by the 3 most significant bits (rather than the least significant bits) of the 6-bits of the DSCP field, the 8 values are not identified as 0 to 7. Instead, the DSCP value corresponding to IP Precedence value n is redefined as $n \times 8$, for $n = 0$ to 7. IP Precedence values 0, 1, and 7, correspond to DSCP values 0, 8, and 56, respectively. DSCP value 40 (decimal) corresponds to IP Precedence value 5. Although the DSCP field supports backward compatibility with the IP Precedence marking, it provides a richer set of traffic classifications and behaviors.

The 8 special values of the DSCP field that are used for backward compatibility with the IP Precedence are called the *Class Selector* (CS) codepoints. IP routers that are DiffServ-aware implement Per-Hop Behaviors (PHBs) [RFC3140], which define how traffic should be treated based on its class, that is, the packet-forwarding properties associated with traffic classes. Different PHBs may be defined in a network to provide, for example, low-loss or low-latency traffic delivery.

DiffServ allows a network to support (in theory), up to 64 (= 2^6) different traffic classes using the full range of markings in the DSCP field. Most networks use the following commonly-defined PHBs:

- **Default PHB**: This is typically used to forward best-effort traffic. Any traffic that does not meet the requirements of any of the other DiffServ defined classes is placed in the Default PHB [RFC2474]. The recommended DSCP for the Default PHB is 000000 (in binary).
- **Expedited Forwarding (EF) PHB**: This is dedicated to low-loss, low-latency, and low-jitter (i.e., low PDV) traffic such as real-time streaming voice and video [RFC3246]. Traffic that conforms to the Expedited Forwarding PHB is admitted into a DiffServ network using a Call Admission Control (CAC) procedure. Expedited Forwarding traffic may also be subjected to traffic policing and other mechanisms to ensure that delay and PDV requirements are not violated. Expedited Forwarding traffic is typically transferred using strict priority queuing with respect to all other traffic classes. The recommended DSCP for Expedited Forwarding PHB is 101110 (i.e., 46 in decimal or 2E in hexadecimal).
- **Voice Admit PHB**: This PHB is defined in [RFC5865] and has identical characteristics as the Expedited Forwarding PHB. Both PHBs allow traffic to be admitted by a network using a (CAC) procedure. However, traffic conforming to Voice Admit PHB is admitted by the CAC procedure that also involves authentication, authorization, and capacity admission control, or traffic is subjected to very coarse capacity admission control (refer to [RFC5865] for description of authentication, authorization, and capacity admission control). The recommended DSCP value for Voice Admit PHB is 101100 (44 in decimal or 2C in hexadecimal).
- **Assured Forwarding (AF) PHB**: This PHB gives assurance of traffic delivery under certain prescribed conditions [RFC2597] [RFC3260]. AF PHB provides assurance of traffic delivery as long as the traffic does not exceed some pre-defined rate. Traffic that exceeds the pre-agreed rate has a higher probability of being dropped if network congestion occurs. Four separate AF classes are defined for the AF PHB group with Class 4 having the highest priority. Packets within each AF class are given a drop precedence (i.e., high drop precedence, medium drop precedence, or low drop precedence). A higher drop precedence means relatively more packet dropping. Also, three sub-classes exist within each class x (i.e., AFx1, AFx2 and AFx3), with each sub-class defining a relative drop precedence which determines which packets should be dropped first if a class queue is full. For example, within class 2, traffic assigned to the AF23 sub-class (i.e., high drop precedence in class 2) will be discarded before traffic in the AF22 sub-class (i.e., medium drop precedence in class 2), which in turn will be discarded before traffic in the AF21 sub-class (i.e., low drop precedence in class 2). The combination of classes and drop precedence yields 12 separate DSCP encodings from AF11 through AF43 (see Table 8.3). During periods of congestion affecting AF classes, the traffic in the higher AF class

TABLE 8.3
Assured Forwarding (AF) Behavior Group

	Class 1	Class 2	Class 3	Class 4
Low Drop	AF11 (DSCP 10)	AF21 (DSCP 18)	AF31 (DSCP 26)	AF41 (DSCP 34)
Medium Drop	AF12 (DSCP 12)	AF22 (DSCP 20)	AF32 (DSCP 28)	AF42 (DSCP 36)
High Drop	AF13 (DSCP 14)	AF23 (DSCP 22)	AF33 (DSCP 30)	AF43 (DSCP 38)

is given priority. AF PHB generally does not use strict priority queuing, instead more balanced queue scheduling algorithms such as weighted fair queuing (WFQ) are used. When congestion occurs within an AF class, the packets with the higher AF drop precedence (AF sub-classes with higher drop precedence) are dropped first. To prevent problems associated with queues using tail drop, AF PH often uses more sophisticated packet drop algorithms such as random early detection (RED) as discussed below. AF PHB in general uses mechanisms that ensure some measure of priority and proportional fairness between traffic in different AF traffic classes.

- **Class Selector PHBs**: This maintains backward compatibility with the IP Precedence field in the old IP header ToS field allowing interoperability with network devices that still use the IP Precedence field. The Class Selector codepoints are of the form "xxx000" (binary), where the first three bits (xxx) are the IP Precedence bits [RFC4594]. Each IP Precedence value can be mapped into a corresponding DiffServ class. For example, a packet with a DSCP value of 111000 or 56 (equivalent to IP Precedence of 7), is provided preferential treatment over a packet with a DSCP value of 101000 or 40 (equivalent to IP Precedence of 5). The different Class Selector PHBs and their corresponding binary and decimal values are given in Table 8.4. If a DiffServ-aware router receives the packet from a non-DiffServ-aware router that uses IP Precedence markings, the DiffServ router can still interpret the encoding as a Class Selector codepoint.

Simply, DiffServ provides a framework that allows classification and differentiated treatment in networks. The standard traffic classes described in the IETF standards only serve to simplify and streamline interoperability between different vendors equipment and different networks. It should be noted, however, that the details of how individual routers deal with the DSCP field and packet markings are operator-specific. Unless, any two network operators coordinate and interoperate their DSCP markings, it is difficult to predict end-to-end CoS behavior. A network operator may choose to implement classification and differentiated traffic forwarding within the local network in whichever way is suitable as long as there are no issues about interworking and interoperability with other network providers.

DiffServ defines the concept of a Bandwidth Broker which is an agent that is supplied with some knowledge of the priorities and policies used by a network operator and allocates bandwidth with respect to those policies [RFC2638]. To facilitate

TABLE 8.4
Class Selector Values

DSCP	Binary	Decimal (n×8)	Corresponding IP Precedence (n)
CS0 (Default)	000 000	0	0
CS1	001 000	8	1
CS2	010 000	16	2
CS3	011 000	24	3
CS4	100 000	32	4
CS5	101 000	40	5
CS6	110 000	48	6
CS7	111 000	56	7

TABLE 8.5
Recommended DiffServ Markings in RFC 4594

Service Class	DCSP	DSCP Value	Examples
Network Control	CS6	48	Routing Updates
IP Telephony	EF	46	VoIP bearer, VCoIP Audio
Signaling	CS5	40	VoIP, VCoIP Signaling
Video Conferencing	AF41, AF42, AF43	34, 36, 38	VCoIP Payload
Real-Time Interactive	CS4	32	Gaming
Streaming Video	AF31, AF32, AF33	26, 28, 30	Internet Video
Broadcast Video	CS3	24	IP TV
Low Latency Data	AF21, AF22, AF23	18, 20, 22	ERP, CRM
OAM	CS2	6	OAM
High Throughput Data	AF11, AF12, AF13	10, 12, 14	Bulk Data, FTP
Standard	DF (CS0)	0	Other Data Apps
Low Priority Data	CS1	8	Applications With No Assured Bandwidth

end-to-end allocation of resources across separate DiffServ domains, the Bandwidth Broker managing a particular DiffServ domain will have to communicate with adjacent Bandwidth Broker peers to allow end-to-end services to be constructed based on bilateral DiffServ interworking agreements.

Table 8.5 provides a summary of the DiffServ classifications for various types of traffic recommended in the IETF RFC 4594 [RFC4594]. The intent of the recommendation is to establish some level of consistency between service provider in regards to DiffServ traffic classifications and PHBs. DiffServ has become the most commonly used mechanism for implementing QoS enterprise and service provider networks.

In a DiffServ domain, the complicated functions such as packet classification, packet marking, policing are typically carried out at the edge of the network by edge DiffServ-aware routers. Generally, no classification and rate-limiting functions such as policing are required in the core router, keeping functionality on these devices simple. The core routers simply apply various DiffServ PHBs to packets based on their markings. The treatment of packets according to the configured PHB is achieved in the routers using a combination of queuing and scheduling policies. The core routers are relieved from the complexities associating with enforcing policies or agreements, and collecting data for billing purposes. Traffic crossing DiffServ domains may be subjected to the more complex functions of packet (re)classification and rate-limiting.

8.4.4 TRAFFIC FILTERING AND DEEP PACKET PATTERN MATCHING

IP filters using ACLs can be used to manage traffic and provide security, by allowing specific actions to be performed when defined criteria are matched. Only data that matches a predefined pattern is allowed to pass through the filter into a device or network. The filters can be used to set traffic priority, drop or allow packets, as well as define the conditions for mirroring traffic.

Traffic mirroring is the process of copying inbound and outbound traffic in a network device to be sent to a collector for being analyzed by network and security analysis tools (e.g., inspect content, traffic profiling, statistics collection, troubleshoot any network issues, monitor potential security threats). Deep packet pattern matching is an advanced implementation of filtering that allows filters to match fields deep within packets by specifying both an offset (from the beginning of a packet) and a value (pattern) to match.

8.4.5 REAL-WORLD IMPLEMENTATIONS OF CLASSES OF SERVICE

The simplest way of managing QoS across a network is to define a limited number of traffic classes (typically, not more than four) (see Figure 8.3). If the classification rules are such that all traffic can be classified into just four (instead of eight) classes, then the system will effectively behave as if it were a four-queue system instead of an eight-queue system.

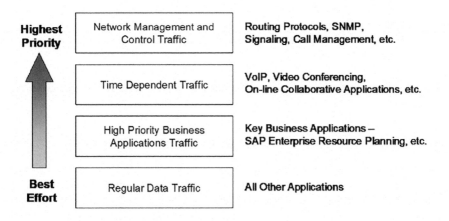

FIGURE 8.3   Implementation of classes of services.

The QoS-based forwarding system then classifies and marks traffic accordingly traffic into a small number of classes. The QoS identifier on each packet (e.g., IEEE 802.1p, IP DSCP) provides the network nodes with specific instructions on how traffic in different classes should be treated. On any node in the network, the configured packet scheduling and discard policies determine how packets in each class are treated.

Typically, when four classes of service are used, their recommended priority values are as follows:

1. **Highest Priority**: IEEE 802.1p =7, IP Precedence = 7, DSCP = 56. Examples: Network management traffic (e.g., SNMP), routing protocol traffic (e.g., RIP, EIGRP, OSPF), Spanning Tree Protocol (STP) traffic, etc.
2. **Time/Jitter Sensitive Traffic**: IEEE 802.1p = 5, IP Precedence = 5, DSCP = 40. Examples: Real-time voice, streaming video, video conferencing, etc.
3. **Expedited Mission Critical Applications (High Priority Data Traffic)**: IEEE 802.1p = 1, IP Precedence = 1, DSCP = 8. Examples: Enterprise Resource Planning (ERP) software traffic (e.g., SAP), Web traffic, etc.
4. **Best Effort**: IEEE 802.1p = 0, IP Precedence = 0, DSCP = 0. All other traffic.

Although it provides coarse-grained QoS, the main reason network operators use a limited number of traffic classes is that it is much easy to implement in practice by defining network-wide traffic classes. Once the traffic classes are agreed upon on an enterprise-wide basis, the network operator uses a policy management tool to set up rules that allow the network nodes to classify and assign traffic to their respective traffic classes. The policy management software translates these rules into specific traffic classification and forwarding instructions used by the network nodes. The rules may be based on the following parameters:

- Source/destination IP address – for user, workgroup, system applications
- TCP/UDP port number – for various end-user applications
- Application Layer headers – for applications, specific content, or context
- IEEE 802.1p priority – pre-classify traffic at end-system or classify traffic at upstream network node
- IP DSCP field – pre-classified traffic at end-system or classify traffic at upstream network

A network operator may use any or all of these classification tools on the devices in the network.

8.4.6 EXAMPLE APPLICATIONS OF PACKET CLASSIFICATION AT LAYER 2

The Layer 2 to 4 classification capability of the switch/router typically includes mechanisms for assigning CoS to traffic with no degradation in device performance. The classification capability allows a network administrator to classify received frames based on specific Layer 2 to 4 information. Classification of frames can be used for various purposes including the following (see Figures 8.4, 8.5, and 8.6):

Packet Classification at Layer 2

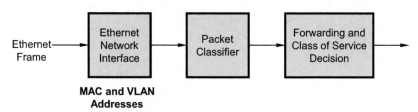

**MAC and VLAN
Addresses**

Layer 2 Classification and Forwarding Decisions

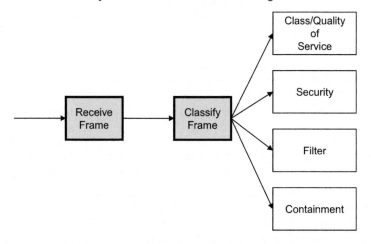

FIGURE 8.4   Layer 2 packet classification.

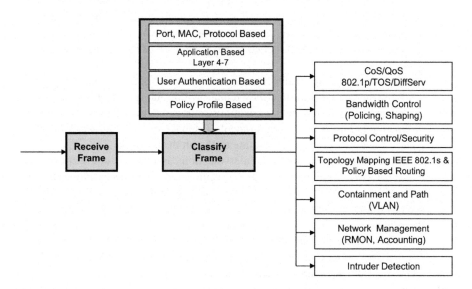

FIGURE 8.5   Layer 2 classification example.

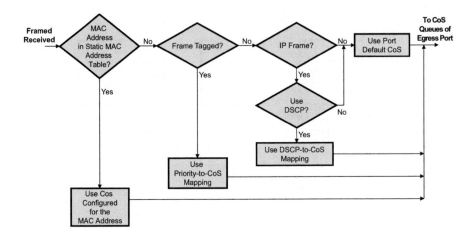

FIGURE 8.6   Determining the class of service of a received frame.

VLAN creation, traffic prioritization and CoS/QoS control, traffic containment, filtering, and security.

- **VLAN creation**: Classification of received frames based on their Layer 2 to 4 information can be used to create distinct VLANs in a network. Frame classification rules can be defined for assigning VLANs to received traffic. A Layer 2 forwarding device can classify frames into VLANs (and priority levels) based on the specific Layer 2 to 4 information in the frame.
- **Traffic prioritization and CoS/QoS control**: The results of traffic classification can be used to prioritize traffic throughout a network. The priority indicator in the IEEE 802.1Q tag of an Ethernet frame allows traffic to be prioritized throughout the network. Classification allows the device to associate a transmit priority to each received frame. Based on the priority-to-queue mapping, each classified frame is mapped to the desired transmit queue.
- **Traffic containment**: Traffic classification can be used for containing or scoping frames to ensure that they remain within the specific boundary of a network (a VLAN). A network administrator can define rules to logically group together users of a given protocol, IP subnet, or application as well as control the flow of the corresponding traffic on the network.
- **Filtering**: The outcome of traffic classification can be used to prevent specific protocols, applications, and/or specific users from accessing certain parts of a network. A network administrator can define rules to filter traffic such as traffic from specific IP addresses from entering designated areas of the network.
- **Security**: Traffic classification can be used for securing certain resources in a network, for example, allowing or preventing certain MAC and/or IP addresses from accessing certain resources.

8.5   TRAFFIC MANAGEMENT

Compared to the traditional data-only networks, networks that transport real-time traffic (e.g., voice, broadband video) generally require significantly more complex and advanced QoS features. Networks that support real-time traffic must be capable of delivering a steady stream of packets for each real-time traffic session without being interrupted by the typically bursty data traffic or other events occurring within the network infrastructure.

For real-time applications, latency and latency variations (or PDV) must be maintained within strict limits to protect the perceived quality of the real-time sessions. In contrast, data applications based on TCP are designed to be quite tolerant of latency, PDV, and packet loss. Data applications that can send very bursty traffic can also adversely affect the transmission and performance of real-time applications over a common link. This is because TCP applications are designed to be capable of expanding their window sizes to maximize bandwidth utilization as much as possible until the link capacity is reached and congestion is detected by the end application via packet losses.

In order to deliver the right level of service to real-time applications, appropriate QoS mechanisms must be introduced to properly manage the shared network bandwidth. The bandwidth of the network must be carefully managed on a short-term or instantaneous basis by incorporating appropriate QoS control mechanisms such as admission control, traffic queuing and scheduling algorithms, traffic policing, traffic shaping, and random packet discard. Bandwidth is managed on a longer term basis through network capacity planning and frequent monitoring of network and application performance.

Given the widely different characteristics of data, voice, and video traffic, the switch/router needs to support a comprehensive array of QoS features that ensure each class of traffic receives the level of service it requires without affecting other traffic types. In cases where the uplink ports are undersubscribed (e.g., for limited 100 Mb/s desktop connectivity and 10 and higher Gigabit Ethernet uplinks), the QoS features in the switch will rarely come into play. However, as more end-systems demand access to the network, the uplinks are more likely to be oversubscribed, the probability of congestion increases, and the QoS features start to play a more significant role.

8.5.1   SETTING THE SCENARIO FOR CONGESTION IN INTERNETWORKS

Currently, enterprise and service provider networks (and the Internet in general) are designed in such a way that virtually all traffic from the end-users and access nodes aggregates at the uplinks to the core networks. In this scenario, as long as the utilization of the uplink is low and the traffic in the core links is low, congestion is infrequent enough to allow delay-sensitive applications to work quite well, even across parts of the network where no QoS control mechanisms are deployed. In many cases, the uplink which links the end-users and access network to the core is where congestion normally occurs.

QoS-related problems that applications experience are primarily due to congestion at these aggregation points in the network. Congestion occurs when the capacity of these uplinks are momentarily oversubscribed even if the core has sufficient capacity for the converged traffic. Congestion is particularly exacerbated when the following conditions occur:

- Access bandwidths are higher than uplinks especially with the trend toward Gigabit Ethernet to the desktop and server connections moving to 10 and higher Gigabit Ethernet.
- Uplinks (as well as access links) are frequently oversubscribed due to economic necessity. It is not uncommon to see expensive leased lines used as uplinks.
- Buffers on links carrying TCP traffic that have capacities less that the bandwidth-delay product resulting in frequent packet losses.
- Aggregate traffic patterns from the access network are less predictable. Traffic patterns depend on a number of variables: number of users, types of applications, time of day, degree of contention for shared resources (e.g., storage or database resources).

The above conditions particularly set the stage for congestion in a network. Now with more multimedia and peer-to-peer applications being deployed in internetworks, and with these applications becoming more bandwidth hungry, it is expected that congestion will continue to be more prevalent in networks. The average access bandwidth continues to increase rapidly making congestion a serious issue that has to be addressed since it severely affects packet loss-sensitive applications and delay-sensitive applications such as real-time streaming voice and video and interactive applications.

8.5.2 OUTPUT QUEUE DROPS

We see from the discussion above that on a network that is well-designed (i.e., during the network planning phase), packet drops are mainly caused by short-term network node interface congestion. We have noted that a common cause of this, is traffic from a high bandwidth link being forwarded onto a lower bandwidth link, or traffic from multiple inbound links with higher aggregate bandwidth being forwarded onto a single outbound link. For example, if a large amount of bursty traffic comes in on a gigabit interface and is forwarded to a 100 Mb/s interface, this can cause significant packet drops on the outbound interface. This is because the outbound interface is overwhelmed by excess traffic due to the speed mismatch between the interfaces

One solution to this problem, which is not often practical to realize, is to increase the outgoing line speed. There are other simpler ways, however, to prevent or control packet drops without having to increase the line speed. It should be note that, one can prevent packet drops only if those drops are a consequence of short data bursts. If packet drops are consistent and are sustained over long periods, then these drops cannot be prevented, but can only be remedied through network redesign or some form of traffic access control (using, for example, rate-limiting mechanisms as discussed in Chapter 9).

The simplest form of dropping packets in an overflowing queue is *tail drop* which involves dropping packets at the tail end of the queue, that is, dropping packets that are just arriving – the most basic congestion avoidance mechanism. Packets just arriving at a congested queue are dropped. Tail drop treats all traffic equally and does not differentiate between CoS when queues begin to fill during periods of congestion. When the queue is full and tail drop is in effect, arriving packets are dropped until the congestion subsides and the queue is no longer full. Tail drop is the most basic type of congestion avoidance and does not take into account any traffic differentiation and QoS parameters. The discussion below presents other more effective ways for better managing or controlling packet drops.

8.5.3 BUFFER SIZING

This section describes the rationale for providing network nodes with port buffers for TCP traffic that are on the order of magnitude equal to the expected bandwidth-delay product (BDP) of the TCP connections. Obviously, the first step to address congestion in a network node is to provide adequate buffering in the system. Note that interfaces that support real-time traffic generally have smaller buffers to control the delay experience by the traffic. Studies of TCP traffic [AWECN2001] [LINSIG1997] [MORCNP1997] [VILLCCR1995] have showed that adequate buffering (proportional to the Round-Trip-Time (RTT)) is necessary to minimize TCP packet loss and maximize the utilization of the end-to-end network.

The rule of thumb for TCP traffic is that a port should have a buffer capacity B equal to the average *RTT* of the TCP sessions flowing through the link times the link bandwidth (*linkBW* in bits per second), the so-called *bandwidth-delay product* (*BDP*):

$$B = RTT \times linkBW = BDP$$

The BDP rule of thumb is based on the desire to keep a link on the end-to-end TCP path as busy as possible in order to maximize the throughput of the network. This typically applies to long-term application flows based on TCP, such as large file transfers with FTP. TCP's congestion control algorithm (as described below) is designed to continually probe the network to find the maximum possible transfer rate. This is done by deliberately attempting to fill the buffer of any port in order to ensure that the link is fully utilized. Buffer overflow causes packet loss that in turn throttles back the TCP sender's data transmission rate.

In general, increasing the buffer size reduces TCP packet losses and increases link utilization but also causes longer queues at the bottleneck link and higher end-to-end delays. In most networks, the congested link will be carrying a combination of different types of flows, including short-term TCP flows and UDP flows in addition to long-term TCP flows (which tend to dominate bandwidth consumption). While large buffers do not adversely affect short-term TCP flows, the resulting queuing delays can have a negative effect on delay-sensitive UDP applications.

We see from the above discussion that while large buffers primarily benefit long-term TCP application flows, they negatively affect other traffic types. This means appropriate QoS measures have to be taken to allocate the required levels of buffering and bandwidth to different classes of traffic and to protect delay-sensitive applications (typically UDP applications) from excessive queuing delays normally associated with large buffers. Where serving delay-sensitive applications is a major concern, appropriate QoS mechanisms must be deployed to minimize queuing delay, especially, for applications such as VoIP. This has to be done while still providing the large buffers needed to optimized bulk TCP data transfers. For jitter-sensitive traffic such as voice and video, the traditional approach is to use small buffers and tail drop.

8.5.4 PACKET DISCARD

In virtually all types of network devices, decisions have to be made about what actions should be taken when processing and memory resources are oversubscribed or fully exhausted. In packet devices such as Ethernet switches, router or switch/router, when buffers are exhausted due to output congestion, packets are dropped. Where these packets are dropped, the circumstances under which they are dropped, in addition to when the packet discard process begins, all can have an impact on network-wide behaviors. We discuss next, the TCP congestion control mechanisms and how they impact network congestion.

8.5.4.1 TCP Slow Start and TCP Window Sizes

Data from applications that require delivery reliability are sent over TCP as discussed in Chapter 3. In TCP, the amount of data outstanding within the network (in transit) before the source stops to wait for an acknowledgment from the receiver is the *TCP window size*. In general, the use of larger window sizes is both more CPU efficient as well as more link utilization efficient, particularly over long (i.e., high latency) links. Without using the TCP window scale option [RFC7323], TCP supports windows sizes up to 65,535 bytes which is the largest window allowed by the 16-bit TCP window field.

A given TCP window size consists of multiple TCP segments. When the transmission of the sender (e.g., a 10 Gb/s server interface) is significantly higher than that of the receiver (e.g., a 1 Gb/s desktop interface) or there is an intervening slow speed network link (e.g., 100 Mb/s), then the potential for losing packets increases. Further, the larger the TCP window size is in this scenario, the more data has to be retransmitted when packet losses occur and the more memory is consumed in the receiver to buffer the window prior to sending an acknowledgment.

In order to mitigate the packet loss behavior of TCP, all TCP implementations use the TCP *slow start mechanism* (or its variant). Studies have found that if TCP starts transmitting its window of data relatively slowly and gradually increases its transmission rate to the full rate of the network path to the receiver, it can achieve a balanced transmission rate that does not overmatch the available network path resources. A TCP sender accomplishes this by doubling its transmission rate with each acknowledgment from the receiver and halving the transmission rate whenever there is data retransmission due to packet loss or receiver window exhaustion. Reference [RFC5681] describes the various TCP congestion control algorithms (slow start, congestion avoidance, fast retransmit, and fast recovery) used in practical TCP implementations.

The typical behavior of a single long-term TCP session with a bottlenecked link can be described briefly as follows (see [RFC5681] for details). At the start of the TCP session, the sender's window size is increased exponentially (*slow start phase*) until a buffer somewhere in the network fills up and multiple packets are dropped. The lost packets provide the feedback that causes the TCP sender to cut the previous window size to zero and recommence a slow start until the current window has attained half of the previous window size. TCP then increments the window size from this point linearly (*congestion avoidance phase*) until some network buffer fills up again and a single packet is dropped. At this point, the sender reduces its window size by one-half and reenters the congestion avoidance phase, resulting in the saw-tooth pattern being repeated throughout the TCP session.

Because of the TCP slow-start mechanism, the use of intelligent packet discard algorithms at network nodes supporting the TCP session can flow control the TCP source. Further, the fact that the majority of Internet traffic such as HTTP, FTP, Telnet, and other higher-layer application protocols use TCP instead of UDP (see Chapter 3) means that intelligent packet discard algorithms such as Random Early Discard (RED) [FLOYACM1993] [RFC7567], Weighted Random Early Discard (WRED), and their many variants [AWECN2001] can be used to cause TCP senders to slow down their transmission rates when short-term congestion occurs in the network. However, it should be noted that selective packet discard only works on TCP traffic, not on UDP. This means, protocols that use UDP, such as Real-Time Protocol (RTP), used by Voice over IP (VoIP), NFS, other applications, are not helped by such packet discard mechanisms.

UDP is an unreliable Transport Layer protocol (as discussed in Chapter 3) and does not support any inbuilt congestion control or data acknowledgment mechanisms. Some protocols such as RTP and NFS have either acknowledgment mechanisms or implicit sequence numbers embedded at the Application Layer instead of the Transport Layer and implement either retransmission or packet loss notification at the Application Layer.

8.5.4.2   Random Early Detection

In the best-effort transmission model (where no QoS mechanisms are enabled in the system), packets are buffered in a single FIFO queue and serviced in the order in which they are received. In the event of congestion, all subsequent arriving packets are dropped causing buffer overflows – a process known as *tail drop*. Buffer overflows associated with tail drop disrupt UDP sessions and can simultaneously drive numerous TCP sessions all the way back to *slow start*, causing a condition called *global synchronization*. This happens when TCP sessions synchronize their window growth rates at approximately the same time due to packet losses in the network and feedback being delivered to all the TCP senders at approximately the same time.

While packet losses due to buffer overflow provide the feedback mechanism that throttles back the rate at which TCP sends traffic, it is preferable to have a less drastic means of providing the needed feedback to the TCP sender. Reducing the number of occurrences of tail drop and global synchronization of TCP sources is the rationale behind congestion avoidance mechanisms like RED and WRED. Buffers sizing based on the BDP together with the use of RED-based congestion avoidance has been shown to be an effective way of optimizing the utilization of congested links supporting TCP applications.

RED [FLOYACM1993] is a congestion avoidance scheme that tries to prevent the oscillations between link under-utilization and over-utilization that occur during TCP global synchronization caused by tail drop. RED tries to control the queue length by randomly dropping TCP packets as buffers fill beyond a configured threshold level. As the buffer fills and packets are randomly dropped, the affected TCP sources will throttle their rates and comparatively a smaller number of the affected TCP sources will decrease their transmission rates at the same time (avoiding global synchronization). If the buffer continues to fill, a higher and higher percentage of packets is randomly dropped. Generally, RED greatly reduces the number of buffer overflows, tries to distribute packet loss fairly among the TCP sessions, and minimizes queue length while still absorbing traffic spikes.

The use of techniques such as RED is commonly referred to as *active queue management* (AQM) [RFC7567]. AQM is the process of intentionally dropping packets randomly at a network node to signal TCP sources to slow down their data transmission rates. TCP is designed with congestion control mechanisms that allow a TCP source to adapt its data sending rate when packets are dropped or lost along the connection to the TCP destination [RFC5681]. Thus, network nodes (i.e., intermediate switches or routers) can exploit this TCP congestion control feature to intentionally drop packets "intelligently" when they experience local congestion or overutilization of local resources (e.g., link utilization, buffer oversubscription, processor overload, etc.).

Because of the randomness of the packet discard process, multiple randomly chosen TCP senders will slow down their transmission rates without causing global synchronization. The randomness of packet drops in RED causes the TCP senders to gracefully adjust their rates to the available congested network resources rather than overly penalizing any single TCP sender as would be the case when a tail drop policy is used which potentially causes application timeouts and application aborts.

Figure 8.7 shows the drop profile of the classic RED [FLOYACM1993]. RED begins dropping packets when the average queue length reaches a minimum

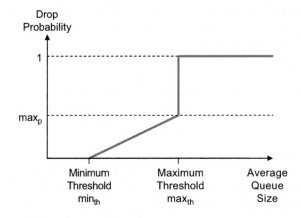

FIGURE 8.7   Classical RED drop profile.

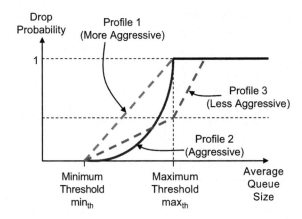

FIGURE 8.8   Other RED drop profiles.

threshold (min_{th}). Beyond this threshold, the probability of a packet being randomly dropped increases linearly up to the maximum threshold (max_{th}), after which all arriving packets are dropped, as in tail drop. The packet drop probabilities are calculated based on the minimum threshold, maximum threshold, and specified mark probability (max_p).

With a given mark probability, the fraction of packets dropped is max_p when the average queue length is at the maximum threshold max_{th} [FLOYACM1993]. With each packet arrival at the queue, the router determines the average queue size based on the previous average and the current size of the queue using an exponentially weighted moving average formula (see details of the RED algorithm in [FLOYACM1993]).

As shown in Figure 8.8, other variants of RED modify the drop profile in order to approach the tail drop point more gracefully. For example, the drop profile could be implemented through a piecewise linear algorithm that approximates an ideal quadratic behavior and eliminates the abrupt step function at the maximum threshold as in RED.

8.5.5   HANDLING CONGESTION WITH MULTIPLE QoS QUEUES

The best effort transmission model assigns one FIFO queue for all traffic – both delay-sensitive and delay-tolerant traffic are lumped in one queue. However, large FIFO queues when full during times of congestion result in long queuing delays. For example, a half full 125 MB buffer on a 10 Gigabit Ethernet port would introduce a 50 ms queuing delay for any new incoming packet. This simple analysis shows that network nodes with large buffers need to be able to identify and protect any critical delay-sensitive traffic (or critical loss-sensitive traffic) by employing buffering that supports multiple QoS queues in conjunction with appropriate congestion avoidance mechanisms.

The multiple QoS queuing approach can be accomplished by configuring the port buffers into multiple individual queues. Then traffic that is sensitive to delay and delay variance (such as VoIP and video packets) are assigned to queues that receive, for example, strict priority scheduling over the remaining queues that carry

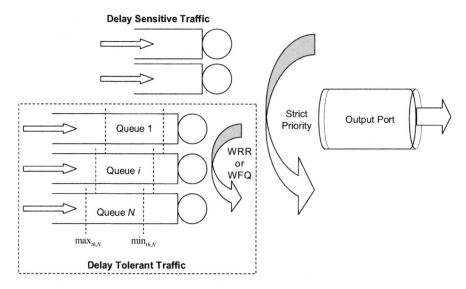

FIGURE 8.9   Example of handling congestion with multiple QoS queues.

delay-tolerant traffic (Figure 8.9). With priority queuing, the bandwidth allocated to low-priority queues is immediately moved to any high-priority traffic that enters the system. Therefore, with priority queuing, the only traffic that can delay critical time-sensitive traffic is other time-sensitive traffic already in the same queue. The probability of this form of congestion occurring can be minimized by proper regulation of the higher priority traffic using rate-limiting mechanisms as discussed in Chapter 9.

The delay tolerant traffic can be classified and assigned to one queue in a group of queues that are managed by a scheduling mechanism, such as Weighted Round Robin (WRR) or Weighted Fair Queuing (WFQ). Each WRR or WFQ queue can be configured with one or more packet drop thresholds. By employing multiple priority queues within a buffer and assigning drop thresholds to each queue, the network node is able to make more intelligent decisions when congestion occurs.

RED (or any of its variants) can be used in conjunction with the multiple priority traffic classes (queues) where each individual queue runs a separate RED instance. Each class (queue) can be independently configured with drop thresholds, particularly relevant for traffic in lower priority classes. In addition, different drop profiles can be applied to traffic within a queue (as discussed in Chapter 9) based on whether the traffic is within committed service parameters (green), within peak service parameters (yellow), or out of profile (red). One packet drop approach is Weighted RED (WRED) where the packet drop probability is weighted to favor high-priority traffic within a single class (queue) that conforms to the service parameters specified; high priority traffic within the single queue experiences relatively lower packet drops during congestion.

Packet loss can be essentially eliminated if buffers are never allowed to fill to capacity. Thus, overflows can be avoided by applying WRED to the lower priority traffic as the buffer fills beyond specified levels. This eliminates the possibility of high-priority packets arriving at a buffer that is already overflowing with lower priority packets.

In the multiple priority queue approach, long-term TCP flows (e.g., large file transfers), for example, can be assigned to a separate traffic class with a large buffer allocation and a large share of the link bandwidth. The RED minimum threshold (min_{th}) for this traffic class could be set equal to the *RTT* multiplied by the minimum bandwidth allocated to long-term TCP traffic class:

$$min_{th} = RTT \times BWtcp\_long\_term$$

The maximum threshold (max_{th}) could be set close to *RTT* × *linkBW*, which would be the full buffer capacity of the port. Note that delay-sensitive traffic has to be isolated and protected from the above assignment. Real-time delay-sensitive traffic like VoIP traffic would be assigned to a strict priority queue given only a small share of the buffer capacity to limit data transfer delay.

8.6 PACKET MARKING AND REMARKING (REWRITING)

To serve the growing numbers of diverse end-users, applications, and their needs, networks have become more complex and often include, at a minimum, Layer 2 and Layer 3 network devices. With this continued growth, network operators must efficiently identify the packets passing through the network and apply appropriate service behaviors before sending them to their destinations.

As discussed above, QoS packet marking is a tool used to differentiate packets by assigning them designated markings in the packet headers allowing the network to allocate resources accordingly. Using marking, the network can assign network resource to the multiple priority levels or classes of service according to their needs and existing network operator policies. Packet marking also simplifies the network design and operation from a QoS perspective. Using packet marking and appropriate QoS configuration tools and mechanisms in the network nodes reduces significantly the overhead of packet classification, priority queuing, and traffic scheduling.

QoS packet marking can be configured on a network node interface, subinterface, or any individual virtual circuit or connection (VC). We discussed above that traffic marking involves setting certain bits inside Ethernet frames and IP packet. Traffic marking can also be done in MPLS packet header fields that are specifically designated for QoS marking. Network devices examine the marked bits, classify, and queue packets based on their marked values. Packet marking can also be used or performed solely internally (within) a network device when resources are allocated and used. In this case, the internal mechanisms used for marking packets and the markings themselves affect only the device's behavior; the internal marking is limited to the local device and not passed on to other network devices.

The following are benefits of QoS packet marking:

- **Network Partitioning and Categorizing**: Packet marking allows the user to partition the network resources into multiple priority levels or classes of service.

- **Layer 2 to Layer 3 or Layer 3 to Layer 2 Mapping**: Packets may need to be marked to differentiate user-defined QoS services in a Layer 2 network domain and mapped to corresponding markings when undergoing Layer 3 forwarding. Also, packets may need to be remapped when leaving a Layer 3 forwarding domain (a router) and entering a Layer 2 domain (an Ethernet switch). The router can set the CoS value of the outgoing Ethernet packet so that the Ethernet switch can appropriately process the CoS markings.
- **WRED Configuration**: Routing devices supporting WRED can be configured to use IP Precedence or IP DSCP values to determine the drop probability of packets. This allows WRED to be used with IP Precedence or IP DSCP markings in a DiffServ network.

Network devices can implement queuing techniques that are based on DiffServ PHBs derived from IP Precedence or DSCP values in the IP packet headers. Based on the IP Precedence or DSCP markings, traffic can be placed into appropriate service classes and allocated network resources accordingly.

8.6.1 IP PRECEDENCE MARKING

A network device may mark the priority of a packet by using the IP Precedence marking mechanism. IP Precedence marking provides the following benefits:

- **Traffic Queuing and Scheduling**: IP Precedence field can be used to determine how to queue and schedule traffic.
- **Avoid Congestion**: IP Precedence field can be used to determine how to handle packets when packet-dropping mechanisms, such as WRED, are configured.
- **Traffic Policing**: Networking devices can use IP Precedence values to determine how to rate-limit inbound traffic.

The Layer 2 media often changes as packets traverse network nodes from source to destination. Thus, a more ubiquitous marking mechanism is one that can take place at Layer 3, that is, using the IP Precedence, for example. A network operator may configure a QoS policy to use IP Precedence marking for packets entering the network. Devices within the network can then use the newly marked IP Precedence values to determine how to treat the packets. For example, class-based WRED in a network node may use IP Precedence values to determine the probability of dropping an arriving packet when resources are oversubscribed. Real-time traffic like streaming voice can also be marked with a particular high-priority IP Precedence value. The network operator may then configure low-latency high-priority queues for all these high-priority traffic.

When IP Precedence-based WRED is configured on an interface and the outgoing packets are MPLS packets, the network device can drop the MPLS packets based on the 3-bit Traffic Class field (previously called the Experimental (EXP) bits) in the MPLS label [RFC5462], instead of using the 3-bit IP Precedence field in the underlying IP packets.

After the IP Precedence bits are set, other QoS features such as WFQ and WRED can then operate on the bit settings. The network can give priority (or some type of expedited handling) to marked traffic through the application of various mechanisms such as WFQ and/or WRED.

8.6.2 DIFFERENTIATED SERVICES CODE POINT MARKING

As discussed above, the DiffServ model allows a network operator to classify packets based on IP DSCP markings. IP DSCP markings allow the network to allocate network resources according to traffic priority levels or class of service. By marking traffic, network devices along the forwarding path can effectively determine the proper CoS policies to apply to each traffic class.

An important aspect of DiffServ is that the packet markings must be consistently interpreted from end to end. All devices in the network path must understand the PHB to apply to a specific class of traffic. If one of the network devices in the path does not interpret and process appropriately the packet markings, the overall service provided to the corresponding traffic will not be satisfactory or may not even be understood.

Because the IP Precedence and DSCP fields are part of the IP header and, therefore, carried end to end, only one of these fields can be marked at any given time. DSCP and IP Precedence values are mutually exclusive. A packet can have only one of these values, but not both.

When a DSCP-based WRED is configured on an interface and the outgoing packets are MPLS packets, the network device can discard the MPLS packets based on the 3-bit Traffic Class (or EXP) field in the MPLS label, instead of using the 6-bit DSCP field in the underlying IP packets. To remark packets from MPLS EXP markings to DSCP markings, a device can shift the 3 EXP bits to the left, making it 6 bits in length. For example, if the value of the EXP bits is 5 (binary 101), the device left-shifts the bits to make them binary 101000, thus making it look like a 6-bit DSCP field. The network device may then drop packets based on the shifted binary DSCP value.

8.6.3 CLASS OF SERVICE MARKING USING IEEE 802.1P

The 3-bit PCP field in IEEE 802.1Q tag of an Ethernet frame (see Figure 8.1) enables a network device to:

- Classify inbound Ethernet packets based on the value in the PCP field
- Optionally, reset (remark) the value in the PCP field of outbound packets

The PCP marking enables Ethernet devices to interoperate to deliver end-to-end QoS. The PCP values can also be mapped to corresponding IP Precedence, DSCP, or MPLS EXP values. The 3-bit PCP field on Ethernet frames provides up to eight classes of service (0 through 7) similar to the IP Precedence, DSCP, and MPLS EXP. Network devices can use the PCP value to determine how to prioritize packets and also to perform Layer 2 to Layer 3 CoS mapping.

To allow IP routing devices to interoperate with Ethernet switches, the PCP values can be mapped to IP DSCP values for packets received on inbound Ethernet interfaces. The DSCP values can also be mapped to PCP values for packets forwarded on outbound Ethernet interfaces. In the inbound direction, the network operator can configure the router to match on the PCP bits and then perform an action, such as setting the IP Precedence or DSCP bits on IP packets. In the outbound direction, the router can be configured to set the PCP bits of outbound packets to user-specified values.

8.6.4   MPLS Traffic Class (or EXP) Marking

MPLS QoS allows a network operator to provide varying levels of QoS services for different types of traffic in an MPLS network. The QoS of a packet can be "tunneled" across an MPLS network using appropriate packet markings. The MPLS network can classify packets according to their traffic type, input interface, and other criteria without changing the IP Precedence or DSCP field values of the packets.

The MPLS EXP field is a 3-bit field within the MPLS label that is used in QoS marking. The MPLS EXP field also allows up to eight different QoS markings that may be mapped to the eight possible IP Precedence or IEEE 802.1Q PCP values. The EXP field is used to support differentiated services and can carry all of the information encoded in the IP Precedence, DSCP Class Selector or PCP fields. In some cases, the EXP bits may be used exclusively to encode the packet drop precedence within a traffic class.

In typical implementations, the IP Precedence field in the underlying IP packet is copied to the MPLS EXP field during label imposition. Using the MPLS EXP field does not modify the DSCP or IP Precedence markings in the ("tunneled") packet's IP header. In an MPLS network, the edge router copies the IP Precedence bits into the MPLS EXP field.

However, based on the service offering, the network operator might need to set the MPLS EXP field to a value that is different from the IP Precedence value. In this case, MPLS QoS allows the IP Precedence or DSCP setting of a packet to remain unmodified as the packet passes through the MPLS network. During congestion, packets receive the appropriate priority, based on their MPLS EXP setting.

The value of the EXP bits determines the PHB for MPLS nodes and is also used as a transparency mechanism when used with MPLS DiffServ tunneling modes such as *pipe* and *uniform* modes. IP marking does not modify an MPLS packet carrying IP data. MPLS marking takes effect only during MPLS label imposition. Marking and policing can be combined to change the DSCP and MPLS EXP values of an IP packet during MPLS label imposition.

A provider edge (PE) router at the edge of the MPLS network can be configured to map the DSCP or IP Precedence field to the MPLS EXP field. The router then uses the value of the EXP field as the basis for IP QoS. As a result, MPLS routers can perform QoS features indirectly, based on the original IP Precedence field inside the MPLS-encapsulated IP packet. The IP packet does not need to be opened to examine the IP Precedence field. When a packet leaves the MPLS network, IP QoS is still based on the DSCP or IP Precedence value in the IP header.

8.6.4.1 MPLS QoS Services

The MPLS EXP field allows the network operator to specify the QoS for an MPLS packet while the IP Precedence and DSCP fields allow the operator to specify the QoS for an IP packet. By setting the MPLS EXP field, the router does not need to modify the IP Precedence or DSCP field of IP packets as they traverse the MPLS network.

MPLS QoS can support the following QoS services:

- **Traffic Policing**: Networking devices within the MPLS network can use MPLS EXP values to determine how to rate-limit inbound traffic.
- **WRED**: The MPLS EXP field can be used to determine how to drop packets when mechanisms, such as WRED, are configured. Here, WRED uses EXP values to determine the drop probability of packets.
- **Traffic Queuing and Scheduling**: After the MPLS EXP bits are set, other QoS features such as WFQ can then operate on the bit settings. The network can give priority (or some type of expedited handling) to marked traffic using WFQ and/or WRED.

As the packet travels through the MPLS network, the marked value of an IP packet does not change and the IP header remains available for use. In some instances, it is desirable to extend the MPLS PHB to the egress interface between the PE router and customer edge (CE) router. This has the effect of extending the MPLS QoS tunnel, allowing the MPLS network to extend classification, scheduling, and packet discard behavior on that final interface.

The network operator can also mark the MPLS EXP bits independently of the PHB. Instead of overwriting the value in the IP Precedence or DSCP field, the operator can set the MPLS EXP field, choosing from a variety of criteria (including those based on IP PHB) to classify a packet and set the MPLS EXP field. For example, the operator can classify packets with or without considering the rate of packet arrival at the ingress PE. If the rate is a consideration, "in-rate" packets can be marked differently from "out-of-rate" packets.

8.6.4.2 MPLS Tunneling Modes

MPLS QoS provides QoS on MPLS packets using the following *tunnel modes*:

- **Uniform Mode**: In this mode, the network has only one layer of QoS, which stretches and reaches end to end. This mode provides uniformity in PHB throughout the MPLS network. In this mode, all customers of the MPLS network use the same IP Precedence (or DSCP) bit settings. Any changes made to the EXP value of the topmost label on a label stack are propagated both upward, as new labels are added, and downward, as labels are removed. The ingress PE router copies the IP Precedence or DSCP bits from the incoming IP packet into the MPLS EXP bits of the imposed labels. As the EXP bits travel through the core, they may or may not be modified by intermediate Provider network routers. At the egress PE router, the EXP bits are then copied to the IP Precedence or DSCP bits of the newly exposed IP packet.

- **Short Pipe Mode**: This mode provides a distinct MPLS PHB layer across the entire MPLS network (which operates on top of a customer network's IP PHB layer). With the short pipe mode, the network customers implement their own IP PHB marking scheme and attach to an MPLS network with a different PHB layer. In this mode, the IP Precedence or DSCP bits in an IP packet from the customer are propagated upward into the label stack as labels are added. When labels are swapped, the existing EXP value is kept. If the topmost EXP value is changed, this change is propagated downward only within the label stack, not to the IP packet originating from the customer. When labels are removed, the EXP value is discarded. The egress PE router classifies the newly exposed IP packets for outbound queuing based on the IP PHB associated with the DSCP value of the original IP packet.
- **Full Pipe Mode**: This mode is similar to short pipe mode, except that at the egress of the PE router, the MPLS PHB layer is used to classify the packet for discard and scheduling operations at the outbound interface. The PHB on the MPLS-to-IP link is selected, based on the removed EXP value rather than the recently exposed IP Precedence or DSCP bit settings in the original IP packet. When a packet reaches the edge of the MPLS core, the egress PE router classifies the newly exposed IP packets for outbound queuing based on the MPLS PHB from the EXP bits of the recently removed label. In this mode, the network schedules and discards packets without needing to know the customer PHB settings.

8.6.5 Tunnel Header Marking

As virtual private networks (VPNs) grow to include data, voice, and video traffic, the different types of traffic need to be handled differently in the network. QoS and bandwidth management features allow a VPN to deliver high transmission quality for time-sensitive applications such as voice and video. Each packet is tagged to identify the priority and time sensitivity of its payload, and traffic is sorted and routed based on its delivery priority. *Tunnel header marking* (THM) allows the network operator to mark the outer IP header's DSCP or IP Precedence value during tunnel encapsulation of the packet.

The outer IP header (DSCP) field of a tunneled packet is typically exposed to a different QoS domain from that of the inner IP header. For example, for Multicast Virtual Private Network (MVPN) packets placed in Generic Routing Encapsulation (GRE) tunnels, the router processes the packet's outer DSCP field based on the QoS services of a common core MPLS network. GRE is a tunneling protocol that can encapsulate a wide variety of protocol packet types inside IP tunnels, creating a virtual point-to-point link to routers at remote points over an IP internetwork.

The router processes the packet's inner IP DSCP field based on the QoS services of a particular VRF (virtual router forwarding) at the ingress of the tunnel. Using tunnel header marking, different traffic streams that are aggregated into the same tunnel can mark their outer DSCP field differently. This enables each stream to receive a different level of QoS processing at the outer DSCP field's QoS domain.

Let us take a GRE tunnel marking example. Tunnel marking for GRE tunnels provides a way to define and control QoS for incoming customer traffic on the PE router in a service provider network. This allows the PE to set (mark) either the IP Precedence value or DSCP value in the header of a GRE tunneled packet. This simplifies administrative overhead required to control customer bandwidth by allowing the provider network to mark the GRE tunnel header on the incoming interface on the PE routers.

Let us assume traffic is being received from a CE1 router through the incoming interface on the PE1 router on which tunnel marking occurs. The traffic is encapsulated (tunneled), and the tunnel header is marked on the PE1 router. The marked packets travel (tunnel) through the provider network core and are decapsulated on the exit interface of the PE2 router. THM is designed to simplify the classification of CE traffic and is configured only in the service provider network. This process is transparent to the customer sites connected to the CE routers.

8.7   THE NEED FOR QoS MECHANISMS IN THE SWITCH/ROUTER

During periods of congestion in network devices, if appropriate congestion management features are not in place, packets will be dropped. When packets are dropped, retransmissions occur for protocols such as TCP. When retransmissions occur, the network load can increase. In networks that are already congested, this can add to existing performance issues and potentially further degrade performance.

With today's networks now serving various types of traffic, congestion management is even more critical. Latency-sensitive traffic such as voice and video can be severely impacted if they experience excessive delays. Simply adding more buffers to the network device also does not necessarily alleviate congestion problems. Latency-sensitive traffic needs to be forwarded as fast as possible. The network device needs to identify this important traffic through classification techniques, and then implement appropriate buffer management and scheduling techniques to prevent higher priority traffic from being dropped or excessively delayed during congestion. The device needs to incorporate scheduling techniques to forward high-priority packets from queues as quickly as possible.

8.7.1   IMPORTANT QoS FEATURES

As the demand for bandwidth in networks continues to increase as discussed above, network devices like switches, routers, and switch/routers must also support capabilities that allow them to use the available bandwidth effectively. Along with advanced QoS features that enable the network devices to maximize throughput, congestion management is particularly important.

Some of the features that can help network devices guarantee a minimum QoS include packet classification, priority queuing, ingress traffic metering and policing, internal packet drops (e.g., via RED), and egress traffic shaping and scheduling. These features are essential for ensuring that a network device can properly prioritize and forward traffic as it traverses the device. Figure 8.10 shows an example of Layer 2 forwarding architecture with CoS/QoS and security control lists. Typically, the Layer

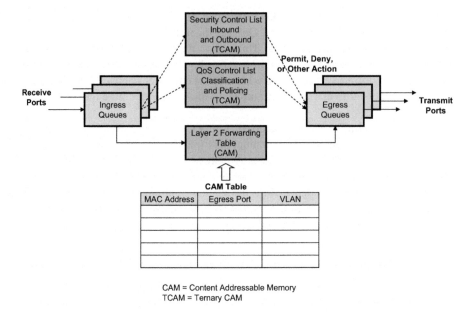

FIGURE 8.10   Layer 2 forwarding with CoS/QoS and security control lists.

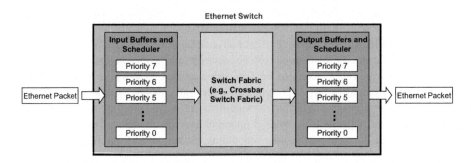

FIGURE 8.11   Priority queueing and scheduling in the generic Ethernet switch.

2 forwarding table is implemented in a Content Addressable Memory (CAM) while the CoS/QoS and security control lists are implemented in Ternary-CAMs (TCAMs).

Figure 8.11 shows the input and output priority queuing mechanisms (using eight priority queues) as packets pass through an Ethernet switch. Figure 8.12 shows the typical QoS features in the line cards of an Ethernet switch. When an Ethernet packet arrives at a switch interface, it is immediately classified according to classification rules as discussed earlier. After classification, the packet is placed in one of the eight priority queues based on the priority determined during classification.

Once the packet is moved from its priority queue and across the switch fabric to an appropriate output priority queue at the destination port, it is scheduled onto the external network using the egress scheduler (which operates while taking into

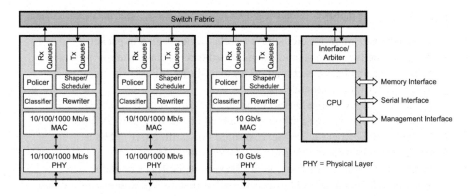

FIGURE 8.12 Classifier, policer, scheduler, and rewriter in the generic Ethernet switch.

consideration the priority of queued packets). The scheduler ensures that high-priority packets traverse the switch with minimum delay and low-priority packets are not totally locked out from proceeding to the external network.

The typical CoS/QoS-capable Ethernet switch with a crossbar switch fabric supports input buffering per port which can be divided up evenly or unevenly depending on the user requirements. The input buffers can be divided on a per-output port basis (with each output port having its own queue, called a virtual output queue (VOQ)), and further on a per-priority basis (e.g., 2, 4, or 8 priority queues depending on user requirements). Input port VOQs are used to prevent head-of-line (HOL) blocking as discussed in Chapter 3 of Volume 2 of this two-part book.

Watermarks can be set on the input queues for low- and high-priority traffic. This allows low-priority traffic to be dropped during extreme congestion while enabling high-priority to pass through. For example, in the input buffer, Priority 0 (lowest priority) may be assigned the lowest watermark and Priority 7 (highest priority) the highest watermark. This allows Priority 0 traffic to be dropped first, followed by Priority 1, and so on until Priority 7 is reached, as congestion in the switch increases. The advantage of this scheme is that the lowest priority traffic is dropped first before it has the chance of causing severe congestion and resulting in higher priority traffic being dropped.

The switch may support the ability to meter and police traffic as it moves into the input buffers (see Figure 8.12 and Chapter 9 of this volume). The metering and policing of traffic can be done on a per-priority basis which allows higher priority, delay-sensitive traffic to move into the input buffers before low-priority, delay-insensitive traffic. Low-priority traffic (Priority 0 to 6) may be subjected to metering and policing while the highest priority traffic (Priority 7) may not be metered/policed at all and allowed into the input buffers unimpeded.

The egress scheduler and the ingress metering/policing functions are both required to maintain a minimum QoS for the user. The servicing of queued packets by the egress scheduler depends on each packet's priority and the scheduling algorithm employed (e.g., Strict Priority, WRR, DRR, WFQ).

REVIEW QUESTIONS

1. What is the standard-based method for classifying Ethernet packets?
2. What are the two main standard-based methods for classifying IP packets?
3. What is the standard-based method for classifying MPLS packets?
4. What is a DiffServ domain?
5. Explain briefly what the DiffServ Default Per-Hop Behavior (PHB) is.
6. Explain briefly what the DiffServ Expedited Forwarding (EF) PHB is.
7. Explain briefly what the DiffServ Assured Forwarding (AF) PHB is.
8. Explain briefly what the DiffServ Class Selector PHB is.
9. What is the purpose of a Bandwidth Broker in a DiffServ domain?
10. What is tail drop in a FIFO queue and what potential problems can it cause for TCP traffic?
11 What is active queue management (AQM) and what are some of its benefits?
12 Describe the five main applications of packet classification at Layer 2.

REFERENCES

[AWECN2001]. J. Aweya, M. Ouellette, and D. Y. Montuno, "A Control Theoretic Approach to Active Queue Management," *Computer Networks Computer*, Vol. 36, Issue 2–3, July 2001, pp. 203–235.

[FLOYACM1993]. Sally Floyd and Van Jacobson, "Random Early Detection (RED) Gateways for Congestion Avoidance", *IEEE/ACM Transactions on Networking*, August 1993, Vol. 1, No. 4, pp. 397–413.

[LINSIG1997]. D. Lin and R. Morris, "Dynamics of Random Early Detection," *Proc. SIGCOMM'97*, Cannes, France, Sept. 1997, pp. 127–137.

[MORCNP1997]. R. Morris, "TCP Behavior with Many Flows," *IEEE Int'l Conf. Network Protocols*, Atlanta, Georgia, Oct. 1997.

[RFC791]. IETF RFC 791, Internet Protocol, September 1981.

[RFC1633]. R. Braden, D. Clark, and S. Shenker, "Integrated Services in the Internet Architecture: an Overview", *IETF RFC 1633*, June 1994.

[RFC2474]. K. Nichols, S. Blake, F. Baker, and D. Black, "Definition of the Differentiated Services Field (DS Field) in the IPv4 and IPv6 Headers", *IETF RFC 2474*, December 1998.

[RFC2475]. S. Blake, D. Black, M. Carlson, E. Davies, Z. Wang, and W. Weiss, "An Architecture for Differentiated Services", *IETF RFC 2475*, December 1998.

[RFC2638]. K. Nichols, V. Jacobson, and L. Zhang, "A Two-bit Differentiated Services Architecture for the Internet", *IETF RFC 2638*, July 1999.

[RFC2597]. J. Heinanen, F. Baker, W. Weiss, and J. Wroclawski, "Assured Forwarding PHB Group", *IETF RFC 2597*, June 1999.

[RFC3140]. D. Black, S. Brim, B. Carpenter, and F. Le Faucheur, "Per Hop Behavior Identification Codes", *IETF RFC 3140*, June 2001.

[RFC3168]. K. Ramakrishnan, S. Floyd, and D. Black, "The Addition of Explicit Congestion Notification (ECN) to IP", *IETF RFC 3168*, September 2001.

[RFC3246]. B. Davie, A. Charny, J.C.R. Bennett, K. Benson, J.Y. Le Boudec, W. Courtney, S. Davari, V. Firoiu, and D. Stiliadis, "An Expedited Forwarding PHB (Per-Hop Behavior)", *IETF RFC 3246*, March 2002.

[RFC3260]. D. Grossman, "New Terminology and Clarifications for Diffserv", *IETF RFC 3260*, April 2002.

[RFC4594]. J. Babiarz and F. Baker, "Configuration Guidelines for DiffServ Service Classes", *IETF RFC 4594*, August 2006.

[RFC5462]. L. Andersson and R. Asati, "Multiprotocol Label Switching (MPLS) Label Stack Entry: "EXP" Field Renamed to "Traffic Class" Field", *IETF RFC 5462*, February 2009.

[RFC5681]. M. Allman, V. Paxson, and E. Blanton, "TCP Congestion Control", *IETF RFC 5681*, September 2009.

[RFC5865]. F. Baker, J. Polk, and M. Dolly, "A Differentiated Services Code Point (DSCP) for Capacity-Admitted Traffic", *IETF RFC 5865*, May 2010.

[RFC7323]. D. Borman, B. Braden, V. Jacobson, and R. Sheffenegger, "TCP Extensions for High Performance", *IETF RFC 7323*, September 2014.

[RFC7567]. F. Baker, and G. Fairhurst, Ed., "IETF Recommendations Regarding Active Queue Management", *IETF RFC 7567*, July 2015.

[VILLCCR1995]. C. Villamizar and C. Song, "High Performance TCP in ANSNET," *ACM Computer Commun. Rev.*, Vol. 24, No. 5, October 1995.

9 Rate Management Mechanisms in Switch/Routers

9.1 INTRODUCTION

With the continued convergence of networks and services to IP- and Ethernet-based technologies, quality-of-service (QoS) has also become increasingly more important. Distinct QoS mechanisms can be applied in different combinations to solve a number of traffic management problems in networks. The QoS features on switches, routers, and switch/routers provide network operators with the means to implement service convergence and at the same time prioritize mission-critical application traffic. More importantly, the availability of a wide range of QoS features on network devices means that changes in QoS management in a network can easily be accommodated, allowing the devices to be used over longer periods of time and to meet new service demands in the network.

Rate management mechanisms are critical to successful network operation because they allow network operators to determine which traffic enters their network, the volume and the rate at which traffic is admitted to the network, and the per-hop packet drop behavior of the network devices as the level of network congestion changes with traffic load. To support a diverse set of users with different QoS requirements across the network, it is critical that the network operator regulate traffic flow to protect the shared resources in the network, and ensure that each user does not consume more than its fair share of bandwidth.

To do this, network operators need tools that allow their networks to determine whether each user is honoring their traffic transfer commitments and what actions should be taken if a user attempts to transmit more traffic than is allowed (i.e., out-of-profile traffic) into the network. This chapter discusses the main approaches to rate management in switch/routers. The discussion is equally applicable to switches, routers, and other network devices.

9.2 TRAFFIC POLICING

Traffic policing and traffic shaping are the two fundamental approaches to rate management in networks. Traffic policing is the process of monitoring arriving packet flows at a network node for compliance with a defined traffic contract while at the same time taking measures to enforce that contract. During traffic policing, packets of a particular traffic flow that exceed the defined traffic contract (i.e., the excess traffic) may be marked as non-compliant, discarded immediately, or admitted into the network as-is. The particular action taken depends on the network administrative policy in place and the characteristics of the excess traffic.

DOI: 10.1201/9781003311249-9

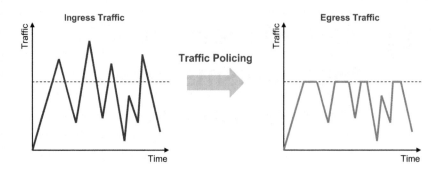

FIGURE 9.1   Traffic policing for bandwidth limiting.

A traffic policer defines a set of traffic rate limits for flows and sets appropriate penalties for traffic that exceed (i.e., does not conform to) the configured limits. When the traffic rate of a flow reaches its configured maximum rate, the excess traffic is dropped, marked, or remarked if it was previously marked by a downstream node. Since traffic on an interface may arrive with a variable rate and in bursts, the result of traffic policing is an output rate that appears as a saw-tooth with crests and troughs as shown in Figure 9.1. The policer may either discard or mark packets in a traffic flow that does not conform to the traffic limits with a different forwarding class or packet loss priority (PLP) level.

Traffic policing may be implemented using, for example, the "*leaky bucket algorithm as a meter*" [ATMFRUNI95] [ITUTTRAFF04] [TURNERJO86] or the "*token bucket algorithm*". The "leaky bucket as a meter" is exactly equivalent to (a mirror image of) the "token bucket algorithm". The "*leaky bucket as a queue*" (discussed below) [TANENBAS03] can be viewed as a special case of the "leaky bucket as a meter" algorithm [MCDSPOH95]. The "leaky bucket as a queue" algorithm [TANENBAS03] can only be used in shaping traffic to a specified rate with no jitter (i.e., delay variation) in the output traffic. The token bucket algorithm is discussed in the sections below.

9.2.1   METERING AND MARKING WITH A POLICER

The two main functions of a traffic policer (as described in [RFC2475]) are *metering* and *marking*. The metering function of a policer measures each arriving packet against the traffic rates and burst sizes configured on the policer. The metering function then passes the packet and the metering result to the marking function, which assigns a packet loss priority that corresponds to the metering result. This process is illustrated in Figure 9.2.

The metering function (the *Meter*) operates in two modes – *color-blind* and *color-aware*. In the *color-blind mode*, the meter treats the packet stream as *uncolored* (i.e., without special interpretable markings understood by network devices). In this mode, any preset loss priorities associated with the packet stream are ignored by the policer. In the *color-aware mode*, the meter inspects the PLP field of packets that may have

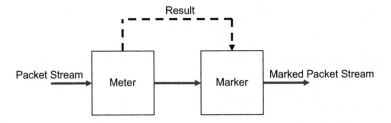

FIGURE 9.2 Tricolor marking policer operation.

been set by an upstream device as high PLP, medium-high PLP, medium-low PLP, or low PLP. In this mode, the marking functions (the *Marker*) changes the preset PLP of each incoming IP packet according to the results provided by the meter.

9.3 TRAFFIC SHAPING

Traffic shaping is the process of smoothing out packet flows by regulating the rate and volume of traffic admitted into the network. Typically, traffic shaping is used to adjust the flow rate of packets when certain criteria are met/matched. The criteria can be, all packets arriving at the shaper, or certain packets identified based on some defined bits in the packet headers (e.g., IP Precedence, IP Differentiated Services Code Point (DSCP)).

Traffic shaping is accomplished by holding arriving packets in a FIFO (First In, First Out) buffer and releasing them at a pre-configured rate. This mechanism can be used to delay some or all arriving packets and release them into the network at a specified rate, bringing them into compliance with the desired traffic profile (Figure 9.3). Packets are stored in a separate FIFO buffer (for each traffic flow or class) and then transmitted into the network such that the traffic rate will be in compliance with the prevailing traffic contract (for that class).

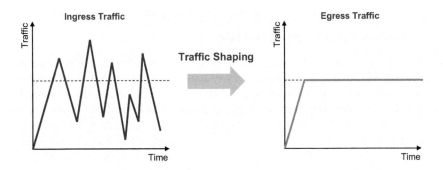

FIGURE 9.3 Traffic shaping for bandwidth limiting.

The following two main algorithms are used to delay the arriving traffic such that each packet complies with the relevant traffic contract:

- **Leaky bucket algorithm as a queue [TANENBAS03]**: This is a traffic smoothing tool that eliminates bursts and presents a steady stream of traffic to the network. This mechanism turns a bursty stream of packets into a regular stream of equally spaced packets (Figure 9.3).
- **Token bucket algorithm**: This is a long-term average traffic shaping tool (described below) that permits traffic bursts of a pre-determined size that satisfies a long-term average transmission rate. This mechanism presents a regulated stream of traffic that can be bursty into the network.

In contrast to policing, traffic shaping stores the excess packets of a flow in a queue and then schedules these packets into the network over increments of time. Traffic shaping when applied to arriving packets, generally, results in a smoothed packet output rate as shown in Figure 9.3.

9.3.1 LEAKY BUCKET ALGORITHM AS A QUEUE

With the leaky bucket algorithm as a queue [TANENBAS03], an unregulated stream of arriving packets is placed into a queue and then transmitted into the network by a fixed-rate traffic scheduler. If the traffic flow presents more packets than the queue can store, the extra packets are discarded. When packets reach the head of the queue, they are transmitted into the network at a constant rate determined by the rate configured on the queue scheduler. This algorithm can only be used for traffic shaping (not traffic policing).

The leaky bucket algorithm as a queue can be used to regulate the flow of packets so that they are not transmitted into the network at a rate greater than the network can or is willing to absorb. The length (or depth) of the packet queue bounds the amount of delay that a packet can incur at this type of traffic shaper. A packet may experience additional delay at downstream nodes if those nodes and their interfaces are not adequately provisioned to support the offered traffic load.

9.4 TRAFFIC POLICING AND TRAFFIC SHAPING USING THE TOKEN BUCKET ALGORITHM

We have discussed above the functional differences between traffic policing and shaping, both of which limit the output rate of a system (e.g., queue, port, link). Though the two mechanisms have important functional differences, both can use the token bucket algorithm (for traffic metering and output rate control functions). We discuss in this section how the token bucket algorithm works for both traffic policing and shaping.

Traffic policers typically use the token bucket algorithm for enforcing the average transmit or receive rate of traffic at a network interface while allowing bursts of traffic up to a maximum value (Figure 9.4). The token bucket algorithm offers more flexibility than the leaky-bucket algorithm as a queue because the policer can allow a specified amount of traffic bursts before it starts to discard packets or apply a penalty to packets (based on the queuing priority or packet drop priority).

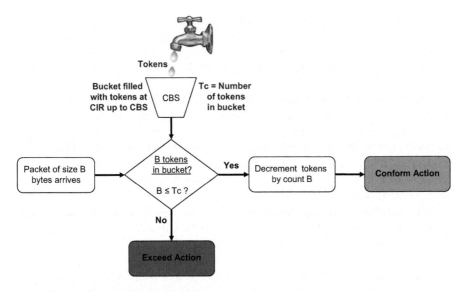

FIGURE 9.4 Token bucket algorithm.

Tokens are added to the bucket at a fixed defined rate (in bytes or bits per second), but only up to the specified depth of the bucket (Figure 9.4). The *Committed Information Rate* (CIR) or the mean rate specifies the average rate at which data can be sent or forwarded into the network. The *Committed Burst Size* (CBS) specifies in bits (or bytes) the maximum amount of data that can be sent within a given unit of time. Tokens are inserted into the bucket at the CIR while the depth of the bucket determines the CBS. The maximum number of tokens that a bucket can contain is determined by the configured CBS.

When the token bucket fills to its capacity, newly generated tokens are discarded and cannot be used by arriving packets. This means, the largest burst a source can send into the network at any given time is roughly proportional to the size of the bucket (i.e., the CBS). This setup allows a token bucket to permit traffic bursts, and at the same time bound them (i.e., limit the length of the bursts). Bounding traffic bursts also guarantees that the long-term transmission rate will not exceed the established rate (CIR) at which tokens are placed in the bucket.

Each token serves as a permission for a source to send a certain number of bytes or bits into the network. For a packet to enter the network, the number of tokens available in the bucket must be at least equal to the packet size; tokens equal to the packet size must be removed from the bucket. If the bucket does not contain enough tokens to send a packet, the packet either waits (i.e., is queued) until the bucket has enough tokens (in the case of a shaper), or it is discarded or marked down (in the case of a policer). Traffic arriving at the bucket when sufficient tokens are available is said to be *conformant*, resulting in the corresponding number of tokens removed from the bucket. If the bucket does not have sufficient number of tokens available, then the traffic is referred to as *excess traffic*.

The token bucket mechanism used for traffic shaping has both a token bucket and a data queue; if a data queue is not present, then the mechanism is referred to as a traffic policer. For traffic shaping, packets that arrive and cannot be transmitted immediately are delayed in the queue. For traffic shaping, the token bucket algorithm is less likely to drop excess packets since excess packets are buffered. Packets are buffered packets up to the length of the queue. Packet drops may occur if excess traffic is sustained for a period of time.

When used for traffic policing, the token bucket algorithm propagates traffic bursts and does not perform traffic smoothing. This mechanism controls the output rate through packet drops but avoids delays because of the absence of queuing. Packets might be dropped or remarked with a lower forwarding class, a higher PLP level, or both. In the event packets encounter congestion at downstream nodes, those with low PLP level are less likely to be discarded than those with a medium-low, medium-high, or high PLP levels.

When sufficient tokens are present in the bucket, traffic flows unrestricted through the interface (as long as its rate does not exceed the CIR). A token bucket itself has no discard or priority policy for the tokens placed in it.

- The rate at which tokens are added to the token bucket, the CIR, represents the highest average data (transmit or receive) rate allowed. The user may specify the CIR as the bandwidth limit of the policer. If the traffic rate at the policer is so high that at some point insufficient tokens are present in the token bucket, then the traffic flow is no longer conforming to the traffic limit.
- The depth of the token bucket in bytes (i.e., the CBS) controls the amount of back-to-back data bursts allowed. The token bucket determines whether an arriving packet is exceeding or conforming to the applied rate. The user can specify CBS as the burst-size limit of the policer. The CBS affects the average transmit or receive rate by limiting the number of bytes permitted in a transmission burst over a given interval of time. Traffic bursts exceeding the CBS are dropped until there are sufficient tokens available to permit the burst to be transmitted.

Though both policing and shaping mechanisms can use a token bucket as a traffic meter to measure the packet rate, they have important functional differences. They differ, however (as described above), in the way they respond to violations:

- Traffic shaping implies the existence of a buffer of sufficient memory size to queue delayed packets, while policing does not. A shaper delays excess traffic by holding packets and shaping their output rate when the arrival rate is higher than expected.
- A traffic policer drops (or remarks) excess packets above the committed rate. A policer typically drops packets, but it may also change the priority setting or marking of excess packets. For example, a policer may drop a packet or rewrite its IP Precedence or DSCP settings if configured to do so.

Traffic policers and shapers are typical deployed today's networks to ensure that traffic flows adhere to their stipulated contracts.

9.5 TRAFFIC COLORING AND MARKING

Packets belonging to traffic flow that is being policed and exceeds the defined limits might be implicitly set to a higher PLP level, assigned to a configured forwarding class, or set to a configured PLP level, or simply discarded. The policer may implement a number of specified actions (which may include setting the IP Precedence or DSCP bits) based on the particular set of traffic limits configured, and the color the arriving traffic belongs to (which can be one of either two or three categories or colors as discussed below).

9.5.1 TWO-COLOR-MARKING POLICER

A *two-color-marking policer* (e.g., using the token bucket algorithm depicted in Figure 9.4) categorizes traffic as either conforming to the traffic limits (*green*) or violating the traffic limits (*red*) [JUNPOLICE]:

- **Green**: A two-color-marking policer implicitly sets the packets in a green flow to the low PLP level.
- **Red**: A two-color-marking policer can be configured to simply discard packets if the traffic flow is red. Alternatively, the user can configure a two-color-marking policer to processing packets in a red flow by setting the PLP level to high, assigning the packets to any forwarding class already configured, or both.

9.5.2 THREE-COLOR-MARKING POLICER

A *three-color-marking policer* categorizes traffic as conforming to the traffic limits (*green*), exceeding the traffic limits but within an allowed range (*yellow*), or violating the traffic limits (*red*) [JUNPOLICE]:

- **Green**: Similar to the two-color-marking policer, the three-color-marking policer implicitly sets the packets in a green flow to the low PLP level.
- **Yellow**: The three-color-marking policer includes a second type of nonconforming traffic – *yellow*:
    - The *single-rate three-color policing* (discussed below) categorizes traffic as yellow when the traffic exceeds the traffic limits but conforms to a second defined burst-size limit.
    - The *two-rate three-color policing* (also discussed below) categorizes traffic as yellow when the traffic exceeds the traffic limits but conforms to both a second defined burst-size limit and a second defined bandwidth limit.

The three-color-marking policer implicitly sets the packets in a yellow flow to the medium-high PLP level so that the packets experience a less severe penalty than those in a red flow.

- **Red**: The three-color-marking policer implicitly sets the packets in a red flow to the high PLP level, which is the highest PLP value. The user can also configure a three-color-marking policer to discard packets in a red flow instead of forwarding them with a high PLP setting.

The main difference between the above two markers from a performance point of view is that, two-color-marking policers allow traffic bursts for only short periods, whereas three-color-marking policers allow more sustained traffic bursts.

9.6   IMPLEMENTING TRAFFIC POLICING

One way to control the effects of network resource over-subscription is to use access control lists (ACLs), policing and shaping to classify and limit the amount of lower priority traffic that can enter the network. Traffic policing as a control tool can be used to limit traffic rates at the ingress (inbound) or egress (outbound) port of network devices. This places limits on the amount of traffic particular users or applications can push into the network.

9.6.1   Single-Rate Two-Color Marker – The Token Bucket Algorithm

A *single-rate two-color policer* (which can be implemented using the token bucket algorithm (Figure 9.4)) can be used to rate-limit a traffic flow to an average rate (as specified by the single bandwidth limit, the CIR). The policer does this while allowing traffic bursts for short periods (controlled by the single specified burst-size limit, the CBS).

This type of policer categorizes traffic as either *green (conforming)* or *red (nonconforming)*. Packets in a green flow are implicitly set to a low loss priority and then transmitted. Packets in a red flow are handled according to the actions specified in the policer configuration. Packets in a red flow can be marked (i.e., set to a specified forwarding class, set to a specified loss priority, or both), or they can be discarded.

The policer can mark packets that exceed the configured rate and burst-size limit with a specified PLP, or simply discard them. A network operator may use this type of policer at the ingress or egress interface of a network device.

9.6.2   Single-Rate Three-Color Marker

The *Single-Rate Three-Color Marker* (srTCM) type of policer is defined in [RFC2697]. This type of policer (Figure 9.5) can be used to rate-limit a traffic flow to a *single* rate and *three* traffic categories (green, yellow, and red). This policer meters a traffic stream based on the following configured traffic criteria:

- **Committed Information Rate (CIR)**: This is the bandwidth limit for guaranteed traffic. This rate determines the long-term average transmission rate. Traffic that falls below this rate will always conform.

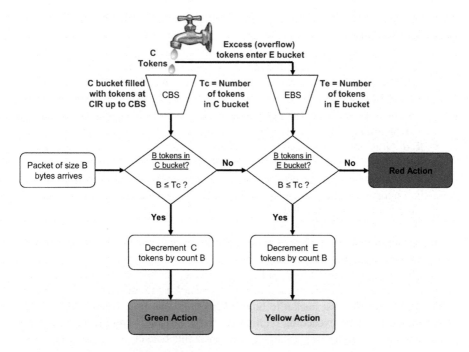

FIGURE 9.5　Single-rate three-color marker (srTCM): Color-blind mode.

- **Committed Burst Size (CBS)**: This is the maximum amount of data permitted for traffic bursts that exceed the CIR. The CBS determines how large traffic bursts can be before *some* traffic exceed the CIR limit.
- **Excess Burst Size (EBS)**: This is the maximum amount of data permitted for peak traffic. The EBS determines how large traffic bursts can be before *all* traffic exceeds the CIR limit. The EBS is greater than or equal to the CBS, and neither can be 0. Extended burst is configured by setting the EBS greater than the CBS.

An srTCM defines a bandwidth limit (CIR) and a maximum burst size for guaranteed traffic (CBS), and a second burst size for peak traffic or excess traffic (EBS). The srTCM is so-called because traffic is policed according to one rate (the CIR) and two burst sizes (the CBS and EBS).

The srTCM classifies traffic as belonging to one of three color categories and performs congestion control actions on the packets based on the color marking:

- **Green**: Traffic that conforms to the limits for guaranteed traffic is categorized as *green*. This is traffic that conforms to either the bandwidth limit (CIR) or the burst size (CBS) for guaranteed traffic. For green traffic, the srTCM marks the packets with an implicit loss priority of low and transmits the packets.

Nonconforming traffic falls into one of two categories and each category is associated with an action:

- **Yellow**: Nonconforming traffic that does not exceed the burst size for excess traffic is categorized as *yellow*. This is traffic that exceeds both the bandwidth limit (CIR) and the burst size (CBS) for guaranteed traffic, but not the burst size for peak traffic (EBS). For yellow traffic, the srTCM marks the packets with an implicit loss priority of medium-high and transmits the packets.
- **Red**: Nonconforming traffic that exceeds the burst size for excess or peak traffic (EBS) is categorized as *red*. For this traffic, the srTCM marks packets with an implicit loss priority of high and, optionally, discards the packets. If congestion occurs downstream, the packets with higher loss priority are more likely to be discarded.

A single-rate three-color policer is most useful when a service is structured according to *packet length* and not *peak arrival rate*.

As illustrated in Figure 9.5, the srTCM uses two token buckets to meter the traffic that passes through the system: *conforming token bucket* and *exceeding token bucket*. The srTCM uses the first bucket to hold tokens that determine whether the CIR is conforming (green) or exceeding (yellow). A traffic stream is conforming when the average number of bytes over time does not cause this bucket to overflow. The *conforming token bucket* can hold bytes up to the size of the CBS before overflowing. The *exceeding token bucket* can hold bytes up to the size of the EBS before overflowing. The srTCM updates the tokens for both the conforming and exceeding token buckets based on the single token arrival rate, the CIR.

9.6.3 Two-Rate Three-Color Marker

The *Two-Rate Three-Color Marker* (trTCM) is a type of policer defined in [RFC2698]. It manages the maximum rate of traffic using two token buckets: *committed token bucket* and *peak token bucket*. This policer (Figures 9.6 and 9.7) can be used to rate-limit a traffic flow to *two* rates and *three* traffic categories (green, yellow, and red).

The trTCM defines two bandwidth limits (one for guaranteed traffic and one for peak traffic) and two burst sizes (one for each of the bandwidth limits) [RFC2698] [RFC4115]. The policer meters a traffic stream based on the following configured traffic criteria:

- **Committed Information Rate (CIR)**: This is the bandwidth limit for guaranteed traffic.
- **Committed Burst Size (CBS)**: This is the maximum amount of data permitted for traffic bursts that exceed the CIR.
- **Peak Information Rate (PIR)**: This is the bandwidth limit for peak traffic. The PIR specifies the maximum rate at which traffic is admitted into the network and must be greater than or equal to the CIR.
- **Peak Burst Size (PBS)**: This is the maximum amount of data permitted for traffic bursts that exceed the PIR.

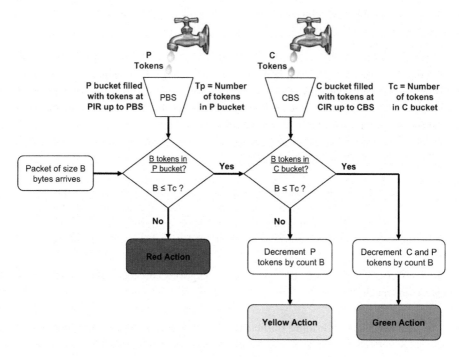

FIGURE 9.6   Two-rate three-color marker (trTCM): Color-blind mode.

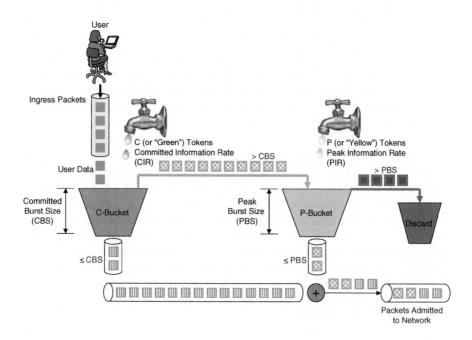

FIGURE 9.7   Another depiction of the two-rate three-color marker.

guaranteed traffic (CBS), plus a bandwidth limit (PIR) and burst-size (PBS) limit for peak traffic. The trTCM is so-called because traffic is policed according to two rates: the CIR and the PIR.

The *committed token bucket* can hold bytes up to the size of the CBS before over-flowing. This token bucket holds the tokens that determine whether a packet conforms to or exceeds the CIR. The *peak token bucket* can hold bytes up to the size of the peak burst (PBS) before overflowing. This token bucket holds the tokens that determine whether a packet violates the PIR.

The trTCM classifies traffic as belonging to one of three color categories and performs congestion control actions on the packets based on the color marking:

- **Green**: Traffic that conforms to the limits for guaranteed traffic is categorized as *green*. This is traffic that conforms to the bandwidth limit (CIR) and burst size (CBS) for guaranteed traffic. For green traffic, the trTCM marks the packets with an implicit loss priority of low and transmits the packets.

Nonconforming traffic falls into one of two categories and each category is associated with an action:

- **Yellow**: Nonconforming traffic that does not exceed peak traffic limits is categorized as *yellow*. This is traffic that exceeds the bandwidth limit (CIR) or burst size (CBS) for guaranteed traffic but not the bandwidth limit and burst size for peak traffic (i.e., the PIR and PBS). For yellow traffic, the trTCM marks packets with an implicit loss priority of medium-high and transmits the packets.
- **Red**: Nonconforming traffic that exceeds peak traffic limits is categorized as *red*. This is traffic that exceeds the bandwidth limit (PIR) and burst size (PBS) for peak traffic. For red traffic, the trTCM marks packets with an implicit loss priority of high and, optionally, discards the packets. If congestion occurs downstream, the packets with higher loss priority are more likely to be discarded.

Token buckets control how many packets per second are accepted at each of the configured rates and provide flexibility in dealing with the bursty nature of data traffic. At the beginning of each sampling period, the two buckets are filled with tokens based on the configured burst sizes and rates. Traffic is metered to measure its volume. When traffic is received, and if tokens are available in both buckets, one token is removed from each bucket for every byte of data processed. As long as tokens are available in the committed token bucket, the traffic is treated as *committed*. When the committed token bucket is empty but tokens are available in the peak token bucket, traffic is treated as *conformed*. When the peak token bucket is empty, traffic is treated as *exceeded*.

The trTCM updates the tokens for both the committed and peak token buckets in the following way (Figure 9.6):

- The trTCM updates the committed token bucket at the CIR value each time a packet arrives at the interface. The committed token bucket can contain up to the CBS value.
- The trTCM updates the peak token bucket at the PIR value each time a packet arrives at the interface. The peak token bucket can contain up to the PBS value.
- When an arriving packet conforms to the CIR, the trTCM takes the *conform* action on the packet and decrements both the committed and peak token buckets by the number of bytes in the packet.
- When an arriving packet exceeds the CIR, the trTCM takes the *exceed* action on the packet, decrements the committed token bucket by the number of bytes in the packet, and decrements the peak token bucket by the number of overflow bytes of the packet.
- When an arriving packet exceeds the PIR, the trTCM takes the *violate* action on the packet, but does not decrement the peak token bucket.

A trTCM is most useful when a service is structured according to *arrival rates* and not necessarily *packet length*. The trTCM is a more predictable algorithm. It improves bandwidth management by allowing the network operator to police traffic streams according to two separate rates. Unlike the srTCM, which allows the network operator to manage bandwidth by setting the EBS, the trTCM allows the operator to manage bandwidth by setting the CIR and the PIR. Therefore, the trTCM supports a higher level of bandwidth management and provides a sustained excess rate. The two-rate policer (trTCM) also enables the operator to implement differentiated services (DiffServ) assured forwarding (AF) per-hop behavior (PHB) traffic conditioning.

The trTCM is often configured on interfaces at the edge of a network to limit the rate of traffic entering or leaving the network. With packet marking, the network operator can partition the network into multiple priority levels or classes of service (CoS). For example, the trTCM can be configured to do the following:

- Assign packets to a QoS group, which the router then uses to determine how to prioritize packets.
- Set the IP Precedence level, IP DSCP value, or the MPLS EXP value of packets entering the network. Networking devices within the network can then use this setting to determine how to treat traffic. For example, a weighted random early detection (WRED) drop policy can use the IP Precedence or DSCP value to determine the drop probability of a packet.

A network operator can utilize the trTCM to provide three service levels: guaranteed, best effort, and deny. The trTCM is useful for marking packets in a packet stream with different, decreasing levels of assurances (either absolute or relative). For example, a service might discard all red packets because they exceed both the committed (CBS) and excess burst (EBS) sizes, forward yellow packets as best effort, and forward green packets with a low drop probability.

9.6.4   COLOR-BLIND VERSUS COLOR-AWARE POLICING

The srTCM and trTCM both provide users with three actions for each packet: a *conform* action, an *exceed* action, and an optional *violate* action. Traffic entering a network interface with any of these mechanisms configured is placed into one of these categories. Within these three categories, the network can decide on packet treatments. For instance, packets that *conform* can be configured to be transmitted, packets that *exceed* can be configured to be sent with a decreased priority, and packets that *violate* can be configured to be dropped.

Both srTCM and trTCM policers can operate in two modes:

- **Color-blind**: In *color-blind mode*, the three-color policer (trTCM) assumes that all packets examined have not been previously marked or metered. In other words, in this mode, the three-color policer is "blind" to any previous color a packet might have had.
- **Color-aware**: In *color-aware mode*, the three-color policer assumes that all packets examined have been previously marked or metered. In other words, in this mode, the three-color policer is "aware" of the previous color a packet might have had. In this mode, the policer assumes that some preceding entity has pre-colored the incoming packet stream so that each packet is either green, yellow, or red. In color-aware mode, the three-color policer can increase the PLP of a packet, but can never decrease it. For example, if a color-aware three-color policer meters a packet with a medium PLP marking, it can raise the PLP level to high, but cannot reduce the PLP level to low.

The *Marker* (Figure 9.2) colors an IP packet according to the results of the *Meter*. The color may be coded in the DSCP field of IP packets in a Per-Hop-Behavior-specific manner.

9.7   CONGESTION MANAGEMENT USING RATE LIMITERS

When a rate-limit profile is configured at an interface, packets may be tagged with a drop preference. The color-coded tag is added to a packet when the committed and peak burst values for the rate-limit profile of the interface are exceeded. For example, an egress packet forwarding on a network device engine may use the drop preference to determine which packets to drop when there is contention for outbound queuing resources. The forwarding engine may use packet drop eligibility to select packets for dropping when congestion exists on an interface (Figure 9.8).

Packets can be categorized as *committed*, *conformed*, or *exceeded* as discussed above:

- Up to the committed rate, packets are considered as *committed*.
- From the committed to the peak rate, packets are considered as *conformed*.
- Beyond the peak rate, packets are considered as *exceeded*.

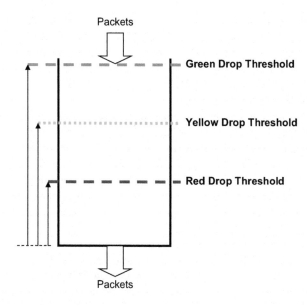

FIGURE 9.8   Color-based congestion management.

A color-coded tag is added automatically to each packet based on the following categories:

- Committed – Green
- Conformed – Yellow
- Exceeded – Red

Each packet queue may have three color-based thresholds as well as a queue limit as illustrated in Figure 9.8:

- Red packets are dropped when congestion causes the queue to fill above the *red threshold.*
- Yellow packets are dropped when the *yellow threshold* is reached.
- Green packets are dropped when the queue limit is reached.

9.8   INPUT AND OUTPUT TAGGING – PACKET RECLASSIFICATION AND REWRITING

A network might have QoS domains within which one either trusts or does not trust the tag values being received in individual packets. An enterprise network may define at least three domains that may have different ways of interpreting and managing QoS tagging. These may be the desktop, LAN backbone, and WAN edge domains.

The network operator may want to manage, for example, four (or fewer) traffic classes, meaning, it may be necessary to perform IP Precedence, DSCP, or IEEE 802.1p remapping into contractually committed service agreements at the WAN edge. The operator may choose to override or re-map the IP Precedence, DSCP, or IEEE 802.1p values based on either application requirements, or contractual commitments from the WAN service provider.

In the enterprise network, the first implementation of QoS tagging may be at the desktop. This may be done for well-behaved applications within the end-system itself. More often, classification rules are used to establish appropriate values for IP Precedence, DSCP, or IEEE 802.1p. Traffic classification at the desktop edge of the network may establish values for IEEE 802.1p for the intervening Layer 2 switches, and IP Precedence or DSCP may be set by the end-user applications themselves. These values may need to be changed to establish the correct handling of packets through the rest of the network.

Modern switch/routers have the ability to read, set, and remap (i.e., rewrite) priorities for Ethernet and IP packets. Rewrite rules determine the information to be written in packets (e.g., a rewrite rule may remark the DSCP bits of outgoing traffic) according to the forwarding class and loss priority of the packets and local conditions in the network device. Traffic conditioning may be based upon srTCM, trTCM, or simple token bucket metering and marking.

9.9   TRAFFIC MANAGEMENT AND QoS FEATURES IN CROSSBAR SWITCH FABRICS

High-capacity, high-performance network devices that support a large number of ports (e.g., 1 Gigabit, 10, 25, 40 Gigabit Ethernet ports) typically have switch fabric architectures that are based on crossbar (also called crosspoint) fabric. To allow high traffic rates through the system, these architectures typically have traffic buffering on both the ingress and the egress ports (Figure 9.9). Ingress buffering (supporting virtual output queuing (VoQ)) is required because of the possibility of resource contention in the fabric which could result in momentary congestion within the switch fabric itself and head-of-line (HOL) blocking.

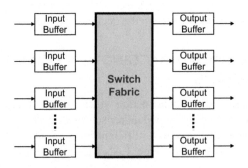

FIGURE 9.9   Input/output buffered switch

The switch fabric also serves as a traffic aggregating mechanism from the ingress ports to each egress port. This makes egress buffering necessary so that an egress port can buffer aggregated traffic from multiple ingress ports in times of congestion. However, because the switch fabric is typically at a higher speed than any port (i.e., a speedup factor of at least 2), fabric congestion is typically very short-lived, which means ingress buffers can be much smaller than egress buffers.

The ingress buffer architecture for crossbar switch fabrics is configured in such a way as to eliminate the HOL blocking [KAROLHM87]. HOL blocking occurs when the packet at the head of an input queue cannot be forwarded immediately because the destination egress port is busy. This is mostly due to some other ingress port in the process of transmitting traffic over the switch fabric to that destination egress port. Karol et al. [KAROLHM87] showed that input FIFO queued switch fabrics can suffer from reduced throughput due to HOL blocking. With FIFO queuing, packets deeper in the ingress queue are blocked in spite of the fact that they may be destined for idle egress ports on the switch fabric.

HOL blocking is eliminated when a VOQ architecture is used at each ingress port of the switch fabric instead of FIFO queuing [TAMIRY88]. In VOQ, each input port maintains a separate queue for each output port (Figure 9.10). VOQ when combined with a suitable scheduling algorithm, has been shown to achieve almost 100% switch fabric throughput [DAIPRAB00] [MCKEOWN96] [MCKEOWN98]. The ability of a switch to achieve 100% throughput is desirable to a network operator, as it assures that all of the (expensive) link capacity can be utilized and real-time traffic in particular will not suffer noticeable performance penalties.

To optimize system performance with VOQ, the ingress packet processing must be fast enough to perform the lookup of the destination port before the packet enters

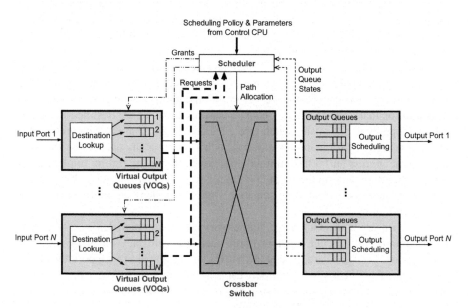

FIGURE 9.10   Crossbar switch fabric with virtual output queues (VOQs) and scheduler.

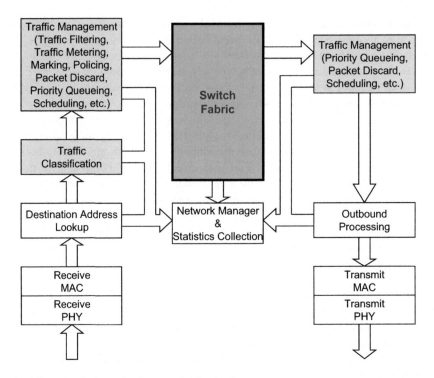

FIGURE 9.11 QoS mechanisms on the data path.

the ingress buffer. This allows the switch to immediately forward any packet in the queue to the correct output port that is not currently busy or full. Suboptimal implementations typically enqueue the incoming packet in the input buffer before the destination lookup has been completed.

As shown in Figure 9.11, an additional enhancement to ingress buffering implementations is to apply ingress QoS policies such as classification, traffic policing, traffic filtering, and packet discard before the incoming packet reaches the VOQ. This ensures that "out-of-profile" packets and packets that are destined to be randomly dropped do not waste switch fabric bandwidth and ingress and egress buffer resources. Ingress QoS is implemented in such a way that any packet loss that does occur at the ingress buffer has minimal effect on high priority traffic. Implementing the QoS function only on the output port could result in ingress packet drops affecting all traffic classes and time-sensitive traffic experiencing more latency and latency variations.

A key concern with VOQ is that, for network nodes with high densities of Gigabit Ethernet and multi-gigabit Ethernet ports, each line card may on its own support several ports. In such a case, if separate buffers are dedicated to each port, the cost of buffering scales linearly with the port capacity of the switch. The overall cost of buffering can be very significant, especially if the buffers are scaled to accommodate high round-trip times of MAN/WAN links.

To address this issue, a switch architecture that provides the needed buffer capacity at a significantly lower cost is required. This can be done by allowing the multiple ports on a single line card to share a common pool of buffer capacity. Each port on the line card can then receive a dynamic allocation of buffer space based on the traffic load and port capacity. A shared pool of buffers is much more efficient than dedicated buffers because no buffer capacity is wasted when ports are idle or lightly loaded.

9.10   ADVANCED QoS FEATURES IN SWITCH/ROUTERS

The main QoS mechanisms in network devices have been discussed above. We now described the advanced QoS features supported in most high-end switch/routers. These features mainly employ hardware-based mechanisms to allow wire-speed processing of traffic:

- **Advanced QoS Control**: These features allow administrators to enforce QoS policies based on port, VLAN, source MAC, ACL rules, IEEE 802.1p priority, IP Precedence, DSCP settings, or rate-limiting parameters. The switch could employ wire-speed Layer 2, Layer 3, and Layer 4 ACLs to control packet forwarding and restricts access to the system's control and management plane.
- **Very Low Latency Forwarding Across all Packet Sizes**: This feature is required to provide consistent low latency for latency-sensitive applications such as real-time streaming voice and video, and high-performance computing. High-end switches would be required to have processing architectures that support non-blocking operation and the ability to maintain full line-rate performance, predictable latency, and minimal latency variations irrespective of the traffic load, QoS features, ACLs, traffic monitoring, or other services that have been configured for traffic control or security.
- **Configurable Combinations of Queuing Disciplines and Congestion Control Policies**: A range of scheduling policies including Strict Priority and Weighted Fair Queuing (WFQ) could be configured to provide flexibility for traffic scheduling. In the event of port congestion, traffic policies can be configured for tail drop or weighted random early detection (WRED) operation.
- **Advanced Bandwidth Management**: This allows intelligent bandwidth control and management using hardware-based enforcement of CIR with excess burst control capabilities. This could be integrated seamlessly with other advanced QoS features including priority traffic marking and tagging. Traffic conditioning can be based on two-rate, three-color token bucket-based metering and marking. Per-port hardware queues can be mapped directly to class-based DiffServ and IEEE 802.1p queuing models.
- **Protection Against Denial of Service (DoS) Attacks**: This feature prevents or minimizes network downtime from malicious users by limiting TCP SYN and ICMP traffic and to protects against broadcast storms. The switch could implement ACL lookups at hardware level, so that QoS and

security control features that are enabled do not adversely affect packet-forwarding performance.

- **Network Evaluation and Monitoring with NetFlow or sFlow (RFC 3176)**: This feature could be used to provide cost-effective, scalable, wire-speed network monitoring to detect unusual network activity. The network evaluation and monitoring tools may need to be capable of measuring the end-to-end network performance parameters (latency, latency variations, and packet loss) that are critical to the successful support of real-time traffic. Traffic monitoring software, such as NetFlow or sFlow, can be used to gather historical data and performance statistics that can be used to help optimize the network for real-time applications. In addition, it may be required to deploy tools that can detect in real-time when the performance thresholds are unacceptable or real-time traffic quality levels have degraded. Continual monitoring of network service quality may be required because of the dynamic nature of both the network application environment and the underlying infrastructure.

High-end switch/routers typically offer superior QoS features designed to ensure high-reliability services in enterprise and service provider networks. The switch can identify, mark, classify, reclassify, and manage traffic based on a wide range of traffic criteria and network operator policies. This enables the operator to classify and manage bandwidth-critical application traffic, while still discriminating among various traffic flows and enforcing bandwidth policies.

REVIEW QUESTIONS

1. Explain the difference between traffic policing and traffic shaping.
2. Explain the difference between traffic metering and marking.
3. Explain the difference between color-blind and color-aware metering.
4. Explain the difference between the "leaky bucket algorithm as a queue" algorithm and the "token bucket" algorithm.
5. Explain the difference between the Single-Rate Three-Color Marker (srTCM) and the Two-Rate Three-Color Marker (trTCM).

REFERENCES

[ATMFRUNI95]. ATM Forum, *The User Network Interface (UNI)*, v. 3.1, ISBN 0-13-393828-X, Prentice Hall PTR, 1995.

[DAIPRAB00]. J. Dai and B. Prabhakar, "The throughput of data switches with and without speedup", *IEEE INFOCOM 2000*, Tel Aviv, Israel, March 2000, pp. 556–564.

[ITUTTRAFF04]. ITU-T, *Traffic control and congestion control in B-ISDN*, Recommendation I.371, International Telecommunication Union, 2004, Annex A, page 87.

[JUNPOLICE]. Traffic Policing Overview, Juniper Networks, *Technical Document*, February 2013.

[KAROLHM87]. M. Karol, M. Hluchyj, S. Morgan: "Input versus Output Queueing on a Space-Division Packet Switch", *IEEE Trans. on Communications*, vol. COM-35, no. 12, December 1987, pp. 1347–1356.

[MCDSPOH95]. David E. McDysan and Darrel L. Spohn, *ATM: Theory and Application*, ISBN 0-07-060362-6, McGraw-Hill series on computer communications, 1995, pages 358–359.

[MCKEOWN96]. N. McKeown, V. Anantharam and J. Walrand, "Achieving 100% throughput in an input-queued switch," *IEEE INFOCOM 96*, pp. 296–302, 1996.

[MCKEOWN98]. N. McKeown and A. Mekkittikul, "A practical scheduling algorithm to achieve 100% throughput in input-queued switches", *IEEE INFOCOM 98*, pp. 792–799, 1998.

[RFC2475]. S. Blake, D. Black, M. Carlson, E. Davies, Z. Wang, and W. Weiss, "An Architecture for Differentiated Services", *IETF RFC 2475*, December 1998.

[RFC2697]. J. Heinanen and R. Guerin, "A Single Rate Three Color Marker", *IETF RFC 2697*, September 1999.

[RFC2698]. J. Heinanen and R. Guerin, "A Two Rate Three Color Marker", *IETF RFC 2698*, September 1999.

[RFC4115]. O. Aboul-Magd and S. Rabie, "A Differentiated Service Two-Rate, Three-Color Marker with Efficient Handling of in-Profile Traffic", *IETF RFC 4115*, July 2005.

[TAMIRY88]. Y. Tamir and G. Frazier, "High-Performance Multi-Queue Buffers for VLSI Communication Switches", *Proc. of the 15th Int. Symp. on Computer Architecture*, ACM SIGARCH, Vol. 16, No. 2, May 1988, pp. 343–354.

[TANENBAS03]. Andrew S. Tanenbaum, *Computer Networks*, Fourth Edition, ISBN 0-13-166836-6, Prentice Hall PTR, 2003, page 401.

[TURNERJO86]. J. Turner, "New directions in Communications (or Which Way to the Information Age?)", *IEEE Communications Magazine*, Vol. 24, No. 10, 1986, pp. 8–15.

Index

A

access control lists (ACLs), 248, 281
adjacency information *see* adjacency table
adjacency table, 126, 135, 209, 219, 264
    adjacency manager, 224
    drop adjacency, 135
    incomplete adjacency, 135
    null interface, 135
    punt adjacency, 135
Address Resolution Protocol (ARP), 45, 55,
        104–105, 126, 166–167, 221, 236
administrative distance, 19–21, 113–114
ARP cache, 105, 126, 129
    ARP aging process, 129
    ARP entry age timer, 129
ARP table *see* ARP cache
availability, 272

C

Class of Service (CoS) *see* Quality of Service
    (QoS)
classless inter-domain routing (CIDR), 14, 16–17,
        72, 109, 129–130, 133, 145
content addressable memory (CAM), 63, 97, 195,
        201, 207, 300
    associated data, 97
    associative memory array, 100
    associative word select register, 100
    comparand register, 98
    hash function, 63, 100
    key word, 97
    label field, 97
    mask register, 98
    matching logic, 97
    mask word, 97
    output register, 100
    search (or lookup) key, 63, 100
    tag bits, 97
    tag memory, 97
control engine *see* route processor
control plane, 14–15, 106–107
control plane redundancy *see* route processor
    redundancy
control processor *see* route processor

D

data plane, 15, 106, 123
device driver, 88, 184, 190
Domain Name System (DNS), 49

E

enterprise network, 2
equal-cost multipath (ECMP) routing, 73, 211,
    219, 220
Ethernet, 43
    Bridge Management Entity, 57
    Bridge Protocol Entity, 56
    broadcast storm, 69, 97, 102
    flapping, 100
    frame, 44
    IEEE 802.3, 43
    Logical Link Control (LLC), 44, 55
    Media Access Control (MAC), 44, 55
    MAC Relay Entity, 56, 58
        Bridge Port States, 59
            Disabled, 60
            Discarding, 59
            Forwarding, 59
            Learning, 59
        Egress Rules, 59
        Forwarding Process, 58, 61
        Ingress Rules, 58
        Learning Process, 58
    Medium Dependent Interface (MDI), 87
    Physical Coding Sublayer (PCS), 87
    Physical Medium Attachment (PMA), 87
    Physical Medium Dependent (PMD), 87
    Spanning Tree Protocol Entity, 56
    transparent bridging algorithm, 65, 67, 70
        filtering, 66
        flooding, 66, 96
        forwarding, 65
Ethernet frame
    baby giant frame, 229
    broadcast MAC address, 66, 124
    EtherType, 44, 124
    globally (or universally) administered MAC
        address, 69
    group (G) MAC address, 68
    IEEE 802.1Q, 225, 226, 274, 275, 295
        Tag Control Information (TCI), 226
            Drop Eligible Indicator (DEI), 226
            Priority Code Point (PCP), 226, 274, 295
            VLAN Identifier (ID), 226, 228
        Tag Protocol Identifier (TPID), 226
    individual (I) MAC address, 68
    jumbo frame, 229
    locally administered MAC address, 69
    multicast MAC address, 68
    NIC-Specific Identifier, 67
    Organizational Unique Identifier (OUI), 67
exact matching lookup, 17, 63, 100, 133, 207

F

fault tolerance, 148
File Transfer Protocol (FTP), 48
Filtering Database, 55, 59
   age, 63
   aging, 63
   aging time (or limit), 62, 63, 101
   aging timer, 101
   dynamic filtering information, 61
      dynamic filtering entry, 61
      dynamic VLAN registration entry, 62, 63
      group registration entry, 62
   Filtering Identifier (ID), 62
   port map, 62
   static bit, 101
   static filtering information, 61
      static filtering entry, 61
      static VLAN registration entry, 61, 63
flow cache *see* route cache
flow/route cache *see* route cache
forwarding architectures
   centralized forwarding, 22–23, 169, 171, 175,
      192, 211, 214, 219, 271
   centralized processor, 22–23, 177, 198, 200
   distributed forwarding, 27, 140–148, 203, 214,
      218, 220, 246, 261
   non-blocking architecture, 271
forwarding engine, 17, 124
forwarding information base (FIB) *see* forwarding
     table
forwarding plane *see* data plane
forwarding table, 16, 18, 26, 72, 122, 134, 192,
     208, 213, 218, 220, 264
   FIB consistency checker, 222
      active consistency checkers, 222
      passive consistency checkers, 222
   FIB maintenance process, 134
   FIB manager, 224
   FIB route resolution process, 134
   resolved routes, 192, 210
   update manager, 224

H

head-of-line (HOL) blocking, 301, 320
hot-swapping *see* Online Insertion and
     Removal (OIR)

I

Internet Control Message Protocol (ICMP), 46,
     125, 132, 136
Internet Group Management Protocol (IGMP), 46
Inter–Process Communication (IPC), 219, 223
IP address lookup, 18

Internet, 2, 42
Internet Message Access Protocol (IMAP), 51
Internet Protocol suite *see* TCP/IP model
IP routing, 1
   default gateway, 167
   path, 14
   route, 14
   routing protocols, 1, 18, 51, 71, 103, 107, 113

L

Layer 2 forwarding table *see* Filtering Database
Layer 3 forwarding engine *see* forwarding engine
Layer 3 route cache *see* route cache
Layer 3 Topology-Based Forwarding Table *see*
     forwarding table
Lightweight Directory Access Protocol (LDAP),
     50
Link Aggregation, 11, 30, 87, 89, 233
Link Aggregation group (LAG), 75, 87
load balancing
   per-destination load balancing, 134
   per-packet load balancing, 133
longest prefix matching (LPM), 17, 23, 72, 130,
     133, 134, 145, 173, 175, 207

M

MAC address table *see* Filtering Database
management
   command-line interface (CLI), 151
   graphical user interface (GUI), 150, 160
   in-band management, 150
   out-of-band management, 150
   management interface, 160
   management module *see* route processor
   management plane, 107
   management and control protocols, 18, 108
   management tools, 18, 108
   network management appliance, 157
Maximum Transmission Unit (MTU), 136, 229
memory, 178
   Flash Memory, 179
      run from Flash memory, 179
   interface FIFOs, 181
   interface receive (RX) rings, 181, 183, 185,
     188, 190, 191
   interface transmit (TX) rings, 181, 185, 189,
     191, 200
   Non-Volatile Random Access Memory
     (NVRAM), 178, 186, 187
      startup configuration, 178, 181
      configuration registers, 178
   Random Access Memory (RAM), 179,
     186, 187
      buffer headers, 182, 185

contiguous buffers, 183
particle buffers, 182
dynamic buffers, 179
input hold queues, 184, 185, 188, 191
interface buffers, 182, 185
internal buffers, 182, 188
Interface Descriptor Blocks (IDBs),
    187, 188
interface queues, 184
main processor memory, 182
output hold queues, 184, 189, 191
private buffer pools, 179, 183, 184, 190
public buffer pools, 179, 183, 184
receive queues, 184
run from RAM, 179
running configuration, 179, 181
shared interface private buffer pools, 179
shared I/O memory, 182, 185, 188, 190
static buffers, 179
system buffers, 179, 182, 188
system interface queues, 184
transmit queues, 184
Read Only Memory (ROM), 178
Boot ROM, 178, 186, 187
bootstrap software (or program), 178, 180
erasable programmable ROM (EPROM),
    178
Power-On-Self-Test (POST) *see* startup
    diagnostic code
ROM Monitor (or ROMMon), 185
RxBoot, 185, 187, 188
startup code *see* bootstrap software (or
    program)
startup diagnostic code, 178, 186
shared I/O memory, 181
multilayer switching, 8, 11, 165

N

network architecture, 9
access layer, 9
aggregation (or distribution) layer, 10
core layer, 10
network (address) prefix, 13, 109
network control protocols, 18
network devices
application switch *see* Layer 4+ switch
bridge (or Layer 2 switch), 55
    promiscuous (or monitor) mode, 66
content switch or content service switch
    see Layer 4+ switch
Layer 4+ switch, 76
    application content inspection and
        switching, 81
    application rate limiting, 82
    application redirection, 82
    application security, 81

global server load balancing, 80
server load balancing, 80
SSL acceleration and compression, 82
system health monitoring, 80
Transparent Cache Switching, 82
repeater (or hub), 53–54
router, 71, 106
switch/router (or multilayer switch), 72,
    165, 241
conventional standalone switch/router
    approach, 242
front-end processor approach, 242
Layer 3 multicast multilayer switching
    cache, 258
Layer 3 (or routed) ports, 234
MLSP messages, 244, 253
MLS Protocol (MLSP), 244, 252, 259
Multicast MLS-Route Processor
    (MMLS-RP), 258
Multicast MLS-Switching Engine
    (MMLS-SE), 257
Multilayer switching aging timer, 252
multilayer switching cache *see* route cache
Multilayer switching fast-aging timer, 251
Multilayer Switching–Route Processor
    (MLS-RP), 243
Multilayer Switching–Switching Engine
    (MLS-SE), 243
switch ports, 234
switched virtual interface (SVI), 234
VLAN interface(s) *see* SVI
Web switch *see* Layer 4+ switch
Network File System (NFS), 50
Network Information Service (NIS), 49
network interface card (NIC), 74, 84
auto-negotiation, 88
autosensing driver, 88
embedded NICs, 84
link aggregation driver, 89
multiport NICs, 85
NIC teaming (or bonding), 75
network interface (or device) driver, 88,
    184, 190
server-Based NICs, 85
workstation NICs, 84
network mask, 14, 109
network reference models, 33–34, 53
communication protocol, 33
cell, 37
datagram, 36, 45, 46
demultiplexing, 37
frame, 36
header, 36
International Organization for Standardization
    (ISO), 33
layers, 33–34, 53
message, 37

multiplexing, 37
Open Systems Interconnection (OSI) reference
        model, 34–35, 53
    OSI Application Layer (Layer 7), 41
    OSI Data Link Layer (Layer 2), 38–39, 55
    OSI Network Layer (Layer 3), 39, 71
    OSI Physical Layer (Layer 1), 37–38
    OSI Presentation Layer (Layer 6), 41
    OSI Session Layer (Layer 5), 41
    OSI Transport Layer (Layer 4), 39–40
packet, 36, 45
protocol data unit (PDU), 37
protocol stack, 34
segment, 37, 46
service data unit (SDU), 37
TCP/IP model, 34–35, 42–43
    fragmentation, 45
    host-to-host communications, 45
    IP, 45
    IP addressing, 45
    Network Layer PDU, 45
    packet formatting, 45
    TCP/IP Application Layer, 48
    TCP/IP Link Layer, 43–44
    TCP/IP Network Layer, 45
    TCP/IP Transport Layer, 46
trailer, 36
network prefix length, 14

O

Online Insertion and Removal (OIR), 79

P

packet classification, 146, 272
packet forwarding
    context switching, 174
    exception packets, 196, 211
    Ethernet Frame Check Sequence (FCS)
        see Ethernet Frame Checksum
    Ethernet Frame Checksum, 22, 106, 123, 124,
        126, 127, 168, 173, 181, 221, 260, 262
    fast-path forwarding, 24, 209
    fast switching (Cisco), 174, 185, 201, 262
    forwarding speed, 168
    frame or packet rewrite, 106
    interrupts, 172, 174, 175, 181, 184, 185, 188,
        190, 191, 204
    interrupt handler, 173, 174, 175, 188
    interrupt level, 172, 174
    IP header checksum update, 22, 106, 123, 125,
        167, 173, 221, 260, 262
    IP header Time-to-Live (TTL) update, 106,
        123, 125, 132, 167, 173, 221, 260, 262
    IP fragmentation, 136
    kernel level, 172

process level, 172
process path see slow-path forwarding
process forwarding see process switching
process switching (Cisco), 21, 171, 174, 262
scheduler process, 172
slow-path forwarding, 24, 108, 171
software processing plane see slow-path
        forwarding
special packets see exception packets
policy database, 145

Q

Quality of Service (QoS)
    active queue management (AQM), 290
    Bandwidth Broker, 279, 280
    bandwidth-delay product (BDP), 287
    color-aware mode, 306
    color-blind mode, 306
    Committed Burst Size (CBS), 309, 313, 314
    Committed Information Rate (CIR), 309,
        312, 314
    Differentiated services (DiffServ), 276,
        279, 295
    Differentiated Services Code Point (DSCP),
        272, 273, 275, 295
    DiffServ domain, 277, 280
    Explicit Congestion Notification (ECN), 275,
        277
    Per-Hop Behavior (PHB), 277
        Assured Forwarding PHB, 277
        Class Selector PHB, 277, 279
        Default PHB, 278
        Expedited Forwarding PHB, 277
        Voice Admit PHB, 277
    Excess Burst Size (EBS), 313
    IEEE 802.1p, 272–274
    IP Precedence, 275, 277, 294–296
    leaky bucket algorithm as a meter, 306
    leaky bucket as a queue, 306, 308
    marker, 307, 318
    meter, 306, 318
    MPLS Traffic Class field (or MPLS EXP bit),
        294–297
    MPLS tunneling modes, 297
        full pipe mode, 298
        short pipe mode, 298
        uniform mode, 297
    Peak Burst Size (PBS), 314
    Peak Information Rate (PIR), 314
    Random Early Detection (RED) see Weighted
        Random Early Detection (WRED)
    packet discard, 180
    single-rate two-color policer, 312
    single-rate three-color policing, 311
    Single-Rate Three-Color Marker (srTCM), 312
    tail-drop, 287, 289

three-color-marking policer, 311
token bucket algorithm, 306, 308
traffic metering, 306
traffic marking, 306
traffic policing, 305
Tunnel header marking (THM), 298
two-color-marking policer, 311
Two-Rate Three-Color Marker (trTCM), 314
two-rate three-color policing, 311
Type-of-Service (ToS) field, 275
Weighted Fair Queuing (WFQ), 292
Weighted Random Early Detection (WRED), 273, 289, 290, 292, 294
Weighted Round Robin (WRR), 292

R

redundancy, 79, 148
Remote Monitoring (RMON), 150, 154
    management console system, 156
    RMON agent see RMON probe
    RMON groups, 158
        Alarm Group, 158
        Event Group, 158
        Ethernet History Group, 158
        Ethernet Statistics Group, 158
    RMON Management Information Base (MIB), 157
    RMON monitor see RMON probe
    RMON probe, 156
    RMON software agent, 157
reverse path forwarding (RPF), 125, 137–140
route cache, 17–18, 24–25, 133, 192, 194–196, 202, 219
    hit ratio, 196, 197
    locality, 194
    spatial locality, 194, 196, 203
    temporal locality, 194, 196, 203
    timers, 206
        Maximum Interval, 206
        Minimum Interval, 206
        Quiet Interval, 206
        Threshold, 207
route preference see administrative distance
route processor, 15, 107, 169, 201
route switch processor see route processor
route processor redundancy, 28–29, 79, 221, 222
routing engine see route processor
routing information base (RIB) see routing table
routing metrics, 20–21, 110
routing prefix see network prefix
routing table, 1, 16–17, 19, 108, 109, 111, 120, 208, 213, 218, 220
    BGP (learned) routes, 115, 117, 120, 133
    default route, 111
    directly attached (or connected) network, 110, 112, 113, 116

directly connected route, 112
dynamic route, 113
filtering and access control, 111
host route, 110
IGP (learned) route, 118
interface, 110
interface (or connected) route, 112
Interface Descriptor Block (IDB), 111, 187, 188
metric, 110
network mask, 109
network prefix, 109
next-hop, 106, 109, 115, 120, 123, 167, 208
Path Descriptor, 111
Prefix Descriptor, 111
Quality of Service (QoS), 111
recursive route, 117, 120, 121, 171, 177, 210
recursive lookup, 120, 121, 171, 177, 197
remote network, 110
Route Resolvability Condition, 114–115
route selection process, 114
static route, 112, 113, 116–118
routing update, 16, 19

S

Secure Shell (SSH), 151
service provider network, 3
Simple Mail Transfer Protocol (SMTP), 50
Simple Network Management Protocol (SNMP), 50, 150, 156
    managed device, 152
    managed object, 153
    Management Information Base (MIB), 153
    Network Management System (NMS), 152
    SNMP agent, 153
    SNMP manager, 152
    SNMP Traps, 154
Spanning Tree Protocol (STP), 37, 56, 70, 102
    Bridge Protocol Data Unit (BPDU), 37, 56, 102
    Multiple Spanning Tree Protocol (MSTP), 70
    Rapid Spanning Tree Protocol (RSTP, 59–60, 70
        Discarding, 60, 71
        Disabled, 60
        Forwarding, 60, 71
        Learning, 60, 71
    STP bridge port states, 60
        blocking, 60, 71
        disabled, 60
        forwarding, 60
        learning, 60
        listening, 60
    spanning-tree algorithm, 102
static routes, 2
Stream Control Transmission Protocol (SCTP), 47

switch fabric module (SFM), 28
switch fabric redundancy, 28
system processor *see* route processor

T

TCP/IP protocol suite *see* TCP/IP model
Telnet, 48
Ternary Content Addressable Memory (TCAM),
    146, 300
traffic classification *see* packet classification
Transport Control Protocol (TCP), 46
    global synchronization, 289
    TCP congestion avoidance phase, 289
    TCP congestion control, 288
    TCP slow start mechanism, 288, 289
    TCP window size, 288
Trivial File Transfer Protocol (TFTP), 49

U

UNIX "r" commands
    rcp (remote copy), 49
    rlogin (remote login), 49
    rsh (remote shell), 49
    rexec (remote execution), 49
User Datagram Protocol (UDP), 47

V

Variable-length subnet mask (VLSM), 14, 17, 23,
    129, 133
virtual output queues (VOQs), 301, 320
Virtual Router Redundancy Protocol (VRRP),
    73, 75
virtual LAN (VLAN), 69, 225
    IEEE 802.1Q VLANs, 225, 230
    inter-VLAN routing *see* inter-VLAN
        communication
    inter-VLAN communication, 70, 229, 230,
        232, 245
        external router, 230
        lollipop routing *see* one-armed router
        one-armed router, 230
        one-arm bandit routing *see* one-armed
          router
        router-on-a-stick *see* one-armed router
        sub-interfaces, 232
        trunking, 230
    IP subnet-based VLANs, 225
    port-based VLANs, 225
    protocol-based VLANs, 225
    Q-in-Q (or QinQ), 229, 275
    stacked VLANs *see* Q-in-Q
    switched virtual interface (SVI), 234
    VLAN interface(s) *see* SVI